PROGRAM EARTH

Electronic Mediations

Series Editors: N. KATHERINE HAYLES, PETER KRAPP, RITA RALEY, AND SAMUEL WEBER

Founding Editor: MARK POSTER

(continued on page 359)

PROGRAM EARTH

*Environmental Sensing Technology and
the Making of a Computational Planet*

JENNIFER GABRYS

Electronic Mediations 49

UNIVERSITY OF MINNESOTA PRESS
MINNEAPOLIS • LONDON

The University of Minnesota Press gratefully acknowledges financial support for the publication of this book from the European Research Council under the European Union's Seventh Framework Programme (FP/2007–2013) / ERC Grant Agreement n. 313347, "Citizen Sensing and Environmental Practice: Assessing Participatory Engagements with Environments through Sensor Technologies."

An earlier version of chapter 1 was published as "Sensing an Experimental Forest: Processing Environments and Distributing Relations," in *Computational Culture* 2 (2012), http://www.computationalculture.net.

Portions of chapter 4 were previously published as "Ecological Observatories: Fluctuating Sites and Sensing Subjects," in *Field_Notes,* edited by Laura Beloff, Erich Berger, and Terike Haapoja, 178–87 (Helsinki: Finnish Bioarts Society, 2013).

Portions of chapter 5 were published as "Monitoring and Remediating a Garbage Patch," in *Research Objects in Their Technological Setting,* edited by Bernadette Bensaude Vincent et al. (London: Routledge, 2016).

An earlier version of chapter 7 was published as "Programming Environments: Environmentality and Citizen Sensing in the Smart City," *Environment and Planning D 32,* no. 1 (2014): 30–48.

Published by the University of Minnesota Press
111 Third Avenue South, Suite 290
Minneapolis, MN 55401-2520
http://www.upress.umn.edu

Library of Congress Cataloging-in-Publication Data
Names: Gabrys, Jennifer. Title: Program earth : environmental sensing technology and the making of a computational planet / Jennifer Gabrys. Description: Minneapolis : University of Minnesota Press, 2016. | Series: Electronic mediations ; 49 | Includes bibliographical references and index. Identifiers: LCCN 2015022473 | ISBN 978-0-8166-9312-2 (hc) | ISBN 978-0-8166-9314-6 (pb). Subjects: LCSH: Environmental monitoring—Remote sensing. | Environmental management—Remote sensing. | Global environmental change—Remote sensing. | Remote sensing—Data processing.
Classification: LCC GE45.R44 .G33 2016 | DDC 363.7/063—dc23
LC record available at http://lccn.loc.gov/2015022473

Printed in the United States of America on acid-free paper

The University of Minnesota is an equal-opportunity educator and employer.

21 20 19 18 17 16   10 9 8 7 6 5 4 3 2 1

CONTENTS

PREFACE AND ACKNOWLEDGMENTS

First sparked through an interest in remote sensing and environments that began with an online project on satellites over ten years ago under the title of Signal Space, the material gathered together for this study on environmental sensing technologies has been a long time in the making. From urban sensing to automated gardens, I have developed an ongoing habit of attending to and working with technologies that would animate and monitor environments. Developed alongside prior work that I have assembled on electronic waste, this work on environmental sensing is also part of a larger project of attending to the environmental and material aspects of computational technologies. Sensor-based technologies are not only environmentally located; they also in-form and "program" environments, have environmental impacts, and take hold in particular environments, whether for managing or monitoring processes. As I outline in the pages that follow, this is an interdisciplinary and even postdisciplinary project of attending to these emerging technical objects and milieus.

As a modest disclaimer, I might also note that this study is inevitably incomplete, since it attends to an ever-growing area of environmental computing. There are many topics I was not able to accommodate fully in these pages. This is not a handbook for understanding the finer details of sensors as technical objects, nor is it a survey of the wide range of citizen-sensing or creative-practice projects using environmental sensors. Despite the earthly expanse of the title, this study is also not quite as "global" as it could be, since inevitably quite incisive discussions could be developed around the commercial, military, and even colonial unfoldings of ubiquitous computing in locations worldwide, with often uneven outcomes. While I engage with citizen-sensing practices here, this is also not primarily an ethnographic or practice-based study, since this work is still in development through a current collaborative research project that I am leading, Citizen Sense.

Instead, what I have developed here is more of a *theoretical and ethico-aesthetic* investigation that emphasizes both the *environmental* and *sensor-based* aspects of ubiquitous computing technologies and practices. As described in these pages, this work has also taken me to many locations where environmental sensing has been in use and under active development. This work would not have been possible if it were not for the generosity and attention of multiple interlocutors who have met up with me to discuss their research, made field sites available for visits, sent along notices for related events, and exchanged ideas about environmental sensors. While this is a far-from-comprehensive account, below are some of the people (and organizations) who have aided in the development of this work.

Thanks are due to the researchers on the CENS sensing project who hosted me during fieldwork conducted in 2008 both at UCLA and the UC James Reserve while writing chapters 1 and 2, including Mark Hansen, Deborah Estrin, Michael Hamilton, Eric Graham, Josh Hyman, Chuck Taylor, Katie Shilton, Hossein Falaki, and Becca Fenwick, among others. I am grateful to Matthew Fuller and Mike Michael for their helpful suggestions for improving and clarifying early versions of chapter 1. I presented a version of chapter 2 at the Emerging Landscapes conference at the University of Westminster, and I am thankful to the organizers and participants at this event for their feedback on this work.

Cecilia Mascolo at Cambridge University helped by meeting with me to discuss her work on monitoring badger movement in 2010 and sent me several key papers and event references to familiarize me with the field of movement ecology, which has informed my work in chapter 3. Erich Berger and Laura Beloff were generous hosts during my residency in Kilpisjärvi in 2012 and in creating a context for experimenting with the topic of environmental computing, which developed into the material written for chapter 4. This text has further benefited from participant and organizer feedback during the presentation given at Sense of Planet: The Arts and Ecology at Earth Magnitude, a National Institute for Experimental Arts symposium at the University of New South Wales (2012).

Chapter 5 would not have emerged if it were not for the generous invitation from Bernadette Bensaude-Vincent to write about the garbage patch for a workshop on technoscientific objects in 2012. Thanks are due to her, Sacha Loeve, Alfred Nordmann, and Astrid Schwarz for organizing this engaging writing workshop, as well as to participants for their feedback on earlier versions of this text. Thanks are also due to Richard Thompson, who provided multiple points of clarification and helpful suggestions on the topic of ocean plastics, and to Charles Moore, who pointed me to useful references on the "patches."

The topic of air pollution developed in chapter 6 has benefited from discussions with Citizen Sense researchers, and thanks are due to Helen Pritchard and Nerea Calvillo for contributing to our first "Air Walk" in 2013, where we tested air-monitoring technologies. Benjamin Barratt and Andrew Grieve from the King's

College Environmental Research Group provided useful input on the details of monitoring air pollution in London. David Holstius, Virginia Teige, and Ron Cohen from the Beacon project at UC Berkeley, and Gayle Hagler and Ron Williams at EPA Research all shared their work on monitoring air pollution, which has provided a valuable resource for this research as well as ongoing Citizen Sense research. Versions of chapter 6 have been presented at the Citizen Scientist on the Move conference at Utrecht University (2012), at the Waag in Amsterdam (2012), at Pixelache in Helsinki (2012), and at the Multimodality Workshop at Cardiff University (2013).

A text that has been through the paces, chapter 7 has benefited from input from Bruce Braun, Stephanie Wakefield, and Natalie Oswin. Versions of chapter 7 have also been presented at the Green Apparatus session of the American Association of Geographers in Seattle (2011), the Platform Politics conference at Anglia Ruskin University (2011), the Digital Media Research Seminar at the University of Western Sydney (2012), the Media Places: Infrastructure, Space, Media symposium at HUMlab at Umeå University (2012), the Speculative Urbanisms seminar for the Urban Salon at University College London (2013), and the Mediating Uncertainty seminar at the London School of Economics (2014).

Chapter 8 has benefited from receptive and helpful comments from audiences at the Digital Citizen events at the Lewisham Library (2013), the Defining the Sensor Society symposium at Queensland University (2014), the Artistic and Cultural Strategies, Technology and Urban Transformations roundtable at Aarhus University (2014), and the Open Data / Smart Citizens seminar for the London Media Cities Network, Birkbeck, at University of London (2015). Thanks are due to the organizers and participants at these many events.

The research presented here has benefited from a long gestation while at Goldsmiths and demonstrates just how much this research environment has informed my thinking. I am thankful to colleagues and students (past and present), as well as Citizen Sense researchers, who have informed this work, whether directly or indirectly. An inevitably incomplete list includes Helen Pritchard, Evelyn Ruppert, Mike Michael, Bill Gaver and the Interaction Research Studio, Mariam Motamedi Fraser, Kat Jungnickel, Anja Kanngieser, Bev Skeggs, Roger Burrows, Tomoko Tamari, Mike Featherstone, Tahani Nadim, Bianca Elzenbaumer, Barbara Neves Alves, Matthew Fuller, Luciana Parisi, Susan Schuppli, Sarah Kember, Nina Wakeford, Rebecca Coleman, Jane Prophet, Lynn Turner, Simon O'Sullivan, and Ele Carpenter, along with students in the MA Design and Environment course who helped me test out ideas and examine environmental-sensing practices at an early stage of this research.

Further afield, and in addition to those mentioned above, thanks are due to many colleagues, friends, and students for intellectual and creative exchanges, as well as logistical support. Again, a partial list includes Gay Hawkins, Adrian

Mackenzie, Lucy Suchman, Clare Waterton, Rebecca Ellis, Blanca Callen, Alex Taylor, Søren Pold, Christian Andersen, Anne-Sophie Witzke, Lea Schick, Winnie Soon, Geoff Cox, Martin Brynskov, Kristina Lindström, Äsa Ståhl, Ruth Catlow, Marc Garrett, Gillian Rose, Sophie Watson, Doina Petrescu, Nishat Awan, Kim Trogal, Rosi Braidotti, Nicole Starosielski, Janet Walker, Lisa Parks, Mel Y. Chen, Kim Fortun, Jennie Olofsson, Finn Arne Jørgensen, Christophe Lécuyer, Mark Andrejevic, Mark Burdon, Douglas Kahn, Jill Bennett, Ned Rossiter, Soenke Zehle, Sarah Barns, Etienne Turpin, Tomas Holderness, Cat Kramer, Zack Denfeld, Cesar Harada, Joel McKim, Scott Rodgers, Kathryn Yusoff, Brandon Labelle, Christian Nold, Muki Haklay, Dennis Quirin, Vivian Chang, Fernando Dominguez Rubio, Amy Zhang, Will Straw, Johanne Sloan, and John Trice. Special thanks are due to the two initially anonymous reviewers of this text, Steven Shaviro and Kevin McHugh, who provided insightful and perceptive feedback. At the University of Minnesota Press, Doug Armato, Erin Warholm, and Danielle Kasprzak astutely guided me through the publication of this text.

The early fieldwork for this research was made possible through seed funding from Goldsmiths, University of London (2007 and 2009–2010). The research leading to these results has also received funding from the European Research Council under the European Union's Seventh Framework Programme (FP/2007–2013) / ERC Grant Agreement n. 313347, "Citizen Sensing and Environmental Practice: Assessing Participatory Engagements with Environments through Sensor Technologies" (2013–2017).

Figure I.1. Scientists at the International Geophysical Year (IGY) 1957–1958 conference viewing *Sputnik* model. Photograph by Howard Sochurek, the *LIFE* Picture Collection. Courtesy of Getty Images.

Environment as Experiment in Sensing Technology

THE EARTH BECAME PROGRAMMABLE, Marshall McLuhan once wrote, the moment that *Sputnik* was launched.[1] Rocketed into orbit on October 4, 1957, and circling around the earth every ninety-six minutes, *Sputnik* was a technological intervention that turned planetary relationships inside out. Inevitably, what springs to mind with McLuhan's easy statement about the transformation of the earth and our relationship to it are the familiar images of *Earthrise* and the *Blue Marble,* which are often pointed to as simultaneously signaling the rise of environmentalism as well as the distancing of the planet through a disembodied space view. And yet, *Earthrise,* an image captured by *Apollo 8,* was not to appear for another eleven years, in December 1968, and the whole-earth view of the *Blue Marble* did not appear until 1973.[2] In contrast, *Sputnik 1* generated not photographic icons of whole or fragile earths, but rather produced a series of inexplicable beeps through a radio transponder, and relayed information about the likely conditions of Earth's upper atmosphere.[3] If *Sputnik* made the earth programmable, it was in part through radio transmissions that encircled the planet and created a live auditory map of a new orbital environment.

Although the twenty-three-inch *Sputnik* did not offer up a *view* of Earth from afar, it did activate a multitude of new experiences for inhabiting the earth. While it sent a signal of Cold War triumph (and even suspected propaganda) for the Soviet Union, in the United States the orbital machine regularly pacing through the skies portended catastrophe, where GDP and money markets as well as science education were all feared to be on the brink.[4] The continual revolutions of *Sputnik* around the earth, which spurred viewing sessions of its orbits in numerous cities, recast spaces of earthly sensibility and began to reshape environments.

Launched during the International Geophysical Year 1957–1958 (IGY), *Sputnik 1* was in many ways a proof-of-concept technology, which contributed to the development of a method for putting a satellite into orbit, while also testing the propagation of radio waves through the upper atmosphere and assessing the endurance of the satellite in space.[5] The testing of *Sputnik 1* further facilitated the development of *Sputnik 2* and *3,* which were launched in 1957 and 1958. *Sputnik 2* was sent into space complete with a dog, Laika, whose heartbeats could be heard through radio transmission. These later satellites were designed to be geophysical laboratories that collected data on the earth's magnetic field, radiation belt, and ionosphere. The remote sensing that the *Sputnik* triad undertook consisted of sending telemetry signals from space to Earth and of experimenting with the conditions necessary for developing a sensing laboratory that could eventually provide data about terrestrial ecologies through further satellite development.

Subsequent to *Sputnik,* satellites such as Landsat became key technologies for undertaking environmental monitoring, whether to detect change or to identify natural resources.[6] As Andrew Horowitz details in a 1973 issue of *Radical Software,* Eastman Kodak Company launched an advertisement in the *New York Times* that promoted the environmental benefits of satellite systems and detailed the endless possibilities for aerial monitoring to aid in the management of the environment, suggesting that this could not only reveal undiscovered dynamics within nature

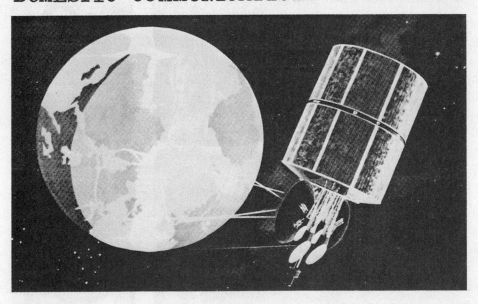

DOMESTIC COMMUNICATIONS SATELLITES

Figure I.2. Domestic communication satellites. Image from article by Andrew Horowitz, *Radical Software* 2, no. 5, chief editors Beryl Korot and Ira Schneider. Courtesy of Radical Software, copyright 1973 by the Raindance Foundation.

but also extend to identifying resources for extraction, and monitoring land use and living patterns.[7] Satellites were promoted as making an easy transition from military research and development to ecological and social applications.[8] Remote sensing developed into a critical technology and method within environmental science and became a crucial way in which to study environmental change on a global scale.[9]

Satellites now regularly monitor environmental change, tracking carbon dioxide in the atmosphere and patterns of deforestation. Satellites are referred to as "eyes in the sky" that communicate to ground stations while relaying data about and through environments, as they watch over earthly spaces and even transform the planet into a digital earth. Our understanding of environmental systems is now bound up with communication technologies that sense earthly processes. Satellites have played an important role in this development. And practices of monitoring environments have further developed from remote sensing to a more distributed array of sensing technologies.

I begin with this discussion of *Sputnik* and the programmability of the earth since this was a moment when a particular approach to sensing emerged that would inform monitoring and approaches to environments. However, what I attend to in this book is not a history of satellites or even Earth as understood from outer space. Instead, I develop an account of more recent developments in sensing technologies through distributed and networked environmental sensors within more earthly realms.

As it turns out, sensing has come down to earth since the time of *Sputnik*. Environments are now monitored not just by satellites but also increasingly by a wide range of sensors that track everything from air quality to traffic levels to microclimates and seismic activity. Such environmental monitoring is a practice that is computational, often networked, frequently automated, and increasingly ubiquitous. Many current scientific initiatives suggest that the monitoring of Earth processes remains one of the core areas of focus and development for the scientific understanding of environmental change. But sensors are also collecting data on any number of environmental processes that include managing cities and facilitating logistics, as well as providing and harvesting a range of data to and from smartphone users. The programmability of environments has expanded from the earth as enveloped in an orbital if experimental technology to a distributed and embedded range of monitoring technologies that inform how environments are sensed and managed. It is this explosion of environmental sensors and environmental sensing operations that I discuss in this book.

While there is much to debate in McLuhan's characterization of *Sputnik* and its relationship to "the natural world," I find the provocation of a planet that has become programmable a key point to take up in relation to the current proliferation of environmental sensors. In his characteristically sweeping essay on media environments, which ranges from the death of Queen Victoria to poetry and newspapers as "corporate poems," as well as Xerox as enabling everyone to become a publisher, and the immersive experiences of electronic "man," McLuhan suggests that *Sputnik* is yet another communications-based revolution that remakes people and environments.

What might he have meant by this rather elliptical discussion, written seventeen years after the launch of *Sputnik 1*? If we take him at his word, *Sputnik* seems to have given birth to a new planet and new environment. As he writes, "Perhaps the largest conceivable revolution in information occurred on October 17, 1957, [*sic*] when *Sputnik* created a new environment for the planet."[10] The usual way to read McLuhan and his sudden leaps of logic would indicate that such a statement runs the risk of technological determinism, and so it might. But I take up in a rather different way the provocation that this proto-remote-sensing device—and that our newer environmental sensing devices—are creating new environments and are programming Earth in distinct ways. I also depart from McLuhan in his understanding of the programmability of the planet, where he goes on to render *Sputnik* as yet another "extension of man," to consider instead how programmability might signal a quite different and distributed way of remaking environments. Programmability, the programming of Earth, yields processes for making new environments not necessarily as extensions of humans, but rather as new configurations or "techno-geographies" that concretize across technologies, people, practices, and nonhuman entities.[11]

Program Earth addresses the programmability of the planet by focusing on the *becoming environmental of computation*. I understand computation to include computationally enabled sensors that are distinct and yet shifting media formations that traverse hardware and software, silicon and glass, minerals and plastic, server farms and landfills, as well as the environments and entities that would be sensed. In other words, I am attending to the extended scope of computation that includes its environmental processes, materialities, and effects. Through discussing specific instances where sensors are deployed for environmental study, citizen engagement, and urban sustainability across three areas of environmental sensing, from wild sensing to pollution sensing and urban sensing, I ask how sensor technologies are generating distinct ways of programming and concretizing environments and environmental relations. I further consider how sensors inform our engagements with environmental processes and politics, and in what ways we might engage with the *"technicity"* of environmental sensors to consider the possibility for other types of relations with these technologies.[12] But before I unfold these concepts and explain how they are important for attending to the specific capacities of these machines, I first provide a bit more background on the growing sensorization of environments.

Figure I.3. Types of sensors. Sensors detect and measure stimuli through a wide range of inputs, including chemical, mechanical, and biological sources. The sensor assemblage typically involves using electronics and software to convert stimuli into electrical and digital signals. Screen capture.

GROUNDING SENSE

While satellites eventually became fully equipped with numerous sensor packages, sensors for use in environmental monitoring on the ground have also proliferated from initial military use to scientific study and commercial deployment. Nonnetworked and analog sensors have been in use in multiple applications for some time, and depending upon how one classifies sensors these could include cameras and microphones, not to mention sensors for use in applications such as radiation detection. For instance, in his work related to the Association for Computing Machinery (ACM) "Working Group on Socially Desirable Applications of Computers" and the "Citizen's Committee for Radiation Information," Edmund Berkeley proposed that radiation sensors could be put to work for political and environmental purposes to better understand radiation hazards and in aid of nuclear disarmament during the Cold War.[13] But the development of one of the first *sensor networks* has been traced to the air dropping of seismic and acoustic sensors by the U.S. military in Project Igloo, where sensors were used to detect movement along the Ho Chi Minh Trail in Vietnam.[14]

Beyond these early instances of sensors and sensor networks, however, the most usual reference for discussing the distributed and networked possibilities of sensors in the form of ubiquitous computing is Mark Weiser's 1991 text, "The Computer for the 21st-Century." Weiser makes the case for computing—and the sensors that would facilitate computational operations—to be distributed in and through environments. Identifying how computers were already present "in light switches, thermostats, stereos and ovens [that] help to activate the world,"[15] Weiser suggested these technologies might allow computing to "disappear" into the fabric of everyday life. Rather than the well-known trope of engagement that involves making the invisible visible, Weiser advocated for further invisibility, to develop computing not as a project principally of cognition and awareness, but rather as something that is integrated into environments and experience.

To this end, Weiser stressed that ubiquitous computing was not simply a project of populating far-flung places with computers. As he writes, "'Ubiquitous computing' in this context does not mean just computers that can be carried to the beach, jungle or airport." Such a strategy would still be focused on the self-contained box-like quality of computing, which would remain a discrete object demanding attention. Weiser emphasizes that ubiquitous computing is not "virtual reality, which attempts to make a world inside the computer." Rather than simulating worlds, he was interested to enhance the world already in existence by making computing an invisible force that runs through the background of everyday life.[16] And he imagined this would take place through networked and computationally enabled sensors.

A growing wave of interest in sensors and ubiquitous computing has occurred on either side of Weiser's proposal, from the 1984 launch of *Sensor Magazine,* to the proposal for technologies such as "smart dust" in 1998 (ambitiously microscaled sensors that were imagined to drift in clouds or swarms and monitor environments), to the coining of the term "Internet of Things" in 1999.[17] "Earth Donning an Electronic Skin," a 1999 article in *Business Week,* made predictions for the imminent encircling of the planet in electronic sensors that would measure and transmit data from millions of points:

> In the next century, planet earth will don an electronic skin. It will use the Internet as a scaffold to support and transmit its sensations. This skin is already being stitched together. It consists of millions of embedded electronic measuring devices: thermostats, pressure gauges, pollution detectors, cameras, microphones, glucose sensors, EKGs, electroencephalographs. These will probe and monitor cities and endangered species, the atmosphere, our ships, highways and fleets of trucks, our conversations, our bodies—even our dreams.[18]

Moving well beyond the singular object of *Sputnik* in space, this article presents a much different vision for a programmable earth, composed of the implementation of "trillions of such telemetric systems, each with a microprocessor brain and a radio,"[19] which would gather and transmit data on the ground by monitoring people, infrastructures, and events. No realm fell outside the reach of these sensor systems, where even dream activity could be surveyed.

A planetary brain, working through parallel and distributed computing, these electronic devices were envisioned to eventually form "a whole ecology, an information environment that's massively connected."[20] Imagined as a "huge digital creature," this ecosystem of electronic sensors, software, and communication networks was intended to be designed to "help human beings, not harm them."[21] With the planetary sensing fabric in place, scientists and technologists could then also turn their attention back to outer space, where this sensory network could spread to Mars and beyond.

While this vision for an electronic sensory network spanning the planet is now nearly two decades past, it continues to influence developments in environmental sensing and the Internet of Things. Today, sensors can be found in traffic infrastructure and ocean buoys, as well as in trees in forests and planted in soil underground. Sensors are used to manage urban traffic flows and to aid in the movement of freight, to signal flooding alerts, and to enable rapid responses in disaster situations. Sensors are located in environments, attached to infrastructure, fixed to vehicles, ported around as wearables, and embedded in smartphones, of which there are now one billion sold every year.[22] As an IBM video pitch for a "Smarter Planet" explains, the increasing instrumentation of the planet is meant to give rise to a "system of systems" that will facilitate heightened levels of observation, responsiveness, and efficiency.[23] "New insights, new activity, new forms of social relations" are meant to come together through an instrumented planet, which as "an information creation and transmission system" becomes newly intelligible. In the aspirations of the Smarter Planet vision, networked environmental sensors make it possible to listen in on a planet that has always been "talking to us," but which we can only now begin to hear.[24]

The drive to instrument the planet, to make the earth programmable not primarily from outer space but from within the contours of earthly space, has translated into a situation where there are now more "things" connected to the Internet than there are people. Some commentators suggest that the defining moment for implementing the Internet of Things was in 2008, when machinic connectivity to the Internet outnumbered human connectivity.[25] Sensing occurs across things and people, through environments and within infrastructures. People-to-people communication is becoming a smaller proportion of Internet and networked traffic in the complex array of machine-to-machine (M2M), machine-to-people (M2P), and people-to-people (P2P) circuits of communication. Cisco has projected, somewhat fantastically, that there could be fifty billion connected objects in circulation by 2020.[26] Many more objects than this could be eventually interconnected, since the IPv6 address system creates 10^{30} Internet addresses per person.[27] Intel ultimately envisions a future where sensors will be monitoring and reacting to us at every second, which would involve "altering reality as we know it."[28]

The basic diagrammatic flow of how sensors are meant to improve environmental understanding and responsiveness goes something like this: Distributed computational sensors monitor real-time events while collecting data on environmental conditions. Data on phenomena such as air quality and temperature, as well as location and speed of bodies and objects, are processed and trigger responses that may be human- or machine-based. These responses are often oriented toward making systems and processes more efficient or "balanced." The real-time "intelligence" provided by sensors is meant to translate into smart systems that continually enable corrective actions. The ambition is that environments and infrastructure can be managed intelligently and cohesively with networked sensor data. Preventative decisions can be taken. And major events such as floods can be instantly reported to ensure intelligent and immediate environmental management.[29]

Sensors are devices that typically translate chemical and mechanical stimuli such as light, temperature, gas concentration, speed, and vibration across analogue and digital sensors into electrical resistors and voltage signals. Voltage signals further trigger digital circuits to output a series of conversions into zeros and ones, which are processed to form readable measurements and data.[30] Data points are captured from a distributed multiplicity of sensors that are often measuring simple variables. By sensing environmental conditions as well as detecting changes in environmental patterns, sensors are generating stores of data that, through algorithmic parsing and processing, are meant to activate responses, whether automated or human-based, so that a more seamless, intelligent, efficient, and potentially profitable set of processes may unfold. Yet what are the implications for wiring up environments in these ways, and how does the sensor-actuator logic implicit in these technologies not only program environments but also program the sorts of citizens and collectives that might concretize through these processes? *Program Earth* takes up these questions and examines the distinct environments, exchanges, and individuals that take hold through these sensorized projects.

THE BECOMING ENVIRONMENTAL OF COMPUTATION

When Weiser made the case that computing should recede into the background, he signaled toward the ways in which environments would become the experiential envelope through which computing would unfold. Computation was to become environmental, or to become even more environmental than it already was. However, at the time of his writing he notes, "Silicon-based technology . . . is far from having become part of the environment," since although "more than

50 million personal computers have been sold," nevertheless, "the computer none-theless remains largely in a world of its own."[31] The environment that Weiser would have computing disappear into was a very particular type of milieu, one of inattention and everyday activity, an automated surround that did not require reflection or focus. But the ways in which computation becomes environmental are not necessarily always a project of disappearing as such, and Weiser articulated a distinct way of understanding what the environment is or would be as computing became more pervasive.

While Weiser suggested that ubiquitous computing would be a way to enhance reality, and Intel goes so far as to propose that sensor networks will create new versions of the real, I take up the work of writers such as Gilbert Simondon and Albert North Whitehead to consider how these technologies are involved in individuating and concrescing environments, entities, and relations. Simondon uses the term "individuate" to describe the processes whereby individuals and collectives take form as they concretize from a "preindividual reserve." For Simondon, not only are individuals not automatically given but also the process of becoming individual is always incomplete and continues to provoke new modes of becoming and individuation.[32] Whitehead uses the term "concresce" to capture ways in which actual entities and actual occasions are realized and joined up as distinct and immanent creatures.[33] In not dissimilar ways, these writers and philosophers are searching for and establishing a set of concepts that help us approach entities not as detached objects for our subjective sensing and contemplation, but rather as processes in and through which experience, environments, and subjects individuate, relate, and gain consistency.

"Environment" as a term has multiple resonances and genealogies. Within this space of examining ubiquitous computing and sensor networks, I consider specifically how environments inform the development of sensor technologies and how these technologies also contribute to new environmental conditions. Not only do computational technologies become environmental in distinct ways, the environments they populate are also in process. The *becoming environmental of computation* then signals that environments are not fixed backdrops for the implementation of sensor devices, but rather are involved in processes of becoming along with these technologies. Environment is not the ground or fundamental condition against which sensor technologies form, but rather develops with and through sensor technologies as they take hold and concresce in these contexts. Distinct environmental conditions settle and sediment along with these technologies as they gain a foothold.[34] These processes involve not just the creation of the entities and environments that are mutually informed but also the generation of the *relations* that join up entities and environments.

As much as computation becoming environmental, this discussion also attends to the ways in which environments become computational, or programmable.

Following Whitehead, this would be a way of saying that environments and entities concresce through processes of relating (as well as excluding) and as units of relatedness and modes of prehension that involve each other.[35] From this perspective, it is possible to see that Whitehead's notion of concrescence does not entail a simple adding together of preformed subjects and objects into an assemblage, but rather articulates the very processes by which entities are parsed, are able to conjoin (or not), and persist in environments. Relations, furthermore, do not precede the acts of relating and are specific to the entities and environments that concresce. Following Simondon and his notion of concretization, this would be another way of saying that how individuals and collectives are individuated gives rise not just to individuals and the environments in which they form but also the relations and potential—especially collective potential—expressed across those entities.

Far from being passive matter upon which human or nonhuman "sense" operates, environments in this way are an active part of how actual entities come to concresce and relate, how organisms endure, and how values—including those values implicit in technology—are expressed.[36] These distinctions and approaches are important since, in discussing the ways in which environments are sensed and monitored by sensor networks, I am bypassing an automatic understanding of sensors as merely detecting preformed environmental data as though there is a world of substantialist phenomena to be processed by a cognizing device. Instead, I consider how distinct environments and environmental relations emerge, take hold, and are programmed with and through these technologies.

Programming Environments, Programming Sense

Sensing is in fact a key part of the way in which computation works, as described in early diagrams outlining the basic components of a computing machine.[37] From input to logic, memory, control, and output, the five basic components of the von Neumann–influenced computer architecture depend upon sensing as part of the process by which computation works in the world. While the modes of input might consist of everything from keyboards to scanners to microphones, the point is that each of these "peripherals" is engaged in a transformation and conversion process. Sensing (broadly understood) in this arrangement has to do with all the ways in which computers input data into internal calculative processes in order to output data in another form. Sensors (as more specific input devices) emerged within this computational arrangement as just one of many possible devices for inputting data into the machine. With this system of input-output, it would seem that you simply need to get a bit of the environment into the machine, process that input, and output the results for onward action.

Environmental monitoring and sensing are inevitably situated within this computational diagram. On one level, environmental sensors are input devices that facilitate monitoring, measuring, and computing. Yet on another level, environmental sensors can be described as engaged in processes of individuating by creating resonances within a milieu, where individual units or variables of temperature and light levels, for instance, are also operationalizing environments in order to become computable. Simondon uses the term "in-forming" as a way to indicate exactly how information-related processes are also ways of giving form, above and beyond an epistemological project, since for Simondon in-forming involves registers of affect and experience as much as cognition and rationality. Sensing is then not just a process of generating information but also a way of in-forming experience.

The title of this introduction, "environment as *experiment in sensing* technology," speaks to the ways in which programming and programmability are approached in this book as experimental engagements in individuating—through sensing—environments. This is not an experimentalism that requires a control subject in order to understand results against a stable indicator. Rather, it is a more speculative way of asking what new entities and environments concresce through computational and distributed sensors. Programmability, in this way, is approached less as an ontology and more as an ontogenesis,[38] where processes of operationalizing environments put dynamic attributes into play rather than simply writing a script against which a workflow is executed.

In her discussion of the "regime of computation," Hayles suggests a salient characteristic of computation is that it is more than a practice of observing and simulating—it is also a process of *generating* new conditions.[39] While one could argue that multiple practices of scientific instrumentation are also generative, Hayles calls attention to the ways in which—within its own inherent logic—computation undertakes generative, rather than merely descriptive, engagements. Computing computes. It processes data to arrive at another point of synthesis. Programming is a way of making operative. In some ways, it attempts to enable processes of self-replication and automaticity. In other ways, it unfurls processes that are potentially open-ended and even speculative.[40] Throughout *Program Earth,* I address these varying ways of understanding programmability in relation to the becoming environmental of computation to consider how environments become programmable and are made to be operational through sensor technologies, as well as the ways in which they might open into speculative engagements and inhabitations.

"Programmability," as I employ the term, has a somewhat wider use than just software or code. Instead, this expanded engagement with programmability considers how code is not a discursive structure or rule that acts on things, but rather is an embodied and embedded set of operations that are articulated across devices, environments, practices, and imaginations.[41] Programmability then exceeds software (and even computation) to encompass the formation of events, spaces, and things. In this study, I open the concept and practice of programmability out into a question of how environments are generated and made operational through sensors and to the ways in which programmability often yields unpredictable (or unscripted) results.

The Multiple Milieus of Environmental Computation

In developing this analysis of the processual environments of sensor technologies, I work across discussions of environments, milieus, technologies, and sensing practices as found in Simondon, Whitehead, Isabelle Stengers, Michel Foucault, and Georges Canguilhem to consider the implications of how computation becomes environmental, and to what a/effects. I start from a point of understanding environments as made up of multiple milieus. "Milieu" is a term with a rich and long history within the history of science and technology, and as Canguilhem draws out in his arresting analysis of milieus, the term has moved from connoting a mechanical-fluid space, to something like the ether, a seemingly necessary binding agent or surround that would bring entities into communication even if not directly connected, an environment influencing genetic adaptation and evolution, and to the contrary, even an environment to which living entities are indifferent.[42]

Both Simondon and Foucault were students of Canguilhem's, and both use milieu as a way to variously describe spaces of transfer, influence, and environmental inhabitation. Foucault's use of milieu often signals the material-spatial conditions in and through which modes of governance may be experienced and lived.[43] Milieus in this respect have relevance for discussions of power and politics. Simondon used multiple terms in his discussion of milieu, including inner milieu, exterior milieu, and associated milieu. These concepts describe the processes whereby environments and entities are formed across individuals (inner) and environments (exterior) through energetic and material exchanges that occur through the transversal field of the associated milieu.[44]

I take up these discussions of milieus to consider how they become situated and multiple zones of transfer and inhabitation within environments. My use of the term "environment" is perhaps closest to Simondon's exterior milieu, which is one milieu of several that designate spaces in communication. I also draw connections across these discussions of milieu to engage with Whitehead's designation of environment as the processual condition and datum influencing the formation of feeling subjects. Environment and milieu are concepts that are threaded throughout Whitehead's and Simondon's approaches to the processual formation of subjects. In varying but not dissimilar ways, for both Whitehead and Simondon there is no such thing as a founding or original subject that cognizes discrete objects. Instead, subjects concresce together with environments to form subject-superjects, where everything—even a stone, as Whitehead would say—counts as an experiencing subject;[45] or where everything is individuated from a shared preindividual reserve, which includes a preformed collective of nonhumans both natural and technical, as Simondon has noted.[46]

In his rather distinct understanding of "mediation," Simondon develops a use of this term that addresses phases of being and becoming that occur through communication. As an example, he describes a plant communicating and mediating between the cosmic and the mineral, the sky and the ground, taking up and transforming energies and materials through its processes. The *associated milieu* operates as this mediatory space, a transversal ground through which transformations play out and new phases of being emerge. Mediation is not, however, a negotiation between two preformed units, but rather is a process in and through which entities transindividuate through communicative exchange. And it is not simply the entities that are individuated but also a milieu with which these entities interact. As Muriel Combes writes in relation to Simondon, "No individual would be able to exist without a milieu that is its complement, arising simultaneously from the operation of individuation: for this reason, the individual should be seen as but a partial result of the operation bringing it forth."[47]

I take up a parallel approach to how sensors harness energies and materials, transforming their own configurations and the environments they would tap into in the process.[48] Sensors are exchangers between earthly processes, modified electric cosmos, human and nonhuman individuals. The environmental computation that materializes here could be described as individual-milieu dyads that become as they communicate, subject-superjects that concresce as entities, and thereby enable particular environments to materialize and sediment. In this way, I am extending an understanding of communication-as-exchange to address the programmability of environments, the conversions across electronics and environments, and the material redistributions of environments and electronics through distinct phases and processes of individuation.

Planetary Computerization and Media Ecologies

The programmability of the earth and its environments as operation-spaces activates distinct ways of approaching the planet as a modifiable object. However, the earth of *Program Earth* is not a stable object undergoing a certain modification. Instead, one could say that, from *Sputnik* to the multiplicity of networked sensors that have since developed, sensing technologies are involved in parsing and making present certain entities and capacities that are bound up in the relational project of programmability. The "earth," as this discussion so far has suggested, is an entity that might be approached as both an antecedent object or datum as well as an entity in process and formed through modes of individuation and concrescence that enable this entity to stabilize and have consistency—as a unit of relatedness, concern, observation, and experience. The planetary then describes processes of individuation and concrescence that in-form the potential of this entity, Earth, to take hold and be experienced in particular ways. Programmability is one way of characterizing a particular process of individuation and concrescence that activates the planet and its entities as an operation space.

Earthly observations can be generative of distinct engagements and relationships. As discussed earlier, *Earthrise* is typically discussed as sparking international environmental initiatives through a "Spaceship Earth" photographic perspective and a counterculture ethos that was simultaneously a sort of neoliberalism in the making.[49] A planetary perspective can at once prove to be limiting and enabling for environmental practices.[50] It can also be the basis for an "infrastructural globalism," as Paul Edwards argues, that binds certain types of scientific practice together in the interest of understanding the planet as a discrete system.[51]

If the satellite view has largely been narrated as a project of making a global observation system and of seeing the earth as a whole object, then the more distributed monitoring performed by environmental sensors points to the ways in which the earth might be rendered not as one world, but as many. Here are multiple earths, in process, programmed and in operation, unfolding through distinct environmental conditions, sites of study, and responsive inhabitations. Where global observation systems might be working toward a planetary-scale project of knowing the earth as an entire system through (ideally) linked-up data sets,[52] in contrast multiple earths are articulated through numerous distributed sensors that as currently implemented rarely form a "system of systems," and more likely produce discrete and localized data sets for particular purposes. What "counts" as an environment—and Earth—then concresces in different ways in relation to the sensors sensing within distinct conditions.

The multiple "views" or "senses" that environmental sensors concretize might be approached through the machinic polyphony described by Félix Guattari in his discussion of "the age of planetary computerization."[53] At the time of his writing, Guattari suggested there was an emerging age characterized by a "polyphony of machine voices along with human voices, with databanks, artificial intelligence, etc." In addition, "New materials made to order by chemistry (plastic matters, new alloys, semi-conductors, etc.)" would take the place of previous materials. In this age, where time and experience were shifting, "the temporality put to work by microprocessors, enormous quantities of data and problems can be processed in miniscule periods of time." With the new machinic subjectivities that he anticipated would arise, as well as the "indefinite remodeling of living forms" that would occur through "biological engineering," he imagined there would be "a radical modification of the conditions of life on the planet, and as a consequence, all the ethological and imaginary references relating to it."[54]

Guattari captures a sense of how the earth and its inhabitants are remade through planetary computerization. In resonance with Simondon, Guattari identifies how the material, energetic, and machinic conditions that take hold and gain consistency become the basis both for remaking environments and remaking the human-machinic subjectivities that unfold in those environments.[55] In this sense, environments are not merely antecedent objects to be translated through informational devices, but rather are entities that concretize along with technologies. Computerization, in Guattari's view, becomes at once planetary and polyphonous, generating new living conditions, subjectivities, and imaginaries.

As the planet becomes a space of newly modified connections and relations, it also joins up and gives rise to new ecologies. McLuhan described *Sputnik* as a machine for generating ecologies.[56] The programmability that he identified as being key to this proto-remote-sensing technology was bound up with notions of what environments are and what it means to monitor and understand them. Contemporaneous with *Sputnik* and the rise of remote sensing, ecology shifted from an embedded field practice to an informational and even cybernetic undertaking, where the earth materialized as an object of management and programmability.

"Ecology" is a term that has multiple resonances and, as discussed throughout this study, also refers to informational or cybernetic management of environments as much as a philosophy of interconnectedness. Moreover, our post–World War II understanding of ecology is predominantly articulated through communication technologies, systems theories, and information science.[57] Donna Haraway has described how ecosystems, similar to immune systems or organisms, materialize through specific technoscientific practices that are in-formed by cybernetic logics. She therefore suggests a project of "probing the history and utility of the concept of the ecosystem."[58] Numerous texts—many of them referring to Haraway's early insights—engage with the question of how information theory and cybernetics have influenced the understanding and practice of ecology.[59] I draw on these informational and cybernetic approaches to ecology to consider how environmental processes and relations are not only increasingly studied through computational technology but also seen to be analogous to computational processes. Read through devices such as sensors and satellites, and assembled into networks and code, ecology is now a shifting entity that typically becomes visible—and manageable—as information. In this way, such ecologies in-form our lived material, political, and ethical engagements, and they contribute to the scope of our environmental practices.

Clearly, in developing these articulations of environment and ecology, I am also situating this work in relation to research on *media* ecologies. "Media ecologies" as a term and area of media research has expanded from its former associations, where ecologies and environments might have been used rather interchangeably to discuss the at-times deterministic effects that media spaces were assumed to have on subjects.[60] Newer work on media ecologies focuses on discussing the material-spatial conditions of media as part of an extended way of understanding what media are and the effects they have—encompassing but also extending beyond devices.[61] More recent approaches to media ecologies also draw extensively on Guattari's notion of the "three ecologies," which makes the case for approaching ecologies across mental, socio-cultural and environmental realms.[62] Within this space, some work on media ecologies goes so far as to even disavow the use of the term "environment" as a problematic term leaning toward unquestioned environmentalisms.[63]

An important point of clarification that is stressed throughout *Program Earth* is that a practice of attending to the milieus of media technology does not automatically translate into an environmental*ist* encounter with media. While these are often discussed in the same space in relation to sensors for environmental monitoring, I make a point of understanding "environment" as not always already environmentalist in order to consider the distinct ways in which environmentalist practices and politics concresce in and through computation technologies as they become environmental. Environmentalism might then be articulated as a response to having monitored environments, for example, in relation to declining habitat or increasing temperatures. Or it might provide the impetus to monitor in the first place, where sensors are tuned to looking for patterns of change or disturbance, and where data is seen as the necessary resource for motivating political action.

Furthermore, "environmental media" as a term often signals a media-based focus on environmentalist topics and environmentalist modes of representation, or alternately points to the "greenness of media." However, I discuss the becoming environmental of computation through the technoscientific processes that environmental sensors enable, rather than assume that this is automatically a project in sustainability.[64] Computational media unfolds not only through the capacities of devices but also through their environmental entanglements and effects, where material conditions such as soil and air together with circuits and screens generate concrete sensor-entities and experiences. With this focus, I am also building on my previous work that has attempted to draw out the environmental aspects of media by, on the one hand, attending to the atmospheric modalities and milieus of media and,[65] on the other hand, by considering the environmental effects of media in the form of electronic waste, which includes disposed gadgets as well as the extended spaces of mining, manufacturing, use, storage, recycling, and decay in and through which electronics circulate.[66] Computational technologies are constitutive of environments, have environmental effects, and also in-form environmentalist practices.

Program Earth then builds on research into media ecologies while making a distinction between environment as referring to *conditions* that form through multiple milieus, in the first instance, and to ecology as articulating the *connections* that take shape within a milieu and across environments, in the second instance, as a way to develop sufficient analytical clarity to be able to discuss both the connections and the conditions whereby the environmental media of sensors take hold. By making this distinction, I am also working in relation to the descriptions of ecology made within scientific literature, which this study draws on in considering how computational sensors are used to study environmental change and advance engagements in citizen sensing.

As mentioned above, the earth in *Program Earth* is not a whole or singular figure. Instead, the earth articulated here is multiple in the ways in which it is put to work, and in the ways it is drawn into experiences of environmental change, practices of environmental citizenship, and optimizations of urban systems. In this sense, I look at this multiplicity not to celebrate the more-than-singular ways in which earth-ness is animated, but instead to consider how a multiplication and accumulation of programming-earth projects shifts the ways in which the practices and effects of digital media unfold. And one of the primary ways in which I take up these environmental sensing practices is by examining the modes of

Figure I.4. One basic example of a "DIY" sensor in the form of Arduino open-source electronics with a carbon monoxide (CO) sensor that would typically be found in a smoke detector. Assembled at a citizen-science workshop in London. Photograph by author.

citizen sensing that are expressed in and through the use of sensors. Since many environmental sensor applications are oriented toward understanding environmental change or managing environments, so too do the ways in which environments come to be articulated through sensing technologies have relevance for the types of environmental politics and citizenship that take hold along with these technologies.

FROM ENVIRONMENTAL SENSING TO CITIZEN SENSING

A key tool within ubiquitous computing, sensors are the technologies that make possible the distribution of computational logics beyond the screen and interface to spatial and environmental applications. While sensors have become embedded in everyday spaces and infrastructures, practices of monitoring and sensing environments have also migrated to participatory applications such as citizen sensing, where users of smart phones and networked devices are able to engage with DIY modes of environmental observation and data collection. Beyond monitoring ecological processes, sensors have then become key apparatuses within citizen-sensing projects that monitor air quality, radiation concentrations, noise levels, and more.

Yet how did the ostensibly technoscientific technology of environmental sensors migrate from computational and scientific uses to more everyday applications? And how effective are practices of citizen sensing in monitoring and addressing environmental issues and in giving rise to new modes of environmental awareness and practice? *Program Earth* examines the migration of environmental sensors from ecological research and commercial applications to a wider array of environmental and "citizen" engagements. By analyzing informational ways of understanding environments, I map the trajectory of the computational and informational approach to environments from ecological sensing applications to more citizen-focused undertakings, and to urban and infrastructural developments that join the objectives of sustainability, intelligent cities, and engaged citizens. I further identify the material, political, and spatial relationships that environmental sensor practices enable; and I ask how a particular version of the "environmental citizen" has become entangled within these relationships and practices. The becoming environmental of computation includes processes of making citizens *and* milieus.

From citizen science to participatory sensing, crowdsourcing, civic science, street science, DIY media, and citizen sensing, a number of widespread practices of environmental monitoring and data gathering are emerging that variously work through ways of democratizing the technoscientific tools and understandings of environments.[67] While these terms are used in different ways to stress the scientific, big data, or civic aspects of these practices, I work with the term "citizen sensing" in order to draw explicit attention to the ways in which computational and mobile practices of environmental monitoring might be discussed as modes of citizen *sensing,* specifically.

Citizen-sensing practices have been described as making inventive contributions to both the research and development of technological tools, as well as to modes of environmental monitoring.[68] These practices range from the use of sensor data to complement other environmental observations, including remote-sensing; ubiquitous-computing approaches that often focus on the capacities and practices of sensor technologies to achieve efficiencies; and engagement with social or civic media projects that emphasize the ways in which social networking can mobilize collected data to influence policy and political action.[69] Citizen sensing as I am defining the practice for the purposes of this study encompasses or refers to those sensing activities that use computational sensing technologies in the form of smartphones, as well as mobile and low-cost electronic devices such as Arduino and Raspberry Pi, and online platforms to monitor and potentially act on environmental events through the collection of environmental data.[70] Such distribution of sensing capabilities across sensor networks and multiple mobile and individualized platforms has become a focused site for environmental and technological engagement.

Citizen-sensing projects are often closely related to citizen-science studies, but differ in the ways in which they seek to enable environmental practice through direct engagement with environmental monitoring technologies. Such citizen-sensing applications, similar to citizen-science, are frequently based on practices of individuals voluntarily tracking and monitoring everything from pollution levels to biodiversity counts.[71] Citizen-science projects are even increasingly transforming into citizen-sensing projects, where digital devices equipped with sensors are used to monitor environments and gather data.

In some cases, sensor technologies have enabled more thorough practices of environmental monitoring and observation that have already been underway through citizen-science initiatives, as in counting and tagging biological activities. In other cases, the capacities of sensor technologies have facilitated more distributed and potentially more accurate collection of data, such as urban air or noise pollution. Some applications extend the scope of citizen sensing not only to encompass sensor data and use of smart phones but also to draw on remote sensing and mapping to enable the tracking of deforestation or animal movements. In still other instances, these mobile sensor applications have sparked new forms of democratic organization and communication about environmental issues by effectively crowdsourcing environmental observations in order to influence environmental policy and action.

Another reason for engaging with these practices as *sensing* practices is then to draw out the ways in which computational devices are at once sensing and actuating technologies, as well as modes for sensing and experiencing environments.

Citizen- or participatory-sensing projects often propose to create "shorter circuits" between environmental information and the observers of that information, and in this way technologists and environmental practitioners have suggested that a more direct line of environmental action may be possible.[72] *Program Earth* specifically charts the ways in which citizen-sensing projects configure environmental practice through data gathering and sensing in order to offer a more in-depth understanding of how environmental practices and politics materialize in relation to observing technologies and communication networks.[73] I consider how environmental monitoring and citizen sensing consist not just of observations of environmental change but also of technical, political, and affective practices that are part of a complex ecology of sensing for environmental action.

What is typically activated in this diverse set of practices is a set of proposals for democratizing environmental engagement and developing other ways of doing environmental science and politics. Yet just as many new questions arise about the ways in which citizen engagement with environments and environmental concerns are in-formed with and through sensing technologies. By using the term "concern," I am here specifically drawing on Whitehead's discussion of concern as an "affective tone" drawn from objects and placed in the experiences of subjects.[74] The becoming environmental of computation includes these ways in which distinct monitoring practices and modes of reporting are enabled—and delimited— through environmental sensors, as well as the citizens and publics that would be activated and affected by these technologies and sensing practices.

Working across citizen-sensing projects that take the form of proposals, experiments, and established practices, *Program Earth* examines the ways in which the distributed and accessible capacities of computational sensors are meant to enable greater engagement with environmental issues. It asks: In what ways do computationally based citizen-sensing engagements influence modes of environmental participation? Citizen-sensing initiatives often depend upon forms of monitoring, reporting, managing, and even self-managing in order to establish environmental engagement. How might the practices of environmental citizens as data gatherers be advanced through a more intensive understanding of these modes of environmental and political practice?

SITUATING THE FIELD

From the Internet of Things to the "quantified self," there is a new set of terms circulating that engage with the ubiquitous aspects of digital media. Within this overarching area there are specific studies that focus on the imagined futures of ubiquitous computing, the distributed and spatial qualities of wireless or pervasive digital technologies,[75] and the ways in which sensor hardware and software move computation out of the black box and into the environment.[76] New texts are also emerging that provide an overview or wide-ranging survey of ubiquitous computing,[77] yet these collections often do not focus intensively on issues of environmental sensing and practice.[78] Other texts engage with the use of ubiquitous computing for social activism, for instance, but the focus on environmental topics is also less intensive.[79] These existing ubiquitous computing texts are useful in establishing context for this emerging area of computing, as well as participatory approaches to digital technology. However, I address sensors explicitly as *environmental sensor* technologies, a function that becomes more evident when devices are used for monitoring environments and collecting environmental data. And although the speculative aspects of computational sensors do influence this study, I especially focus on the ways in which sensors are actually being used and deployed.

Program Earth considers environmental sensing as a technological practice that spans environmental studies, digital culture and computation, the arts, and science and technology studies. As discussed above, the becoming environmental of computation includes considering not just how environments concresce along with individuals and objects but also how distributions of experience might be recast in and through environmental processes. Environmental sensing technologies open up new ways of approaching digital technology as material, processual, and more-than-human arrangements of experience and participation.

While there is comparatively less research within digital media studies that focuses specifically on the environmental articulations and capacities of sensors, there is a significant body of literature dealing with social media and the participatory aspects of digital devices, typically in the form of the mobile and online platforms.[80] Research into social and participatory media is a rapidly burgeoning field, where social media are often analyzed through considerations of alternative content generation, community formation, or social change,[81] as well as the politics and practices of observation, control, and tactical intervention.[82] This work forms an important reference point for understanding the rise in participatory engagements with digital media. However, *Program Earth* is situated somewhat obliquely to studies of participatory and social media, in that while it is focused on the political and participatory enablement of environmental sensing, it is primarily oriented toward more-than-human, environmental, and distributed analyses of how citizens and citizen-based engagements are expressed through this distinct set of technologies. At the same time, this research focuses on the ways in which computing has not only moved beyond screens to environments but also given rise to new imaginaries for how to program environments for digital functionality and participation.

Rather than focusing primarily on individual use or content generation for human-led manipulation of Internet- or screen-based media, I consider how environmental sensors variously articulate practices constitutive of citizenship in and through sensed environments that come into formation through an extended array of technologies and practices. Participation, as I engage with the concept and practice, is also a more-than-human undertaking.[83] I investigate how machines, organisms, energy, networks, code, and atmospheres in-form how distributed and environmental computing materializes and operates. Taking up the more-than-human, machine-to-machine, and algorithmic operations of wireless sensor networks, *Program Earth* addresses the proliferation of environmental and computational entities that concretize and participate in wireless sensor networks.

While this research synthesizes and draws on emerging media theories that deal with ubiquitous and participatory computing applications, it also seeks to develop a new terrain for thinking through distributed sensing technologies as articulating distinct modalities of environmental politics and practices. This book makes the case for different approaches to "sensing" within digital media studies, arguing that distributions and processes of sensation might be more effectively understood by not simply collapsing sensation into fixed sensing categories such as sight or hearing. Environmental sensing technologies entail a transformation of the "objects" that are turned into information; to produce information is a technological intervention that generates distinct types of realities, rather than simply mirroring them. With these insights in mind, it is possible to move beyond the notion that environments are something "out there" to be studied and acted upon by citizen sensors with their sensing devices and instead to look specifically at how the spread of informational techniques co-constitute monitored environments and informed environmental citizens. I draw on the work of Stengers specifically as she discusses the philosophy of Whitehead to develop a constructivist approach to environmental sensors to suggest not that environments are "constructed" (in the sense of being concocted) through sensing technologies, but rather that distinct capacities for *feeling the real* are articulated through these monitoring practices.[84]

It is important to note that in focusing on environmental sensing, *Program Earth* is working in a register that is not a phenomenological treatment of sense and sensation. In existing literature, sensemaking aspects of media technologies are often discussed through theories of mediation or individual attention and embodiment.[85] In related approaches, "sensing" is focused on a human subject and often rendered through theories of phenomenological or prosthetic engagement. The difference in this approach pertains to how environmental sensors are not simply providing access to new registers of information for established subjects but are changing the *subjects of experience* as well as the sensing relationships in which subjects are entangled and through which they act. Hence, vis-à-vis Whitehead's notion of the superject, we could say that the *superjects of experience* are also changing. *Program Earth* works to develop new theories of sensing that do not rely on an a priori human-centric subject or mediated subject–object relationship. Sensing here is not primarily or exclusively about human modalities of sensing, but rather has to do with distributed formations and conjunctions of experience across human and nonhuman sensing subjects, in and through environments. Sensing, in this respect, is understood as a multifaceted process of participating, individuating, and concrescing.

Methods and Chapters

Program Earth examines the monitoring and sensing of environments to question how sensing technologies give rise not just to new modes of environmental data gathering but also to new configurations of citizen engagement, environmental relationality, sensing, and action. Along the way, this work raises questions about the politics and practices of sensing that concresce at the intersection of sensor technologies, citizen participation, and environmental change.

Methods used in developing this material include fieldwork at sites of environmental sensing and testing, interviews with scientists and creative practitioners who have developed environmental sensing applications and devices, residencies and fieldwork at scientific field stations and sensing laboratories, ethnography at creative and scientific conferences and events where sensors for environmental monitoring have been exhibited or under active development (e.g., urban prototyping festivals), a visit to a sensor factory, attending and developing events for using sensing equipment to monitor environments, inventories of sensors and tests with sensor toolkits, virtual ethnography of online sensing communities, and an extensive review of environmental sensing literature, media, and practices.

This book works across this research material while developing a theoretical account of how sense, environmental participation, and politics shift through ubiquitous computing and environmental sensing technologies. Working within a radical empiricism modality, I do not "apply" theory to empirical material, but rather attend to the emergent intersections across theory and practice in order to create openings to inventive encounters with environmental sensing, as well as to enable propositions for practice.

In order to undertake this study of environmental sensing and its migration to more participatory applications, I have divided this text into three main sections that address key aspects of environmental sensing, including "Wild Sensing," "Pollution Sensing," and "Urban Sensing." The first section, "Wild Sensing," discusses the development of sensor networks within ecological applications to track flora and fauna activity and habitats and the ways in which these technologies have moved "out of the woods" to be deployed in more urban and citizen-focused applications. Within this section, chapter 1 focuses on fieldwork conducted at one sensor test site, the James Reserve in California. This chapter suggests that these experimental environmental sensor arrangements mobilize distinct sensing practices that are generative of new environmental abstractions and entities, which further influence practices such as citizen sensing. In chapter 2, I discuss two webcams, a Moss Cam and a Spillcam, to consider how images now operate as sensor data, even more than as stand-alone "pictures." Often located as one mode of input within sensor networks, images are generated from webcams and translated into data that can be parsed through image analytics while also drawing citizens into distinct practices of watching and reporting. Chapter 3 examines how the movement of organisms has become a key site of study facilitated through sensors. Migration-tracking sensors provide new data about the movement of organisms while also indicating the distinct environments and environmental relations in and through which organisms are living.

The second section, "Pollution Sensing," addresses the use and adaptation of environmental sensing technologies to monitor pollution, specifically focusing on the use of sensors as tools within creative practice and citizen-sensing projects. This is an area of sensor development that continues to trigger new proposals for citizen engagement in environmental issues. Yet how do these applications influence the becoming environmental of computation, as well as the concrescence of distinct environmental practices and politics? Beginning this section with chapter 4, I discuss fieldwork and observations gathered from my time spent at the Kilpisjärvi Biological Station while participating in an art-science residency. This field station is located within observation networks focused on studying the Arctic and environmental changes that are primarily influenced by climate change. I relate this site to a discussion of carbon and other environmental monitoring projects underway to consider how climate change is sensed and expressed across arts, sciences, and community monitoring practices. Chapter 5 shifts from carbon sensing to garbage sensing, more specifically in the form of plastic debris in the oceans in the form of "garbage patches." In this chapter, I look at the different ways that forms of marine debris that are relatively amorphous and invisible are brought to sense (or not) across Google Earth platforms, iPhone apps, and drifting ocean sensor floats. Moving from carbon sensing to ocean debris sensing, chapter 6 considers the numerous projects and sensors engaged with sensing air pollution. I develop a discussion of how air pollution is distributed across an array of devices and environments and how the data that are generated through pollution monitoring technologies and practices operate as distinct "creatures" of sense.

The third section, "Urban Sensing," looks at the ways in which citizen-sensing applications have become central to the development of the latest wave of smart city proposals that focus on urban sustainability. Smart cities proposals are developing apace, from IBM to HP and Cisco, and new projects are spun out that address not just the development of new intelligent infrastructures but also the compatible inhabitations of smart and connected citizens. How do these imaginings and deployments of environmental sensing technologies across infrastructures and citizens influence urban environmental politics? And in what ways do new versions of digital technopolitics take hold that potentially limit democratic urban ways of life? Chapter 7 takes up these questions by focusing specifically on governance and power as they are distributed through the new milieus and environments of smart cities. Using the notion of "environmentality" to describe these spatial power dynamics, the chapter asks how visions for an efficient and sustainable city might restrict urban practices and modes of citizenship. Chapter 8 considers the versions of participation that are enabled within DIY digital urbanism projects and platforms and explores the ways in which the "idiot" operates as a noncompliant digital operator and figure that does not participate as intended. Chapter 9 takes up the speculative aspects of smart cities and digital infrastructures as they are built to ask how sensor networks enable distinct types of withness in these new urban environments. I conclude *Program Earth* with a reconsideration of how planetary computerization might point toward expanded ways of engaging with sensor networks as generative of experimental worlds and speculative practices.

Each of the chapters within these sections deals with distinct deployments of and participations within sensing systems while engaging with these technologies as actual concrescences of computation and distributed arrangements of environmental sensing, practice, and politics. Across the chapters, there are multiple grounded instances of sensor technologies used in environmental projects that organize monitoring, facilitate participation, and manage urban processes. In each of these examples, the political implications of how sensing systems inform environmental practice and participation, as well as become enrolled in distinct ways of life, are articulated and addressed. While I have emphasized the becoming environmental of computation, the aspects of citizen sensing as they are expressed within environmental sensing projects are no less important, as they are intimately connected to the ways in which environments and environmentalisms materialize. Moreover, citizen engagement is a recurring lure and organizing device for enfolding people into sensing projects.

Ultimately, this book sets out to advance the conceptual understandings of environmental sensing—and the possibilities of sense—through a theoretical and empirical engagement with technologies and practices of sensing, citizenship, and environmental change. *Program Earth* explores the assertion made by a diverse range of researchers and practitioners that distinct practices of observation connect up with and enable distinct political possibilities.[86] It asks: How do different sensing practices operationalize distinct affective and political capacities? And what are the ways in which these computational sensors *become environmental,* as they take hold and create new feelings for the real?

I

Wild Sensing

Figure 1.1. Wired woods at James Reserve. A diagram showing the sensor ecosystem developed and tested through the Center for Embedded Networked Sensing (CENS) research project. Illustration Copyright 2005 by Frank Ippolito

1

Sensing an Experimental Forest

Processing Environments and Distributing Relations

Surrounded by the San Bernardino National Forest and situated within the San Jacinto Mountain Range in California, there is one particular patch of woods that is distinct in its ecological processes. This forest is equipped with embedded network sensing that digitally detects and processes environmental phenomena, from microclimates to light patterns, moisture levels and CO_2 respiration in soils, as well as the phenology, or seasonal timings, of bluebirds and auditory signatures of woodpeckers. These multiple modes of experimental forest observation are part of a test site for studying sensors in situ. A "remote sensing lab," the University of California James Reserve is an ecological study area that has hosted field experiments since 1966. The use of this ecological study area to test electronic sensors developed through the Center for Embedded Networked Sensing (CENS) research project is at once a continuation of experimental ecological practices in this area, as well as a shift in the technologies and practices for studying environmental processes. The question that arises here is: When the ecological experiment changes, how do experiences also change?

This notion of experimenting and experiencing as springing from a shared modality is put forward by Stengers in her discussion of Whitehead, where she uses "a (French-inspired) neologism" that does not draw a "clear distinction between the terms 'experience' and 'experiment' as there is in English." This merging of terms is also a critical way for describing the speculative approach of Whitehead, which might be characterized by a crossing-over of experience and experiment, where experimenter and experiment are part of a unified and concrete occasion.[1] This point of entry is important for this discussion, as it immediately points toward a consideration of sensors not as instruments sensing something "out there" but rather as devices for making present and interpretable distinct types of ecological processes. These processes are articulated computationally,

Figure 1.2. Ecological study area, James Reserve. Sign indicating the edge of the study area. Photograph by author.

and they draw together a wide range of experiencing entities that begin to inform new arrangements of environmental sensing. The becoming environmental of computation extends to the experiencing entities that sense and express ecological processes.

The use of wireless sensor networks to study environmental phenomena is an increasingly prevalent practice. Sensing projects encompass studies of seismic activity, the health of forests, maps of contaminant flow, and the tracking of organisms from dragonflies and turtles to seals and elephants. These projects generate sensor data that are meant to provide greater insights into environmental processes. At a time when ubiquitous computing is extending to multiple aspects of everyday life, where the Internet of Things promises to have your refrigerator communicating with supermarkets, and smart city designs propose to harvest your location data to ensure your roast-chicken dinner is prepared on time, sensing environments for ecological study is just one set of practices within a larger project of programming environments through distributed modes of computation. Sensor networks arranged over static and mobile platforms and widely distributed throughout environments are the common thread throughout these projects, but the deployment of sensors within ecological study sites has been one of the key and ongoing areas for early sensor research and development.

As discussed in the introduction to this study, although a range of research has been conducted on ubiquitous computing and the Internet of Things,[2] less has been written in the context of digital media theory or science and technology studies about the ways in which understandings and practices of environmental science have shifted through sensor systems and how these shifts have also had ongoing effects on more "participatory" sensor projects. While sensors and sensor systems were initially developed for use in military contexts, wireless and embedded sensor systems have further developed through ecological study, which has in turn provided an additional basis for deploying sensor systems within social media and citizen-sensing contexts.[3] This chapter focuses on the use of sensors for study in environmental science in order to consider how these science-based sensing practices might influence practices in expanded areas such as citizen sensing.

Situated within the context of these ubiquitous computing developments, this chapter specifically focuses on the distinct forms of sensing that concresce in relation to the monitoring of environmental phenomena. One key advantage that sensor systems are meant to provide is the ability to understand the complex interactions and relations within ecosystems in greater detail. Ecological relations are meant to appear in higher resolution because sensors monitor and make available aspects of environmental processes as they unfold over time rather than as more discrete moments—and because more data are available for generating models of complex interactions. This study asks how the ecologies that materialize through more continual sensor observation are not simply the result of increased data output and processing, but might also be understood as generative sensory relations articulated across humans, more-than-humans, environments, and devices. In what ways do distributed sensor technologies contribute to new sensory processes by shifting the relations, entities, occasions, and interpretive registers of sensing? How do the interpretative practices that are individuated in experimental environmental sensing test sites in-form attention to environmental problems? And what are the implications of these experimental environmental sensing arrangements as they migrate into policy and influence participatory sensing processes?

In order to consider these questions, I first give an overview of the increasing use of sensors for monitoring environments and studying environmental change. The generation of more and higher-quality data is seen as critical to developing more advanced insights into how environments are transforming, and so the sense data produced through these projects are often gathered for the purposes of advising science and policy, in addition to testing prototype computational technologies in the field. Environmental monitoring can bring with it a sense of increased responsibility; and the commonly used phrase, "all eyes on earth," is a way of articulating the watchful concern that sensors embody and operationalize through the continual observation of environmental processes.

But sensors connect up more than just a network of human- or sensor-based eyes. This chapter draws on more-than-human theory to move beyond human-centric interpretations of computational sensing technology and engages with Whitehead's approach to experience as something that concresces across human and more-than-human subjects. As Whitehead suggests, perceiving subjects are neither exclusively human nor pregiven, but combine as feeling entities through actual occasions.[4] In this way, sensors might also be understood not as detecting substantialist external phenomena but as contributing to inventive processes for making interpretive acts of sensation possible—and for articulating environmental change and matters of concern. This is a way of saying that interpretation matters, and that experience to be interpreted concresces across multiple registers and entities. In addition, interpretation is integral to processes whereby things come to take hold as objects of relevance.[5]

Based on a consideration of the distinct articulations of sense across more-than-human and environmental processes, this chapter moves to focus specifically on the use of embedded networked sensors at the James Reserve ecological study site. Drawing on fieldwork carried out at this site where sensors were tested in situ, as well as a review of scientific papers and online records of sensor data, I discuss new formations of distributed sense that concresce through these experimental forms of environmental sensing. Part of the way in which sensors might be understood as operative within distinct registers of experience is as distributed computational technologies. Sensors are distributed in at least two ways: in terms of their spatial distribution, by monitoring environments in a widespread and localized way; and in terms of the distributions of experience that generate sense data and interpretations.[6] If we take seriously Whitehead's suggestion that sensing entities concresce through experiences (or prehensions) and that they are inseparable from occasions of experience, then how do experimental environmental sensor arrangements mobilize distinct sensing practices that are creative of new environmental abstractions and entities?[7]

As Whitehead suggests, abstractions are not separate from concrete things, but rather influence "the process of concrescence" and provide a "lure for feeling."[8] The concrescences that come together here might be understood not just as scientists-devices-flora-and-fauna but also as relations that individuate and are individuated through data sets and algorithmic processes, across sedimented environmental effects, and through responsive modes of environmental action. The coming together of an experiment presents the possibility for distinct experiences and subjects to concresce. Sensing an experimental forest is not about detecting information "out there" but about "tuning" the subjects and conditions of experience to new registers of becoming. Tuning is a way to describe the co-creation and individuation of agencies within experiments and the complex process of developing facts or matters of concern within such experiments.[9] This chapter sets out to provide an understanding of the dynamic, distributed, and multiple modes of computational sensing environments that might also provide insights for more "cosmopolitical" participation, where sensing is a process of multidirectional tuning and experiencing.[10] The becoming environmental of sensor-based media is then distributed to include multiple subjects, organisms and technologies, as they process their environments.

INSTRUMENTING THE EARTH

The use of instrumentation within ecological study, from bird ringing to anemometers, has a longer history than the more recent use of networked sensor systems.[11] However, the miniaturization and faster processing speeds of sensors have contributed to their increasing use as instruments within ecological study.[12] Sensor systems—composed of relatively small-scale in situ sensors and actuators that are able to collect and transmit data through networked connections, as well as undergo remote reprogramming—have been described as nothing less than another "revolution" comparable to the rise of the Internet.[13] These imagined and actual transformations involve extending computational capacities to environments through

Figure 1.3. Monitoring station with Bird Box Cam at James Reserve. This CENS monitoring station included weather observation as part of its sensing kit and was a test-bed that contributed data to the U.S. National Ecological Observatory Network (NEON). Photograph by author.

sensors, where objects and phenomena are transformed into sensor data and made manageable through those same computational architectures.

In related literature, sensor networks have also been described as a revolution in scientific instrumentation, similar to the telescope and microscope, where a new order of insights might be realized. But instead of probing outer or inner space, sensor networks operate as "macroscopes," which enable a new way "to perceive complex interactions" through the high density and resolution of temporal and spatial monitoring data.[14] While issues related to providing a reliable power source, ensuring the robustness of hardware, and maintaining the validity and manageability of large data sets remain, sensor systems present the possibility for understanding environmental processes and relations more thoroughly by providing real-time data that are more detailed than existing modes of data collection, including remotely sensed and manually gathered data that may exist at a much larger scale or more discrete moments in time. The hope is that a background of new and undiscovered relations may be connected up and made evident through these sensing devices.

A wide range and number of projects now employ sensors for environmental monitoring, from bird migration and nesting to the social life of badgers, to water quality monitoring, phenological observations, the acoustic sampling of volcanic eruptions, and the monitoring of microclimates in redwood forests.[15] One of the key projects within sensor-systems development—a 2003 study of Leach's Storm Petrels at Great Duck Island, a wildlife preserve in Maine—established that "habitat and environmental monitoring is a driving application for wireless sensor networks."[16] This sensor project employed static sensor nodes and patches, with "burrow motes" and "weather motes"—or sensor nodes—to study the underground nesting patterns of migrating birds.

As with many similar and subsequent sensor deployments, this project produced more detailed data on previously unobserved ecological phenomena and relationships while also providing a test-bed for experimenting with the system architecture of sensor networks. The ecological relationships observed—or sensed—are in many ways coupled with the capacities of sensor networks, which similarly are adapted to and "learn" from the processes under study. The "tuning" of sensor networks may take place not just between scientists and devices but also across devices, code, and ecological processes. In this way, sensors become environmental by tuning in and developing along with the phenomena and organisms under study.

At the same time that sensor observations are intended to provide more detailed accounts of environmental phenomena on the ground, they also contribute to the building up of multiscalar and widely distributed approaches to environmental sensing, including remote sensing by satellites and airborne observations. These data are often generated across scales and derived from diverse modes of sensor input for wider and more detailed views on environmental processes and to study the effects and possible impacts of environmental change.[17] Multiple "observatories," together with long-term ecological research sites (LTERs), and the U.S. National Ecological Observatory Network (NEON), attempt to collect and synthesize sensor data across the United States. While a site-specific sensor project may study the detailed relationship between birds' nesting behavior in relation to microclimate and multiple other environmental factors, this same study may benefit from climate data resources or may contribute to climate monitoring programs. In other words, the sense data gathered may have the potential to elucidate environmental relations within a particular area of study, as well as across expanded and yet-to-be-gathered data sets—as long as the data to be compared are of compatible formats.

Just as sensing systems are proliferating, numerous attempts are underway to amalgamate and make sense of the many forms of data—a key "cyberinfrastructure" task—since the multiple formats and provenances of data may mean that they are rendered meaningless for ongoing use and study if not consistently handled.[18] Sensor-gathered data sets, which are typically "heterogeneous," are increasingly brought together not only in larger data networks but also in mapping platforms where fine-grain sensor data provide a real-time "ground-truth" to coarser remote-sensing and field-gathered data. From Microsoft's SenseWeb to the former DIY-sensing platform Cosm, such platforms intend to consolidate environmental sensor inputs.[19] The range of possible sensor inputs is illustrated by one Microsoft diagram, "Instrumenting the Earth," which outlines twenty different modes of sensor input, from snow hydrology and avalanche probes to citizen-supplied observations and weather stations.[20] Innumerable potential points and processes in the environment become the basis for sensor input, and it is from these delineated sites of input that newly observed relations might be studied, articulated, or managed.

While these sensing projects and networks have been under development within universities and public institutions, technology companies working individually or often in collaboration with universities are also developing multiple sensor network systems for environmental observation. These projects range from Nokia's "Sensor Planet" to IBM's "A Smarter Planet," HP Labs' "Central Nervous System for the Earth" (CeNSE), and Cisco's "Planetary Skin" (in collaboration with NASA, the University of Minnesota, Imperial College, and others).[21] Governments and their militaries are also investing in the development of sensor networks, with white papers and research issuing from the EU, China, and the United States DARPA, among others.[22] Many of these sensing projects raise ethical issues related to surveillance, while still other projects are enabling new forms of resource exploitation. The project of monitoring and managing environmental relationships continues to be a way in which the governmentality—and even environmentality—of sensor systems unfolds, where sensor capacities may point toward particular relations to manage or sustain in distinct ways.[23]

All together, these environmental sensing systems variously undertake a project of instrumenting or programming the earth.[24] Within a sensor-ecology imaginary, the planet might be understood as an entity to be sensed and transformed into data. Improved sensing capabilities are critical to advancing understandings of environmental change while also indicating ways of acting (whether through automated systems or environmental policy) in response to that data. With small-scale, distributed, and pervasive computation embedded in environments, new relationships emerge not just to studying but also to managing environments, since sensor systems computationally describe and capture environmental processes while also providing the promise to "design and control these complex systems."[25]

In many ways, the notion here is that increased amounts of environmental data allow for the improved management of environments. Data are descriptive indicators capturing environmental processes. But from a Whitehead-influenced perspective, it could be argued that sense data are less descriptive simply of pre-existing conditions and more productive of new environments, entities, and occasions of sense that come to stabilize as environmental conditions of concern. The

Figure 1.4. Nesting box with interior electronics at James Reserve. This prototype bird box with camera captured and logged still images of bird activity every fifteen minutes, twenty-four hours per day. Photograph by author.

ways in which phenomena are tuned into as sense data are one part of this opera-tion of the becoming environmental of computational sensors; but the ways in which sensory monitoring gives rise to new formations of sense within and through data, computational networks, humans, more-than-humans, and envi-ronments also in-form distinct sensing practices. Since sensor networks offer dis-tinct insights into the complex interactions and processes within environments, then the ways in which these relationships are joined up, articulated, and trans-formed into new observational capacities also matters.[26]

SENSING AN EXPERIMENTAL FOREST

Turning now to a more detailed discussion of one embedded sensor network project, the CENS sensor installations at the James Reserve forest, I consider how the rise of distributed sensing might be looked at more closely in the context of this experimental project and test site. The CENS initiative is one of many sen-sor developments as discussed previously, and it is a well-known and frequently cited project for sensor research. Established in 2002 as a National Science Foun-dation Science and Technology Center, the CENS project was a collaboration between several California-based universities. The project, which finished in 2012, focused on four key areas of research: Terrestrial Ecology Observing Systems (TEOS), Contaminant Transport and Management, Aquatic Microbial Observing Systems, and Seismology. A fifth area of research, Participatory Sensing, grew out of the project research into ecology and focused on how sensor applications may be used for citizen engagement in environmental and social issues.[27] This discus-sion focuses on the TEOS sensing deployments, which were primarily situated at the James Reserve (while the other study areas were located in a diverse range of sites). Participatory Sensing is a further project research area that I briefly address in the conclusion to this chapter.

The James Reserve ecological study site is in many ways an environment for developing experimental practices as well as for transporting laboratory tech-niques into the "wild." The fieldwork that I conducted at the James Reserve also moved from the laboratory to the field, as I first visited the CENS laboratory at UCLA where most of the sensor prototypes were developed, and then observed the sensors at work in situ at James Reserve. I held informal interviews with researchers involved in the CENS project, mapped the different locations and functions of sensors in the field at James Reserve, and compared the online records of sense data with the sites where sensors were installed. However, this is not a project of "following the scientists," which is by now a well-established area within science and technology studies.[28] Instead, through a discussion of field-work conducted at the site, I attempt to understand processes and sites of sensing as they intersect with ecological practice and cultures of computation. Rather than focus exclusively on how ecologists use sensors to obtain scientific meaning or generate data or facts, I concentrate on James Reserve as a particular ecological research site that concresces through a distribution of sensing processes across organisms, ecological processes, and sensing technologies in the form of computational hardware and software, online interfaces, conservation infrastructures, resident scientists, environmental change, citizen scientists, publics, and visiting researchers. In other words, I attend to the becoming environmental of sensor-based media as a concrescence of these experiencing entities.

The nearly 12-hectare and 1,640-meter-high site is characterized by a complex intersection of ecosystems, "including montane mixed conifer and oak forest, montane chaparral, wet and dry meadows, montane riparian forest, a perennial stream, and an artificial lake."[29] Since James Reserve is located in a relatively remote wilderness setting, it is effectively "off the grid," and is a study area that generates its own solar power and has its own well for water. In this sensing lab or experimental forest, infrastructures are realigned, not as obvious allocations of roads, electricity, and water, but rather as new arrangements of energy, sensation, and observation.

Sensing in the James Reserve is distributed not only across this experimental site (and at distinct locations for the study of ecological processes) but also across larger sensor networks. Many of the CENS James Reserve sensors measure phenomena over time, which is meant to enable researchers to study sequences of data that are fine-grained and relatively continuous in comparison to more discrete data sets, with data captures taking place in localized settings as frequently as every fifteen minutes or more. Still other sensor test beds are in place that connect up to larger networks, including national observatories such as NEON. Observations are successively gathered and joined up in far-reaching networks, so that sense data becomes an amalgamated and comparative networked infrastructure of ecological observatories for studying environments and environmental change.

CENS sensor systems have been developed and deployed within a larger project that seeks to collect data in order to respond more effectively to environmental challenges. Higher-resolution data promise to create more effective models for predicting and managing environmental events. This "new mechanistic understanding of the environment" involves a near-future commitment to developing a "critical infrastructure resource for society" in the form of detailed environmental monitoring.[30] The promise to respond to crises more effectively develops not just through larger data sets but also through more extensive data gathering that is better tuned to detecting anomalies and extreme events, since most ecological data have largely consisted of documenting ecological conditions within a logic of averages and generalities.

However, data expressive of average conditions do not capture the effects that major if singular disruptive events have on environments and rapidly shifting ecological relations and processes.[31] CENS and related projects such as NEON are oriented toward the objectives of monitoring changing environmental processes, where an increasing number of disturbance events are contributing to the perceived need to develop different practices and technologies for sensing environments. The expression and agitation of environments (which, as Whitehead suggests, "seep" into all things) also turn up in and transform the sensing practices and technologies that monitor them. Instruments for capturing sense data are here specifically honed toward disturbance, since environmental change becomes more of a matter of concern within ecological study. At the same time, disturbance detection rather than observation of norms begins to influence what counts as relevant sense data.

Machine Ecologies

The sensors at work in the James Reserve within the TEOS group of research projects consisted of everything from soil sensors that detected moisture levels, a Rhizotron installation of tubes that allowed robotic cameras to capture images of root growth and CO_2 sensors at three different soil depths to estimate soil flux, a bird-audio system involving sonic booms triggered by camera activity to capture woodpecker auditory data, weather stations for gauging microclimatic conditions,

Figure 1.5. Sensors measuring the flow of tree sap throughout different seasons as part of the CENS sensor ecosystem at James Reserve. Photograph by author.

tree-sap flow sensor systems, nest boxes with cameras and audio installed within bird boxes, pan-tilt-zoom tower cameras on thirty-foot-tall poles, and a Moss Cam web camera. At the time of this fieldwork, there were over 550 connected and untethered sensor nodes, as well as reconfigurable robotic mobile sensors working above and below ground, within waterways and across tree canopies, capturing data on plants, animals, birds, soil, microclimate, and more.[32] Sensor observations provided the ability to observe fungal growth patterns, soil CO_2 production, the times at which plants shut down their CO_2 fixing, and all manner of activity that typically takes place outside the scope of direct human observation.[33]

The initial proposal for this project made a bid to develop "distributed sensor/ actuator networks [that] will enable continual spatially-dense observation (and ultimately, manipulation) of biological, environmental, and artificial systems."[34] Midway through the project, many of the initial proposals for comprehensively distributing a large number of small sensors within an area of study shifted to a practice of strategically deploying sensors in precise locations to study specific ecological activities and to develop a hierarchy of sensing platforms that could span from small-scale motes to larger sensors such as imaging robots on cables.[35] The sensor practices and arrangements developed in the James Reserve context were specific responses to site conditions and processes, so that phenomena to be observed in-formed which sensors would be used and how. At the same time, the difficulty of creating a pervasive sensor network led to a focus on specific sites of study as a more feasible test of the technology. This points to a key aspect of the sensor systems: they were almost always physically proximate to that which they monitored. Sensors were distributed in the environment, and networks were developed and paired with those environments.[36] Sensors in the field at James Reserve were wrapped around tree trunks in a loop of foil and cables; they were interspersed in the ground as arrays at regular intervals; and they were clustered at bird boxes to cross-correlate microclimate in relation to nesting at distinct locations.

The ways in which sensors were paired with environments was not a simple mirroring, however. Sensors proximate to roots and soil, for instance, did not stream all possible data all the time. Instead, sensor motes within a network talked to each other to coordinate data detected, processed, and sent according to distinct algorithms. Part of this configuration had to do with energy efficiency, where motes were triggered to record events only at select times and were turned off during times of inactivity to save energy. Indeed, a key aspect of imagining the possibilities of sensors as environmental systems involved thinking through how it may be possible to realize "pervasive sensing" without "pervasive infrastructure,"[37] which primarily meant not requiring a central electrical grid for power. The sensors at James Reserve were in part powered by a solar array that was the primary source of energy to power this elaborate sensing lab, which was supplemented by batteries, including motorcycle batteries, for distinct devices to transmit their sensory data via wireless and networked connections.

Part of the algorithmic processing of sensor data involved setting sensors to pick up, filter, and amalgamate data within established ranges. The processing that sensors undertook was ad hoc and in situ, rather than a continual capturing and streaming of environmental activity. Each mote within a network was already set to detect some things and not others, to make correlations among certain data criteria, and to discard anomalies and redundancies according to predetermined phenomenal ranges. Sensor motes detected events within a specific range, and processed and communicated this data across short distances or hops to other sensors within the network for collection at sensor nodes. Data were typically fused and processed at each individual mote in order to make real-time streaming more efficient and effective.

While sensors were physically proximate to what they sensed, that which was sensed and communicated traveled through channels of algorithmic detection and processing. While sensor applications are intended to record extreme events and anomalies, the algorithms that capture data have a tendency to smooth and fuse data at the source in order to conserve energy and generate manageable quantities of data, which even with these filtering mechanisms can easily run to several million records per year per sensor patch. These syntheses are intended to turn data into "high-level information," where the multitude of records and raw data transform into something like observations or experience.[38] This transformation required "data reduction" in the form of "in-network processing" that aggregated similar data and filtered redundant data.[39] As CENS researchers Jeremy Elson and Deborah Estrin write,

> For example, emerging designs allow users to task the network with a high-level query such as "notify me when a large region experiences a temperature over 100 degrees" or "report the location where the following bird call is heard."[40]

In this way, processes of filtering, aggregating, and selecting have already been put in place to turn sense data into relevant information. At the same time, these filters may not always capture intended phenomena. A researcher walking through the James Reserve forest might create noise that is picked up on sonic booms, which through algorithmic parsing activates cameras to record activity. In this field of environmental sensing, researchers might fall within the data event-space of motion detection, but inaudible birds traveling in a different column of air might not be detected.

Processes of producing data are also processes of making sense: the experiment is generative of modes of experience. These processes include how sensors are developed in the lab, tested in the field by technologists and scientists, merged with historic ecological study practices, and read across new data sets, while also producing distinct insights into ecological relationships by connecting up multiple experiencing subjects. The architectures and algorithmic processes for relating sense data are a critical part of how sensor systems operate. They articulate how sense data will come together into arrangements indicative of environmental and planetary processes.

Inevitably, the focus on gathering massive amounts of sense data raises issues related to data ontologies. Sensor networks provide the basis for monitoring and acting upon environments, and yet the data and connections made across sensors are selectively captured and joined up, and are also subject to failure and incompatibility of data.[41] Different data standards, classification techniques, and dispersed practices in-form the content and processing of dataspaces.[42] Databases and dataspaces are more than collections of objectively observable facts—they are embedded within and performed through infrastructures of science, governance, and public outreach. On the one hand, there are issues related to how an entity becomes data, as Wolff-Michael Roth and G. Michael Bowen have discussed in relation to the digitization of lizards.[43] On the other hand, there are questions about what constitutes data (a lizard may seem to be a clear artifact of digitization; but when its habits and habitat become part of the sensed data, where does the

Figure 1.6. An array of sensors at James Reserve measuring moisture and respiration of CO_2 throughout the soil. Photograph by author.

organism begin and the environment leave off?). Data ontologies in-form which data are collected, but they also in-form possibilities of sense by giving rise to new actual entities and occasions for articulating and experiencing relevant sense data.

System as Sensor and Proxy Sensing

In order to create a more effective parsing of environmental phenomena, sensors are not just used as individual devices that simply generate discrete sense data at the James Reserve. Instead, multiple sensors and sense criteria within a sensor network are often also brought together to form a composite picture of a distinct environment under study. Chemical analysis of pollution may provide readings on contaminant concentration levels, but additional sensors may also work out the direction and speed of contaminant travel, as well as the size of an affected area, by cross-correlating multiple sensor data. In this process of data fusion, the "system is the sensor."[44] Sensors working together within a network establish a computational pattern of correspondences, where the physical sighting, sensor type, coding, and correlating of data coalesce into an environment of sensor data that in-form observations. When the "system is the sensor" and the network operates as a sort of distributed instrument,[45] it might be possible to create models and forecasts of ecological processes and, through these sensor systems, act upon environments.

Sensor systems may also act as proxies for the environments they sense. Sensors as proxies are not standing in for a more-real version of environments, but rather are sensory operations that mobilize environments in distinct ways. Sensor networks perform—and so transform—environmental systems. Data may be correlated across sensor types, or sensors may trigger other sensors to capture phenomena, or trigger actuators to collect samples for later study.[46] Inferences can be made about phenomena through sensors and actuators, and sensors can be arranged through flexible, multiscalar platforms that investigate particular sensing relationships.

As a CENS "Distributed Sensing Systems" white paper notes, "Embedded sensing can involve a mix of observations with inherently different characteristics. For instance, it is common for systems to include multiple sensors, each with a different form of sensory perception or modality."[47] This is the case in James Reserve, where seemingly traditional image and audio technologies provide a new way to "sense" phenomena in the absence of direct biological sensors. While the majority of sensors now available are capable of detecting physical and chemical attributes, devices such as cameras become newly deployed as biological sensors in the absence of direct biological sensing capabilities, where physical and chemical sensors algorithmically set to filter for event detection can automatically trigger cameras to record biological events.[48] Imager and audio modes of sensing are activated within a computational network that mobilizes these forms of sensing as distinct and often proxy operations. The possibility to articulate relationships and interactions within environments to a higher fidelity is something that is meant to be generated through sensor applications that join up environments across sensor system hardware, software, databases, and cyberinfrastructures, as well as distinct sites and the more-than-human processes.

Proxy modes of sensing do not just extend to sensors triggering other sensors or actuators to perform sensing operations but also include proxies that become apparent vis-à-vis more-than-human processes. A not-uncommon technique within environmental study, where climate change in deep time may be studied through ice cores as proxies for past climate events, proxies within sensor-based environmental monitoring are mobilized to infer and detect traces of ecological processes. In the James Reserve, for instance, phenology is a central area of study. In order to capture seasonal relationships, organisms may be observed for the ways in which they "process" environments.

The perceptive capacities of Violet-Green Swallows and Western Bluebirds, in addition to Star Moss and other organisms, are placed under observation through webcams and Cyclops networked image sensors, which capture images and data related to these organisms often at least every fifteen minutes per day, if not more frequently.[49] The bird cams and Moss Cam, or web camera specifically monitoring the growth of Star Moss, generate a store of image data that can be compared to microlocal temperature and related data, as well as data captured throughout the James Reserve site. The birds' choice of a nesting location, or the failure to raise chicks due to absence of food or low temperatures, can be captured in this context where the birds' activities are made available as a sort of proxy sensor of phenological processes. Birds may provide key environmental sense data through computational networks that make sensible these registers of more-than-human experience. What is clear is that sensors do not just capture data, they shift the processes of sense across these multiple registers, so that more-than-human perceptive processes concresce in newly relevant arrangements.

Similarly, the Moss Cam generates images and daily records that contribute to a picture of seasonal patterns and "event effects." These effects might include lack of moisture in the summer, which contributes to mosses "burning through" their CO_2 reserves—in other words, higher temperatures can correlate to an increased release of CO_2 by mosses, as they consume stored energy and move toward states of dehydration and dormancy. Here, what counts as "sensing" is not a simple matter of observing mosses through a web camera over time, but instead involves observing how the moss is a sensor, or a biomonitor that is itself detecting and responding to changes in the environment.[50] The mosses' morphological changes to local conditions are an expression of an ecological relationship that is further entangled in the complex shifts of climate change. In this respect, the mosses may be expressing sensory responses to human-altered worlds, yet to understand more fully what those alterations involve, it is necessary to observe sensing organisms in order to register the effects of increasing carbon and temperatures. The delay and resonance within these environments is not as immediate as a typical sensory example might assume. Yet in this study, the ways in which sensing organisms "take account" of environments multiply, where the sensory input and means of detection are distributed and computational. The becoming environmental of computational media then further takes place through organisms and their processing of environments.

In a sensor-based study of phenology, sense operations are distributed and collaborative. In these forms of collaborative sense, sensors experience and provide proxy experiences across a sensing system that generates distinct occasions of sense. But these collaborative qualities of sense concresce not through researchers primarily but through the dynamic responses of organisms to environments and the sensors that collect data in relation to which algorithms query, filter, and record these changes. The more dynamic sensory modalities that concresce in this relationship are examples of inventive ecological experiences and subject-superjects, as discussed earlier. The timings at which plants leaf out, for instance, might even begin to disrupt and alter scientific models that expect seasonal timings to unfold at times established through prior empirical study. In these encounters and formations of sensory practice across organisms, ontologically prior categories of sense become more mutable and ontogenetic, where more-than-human modalities of sense indicate the shifting encounters of sense in which we are engaged. Sensor systems mobilize multilocated and multispecies processes of sensing, which in part enable the development of distinct capacities to sense change, where the scope of computational sensing and proxy sensing expands to include more-than-technological perceptual processes.

In an account of ubiquitous computing as distributed cognition, Hayles suggests that distributed computation could operate as machines for aiding, and so enhancing, human perception.[51] Here, however, computational devices are not augmenting human perception as such, and humans are not even the central perceptual processors toward which distributed sensation and computation might be directed. More-than-human proxy sensing points to the ways in which sensor technologies, instead of providing supersensing or cognizing capabilities to supplement human modalities, filter, connect up, and in-form environmental relations in distinct ways, and so change what modes of sense humans may even experience. New ecological arrangements of subjects—and superjects—concresce through these sensory processes.

Environmental monitoring through sensor networks is a practice of making—and not just capturing—environments as process. Sensor networks are tuned to distributions of relations. They tune into discrete sense criteria and amalgamate these across sensor networks and through proxy modes of sensing to make

Figure 1.7. Detailed view of soil moisture sensor at James Reserve. Photograph by author.

particular environmental relations more evident and sensible. Environmental monitoring through sensory networks mobilizes and concretizes environments in distinct ways by localizing computational processes of sensing within environments and across more-than-human experiences while also articulating those relations through algorithmic processes for parsing data. As these processes inevitably compose the possibility of sensing environments in particular ways, they also in-form which participants and participatory modes of sensing register in the perceptive processes of sensor technologies. Such sensing practices, moreover, are replete with political effects. Within the context of sensor networks, the sensory arrangements that are identified within data may become the basis for identifying and protecting matters of concern. Yet they might also overlook those "non-sensuous" background events that may still generate new sensing arrangements but which are not interpretable within present modes of sense data.[52]

DISTRIBUTING SENSE

As discussed in the introduction to this study, the initial developments of ubiquitous computing are often attributed to Mark Weiser's 1991 suggestion for computation to move from desktops to the environment, so that computational processes would become a more integrated and invisible part of everyday life.[53] Yet another possible reference point could be Alan Turing's 1948 ruminations on how to build "intelligent machinery" with sensing capacities on par with humans. Turing reviews the options for such a project, first considering how to atomize every part of the human sensing ensemble and replace it with equivalent machinery. Emulating human vision, speech, hearing, and mobility, such a contraption "would include television cameras, microphones, loudspeakers, wheels and 'handling servo-mechanisms' as well as some sort of 'electronic brain.'"[54] This project would inevitably be "of immense size," Turing notes, "even if the 'brain' part were stationary and controlled the body from a distance." But data would not enter the thinking machine through its remaining static, and so "in order that the machine should have a chance of finding things out for itself it should be allowed to roam the countryside." But in such a scenario "the danger to the ordinary citizen would be serious." Add to this the hazards of such a machine taking up all of the usual activities of human interest, and this contraption would be altogether unwieldy. Turing's more practical recommendation is to behead the body, to work with the brain as the critical site of processing, and later attend to the sensory apparatus as a secondary concern.[55]

Even if Turing's proposal does consolidate the "thinking machine" into a central and seemingly Cartesian apparatus, his thought experiment on the sensing body in pieces and distributed throughout the countryside remains a potent figure for ubiquitous computing. What is striking about Turing's example is the way in which the thinking machine, even when distributed, would emulate the human body, which serves as a template for understanding how sensory data would be captured and centrally computed. While computational sensing technology can now be understood as more than a double of or prosthesis for human sensing, Turing's figure of the body in pieces raises questions about how particular distributions of sense might reconfigure environments and processes of sensation.

Could such distributions of sense point toward modes of sensation where computation reassembles not as a singular sensing subject but rather as a processual and multilocated experience comprised of numerous sensing entities? How are sensing practices individuated, and how do they concresce, across potential sensor networks? In this way, sensing also assembles not as a mental or cognitive operation but as an environmental and relational articulation across multiple bodies and sites of sensing.[56] Within Turing's example of the sensing body in pieces, this could mean that we attend not to how the body might reassemble toward human perception and functionality but rather to how the "countryside" and the many inhabitants, processes, and processors of this distributed and distributive milieu begin to rework how the thinking-sensing machine captures, configures, and acts upon its inputs.

Perception in the World

Turing's sensing apparatus points to the distributed processes that make sensing possible, even if the sites of sensation do not return to a coherent human processor. Indeed, as Whitehead suggests, perception might be understood to be in the world and distributed through more-than-human processes—it is not the special preserve of a human decoding subject. Instead, multiple participants express and unfold a distinct experience of the world, independently but contemporaneously within an immanent series of events.[57] At the same time, the excitations of environments are fused to all modes of "matter," where "the environment with its peculiarities seeps into the group-agitations which we term matter, and the group-agitations extend their character to the environment."[58] "There are numberless living things," Whitehead writes, that "show every sign of taking account of their environment."[59] This taking account of environments is a way of capturing what is relevant, and—through being affected—of transforming environments and relations.

Sense data might be seen as a concrescence of multiple ways of *taking account* of environments, whether through researchers or devices or environmental events. But these data are necessarily articulations of the ways in which environments are gathered and expressed through varying subjects—here, with subjects understood in the broadest possible way. Sensing systems generate and concresce distinct articulations of environmental relations within and through data and across sensing "subjects/superjects." Rather than take on a Kantian view of how "the world emerges from the subject," Whitehead, with his "philosophy of organism," seeks to understand how "the subject emerges from the world," thereby constituting a "superject," or a subject that is always contingent upon actual occasions and experience.[60] As Shaviro notes in relation to Whitehead:

> There is always a subject, though not necessarily a human one. Even a rock—and for that matter even an electron—has experiences, and must be considered a subject/superject to a certain extent. A falling rock "feels," or "perceives," the gravitational field of the earth. The rock isn't conscious, of course; but it is affected by the earth, and this being-affected is its experience.[61]

Sensor technologies are constitutive of sense—they too "experience" the world and generate perceptive capacities.[62] Sensors that map in real time a greater density of ecological relations might generate a processual approach to environments by focusing on interactions and even multiple modes of perception. At the same time, to identify a phenomenon as constituting sense data is to make a commitment to distinct "forms of process," so that environmental processes are selected and concretized in those forms. The process of selecting sense data involves capturing a moment in time, an "instant," that is re-sutured with other data to form a pattern of ecological processes. While approximating a more process-based and even real-time monitoring of environments, sensors are also productive of practices of selecting and interrelating discrete observations in order to arrive at an understanding of ecological processes. The selection of temperature, vibration, light levels, humidity, and other measurements across primarily physical (although to some extent chemical and biological) criteria in-forms the instants that are sensed, the forms that are documented, and the processes that might be reconfigured.

The basis for developing "facts" within the sensing experiment then directly pertains to the forms and processes of experience that are generated and connected up across sensing subjects.[63] The concrescence of data also requires subjects that can prehend and experience the data. Subjects may be attuned or resistant to receiving data based on prior or concrescent experiences. But the means of gathering data might also contribute to the possibilities for processing and integrating data. In this way, sense data as experienced by subjects may be generative of superjects where the experiences and perceptions generated are in turn formative of the subjects that experience. This runs counter to the notion that a founding subject is the entity that experiences. If, as Whitehead suggests, subjects are always superjects, then subjects are always necessarily distributed and concrescent in relation to actual occasions.[64] Subjects, whether stones or sensors or humans, become environmental in this way since they are involved in feeling and concrescing actual worlds.

Approaches to media and sensation often focus on the ways in which technologies train or otherwise attune the human senses within a mediatory or prosthetic relation. But the interactions and processes of sense are arguably not fixed within sensory organs or technologies through which mediations are typically understood to occur. In this way, sensation is not primarily an inquiry into relations between human subjects as they perceive more-than-human objects. Instead, the sensory relations within which sensors are mobilized give rise to a more ontogenetic understanding of perception, where sense and expressions of perception are articulated processually and across multiple sites and subjects of inventive sensation.[65] In this way, new perceptual engagements are distributed across sensing capacities and engagements (perhaps similar to what Luciana Parisi has called "technoecologies of sensation"), which give rise to distinct sensory processes, informational-material arrangements, and ethico-aesthetic possibilities.[66]

Such a condition resonates with what Patricia Clough refers to as the importance of focusing on "an empiricism of sensation" rather than "an empiricism of the senses."[67] Technologies, including sensor systems, can be understood as generative ontologies that in-form the experience and conditions that make sensation possible and changeable. Rather than studying "the senses" as given, it may be more relevant to study experience and how distinct types of sensation become possible, and to consider further what modes of participation and relation these processes of sensation facilitate or limit. To bring this analysis back to sensor technologies, sense data are not simply items to be read and gathered as machinic observations of environments that scientists process. Instead, sense data are indications of a process of becoming sensible, where environments, humans, and more-than-humans are individuated as perceiving and perceivable entities.

Collaborative Sensing

The modes of sensing that concresce within the context of ecological sensor applications might, as discussed earlier in this chapter, begin to be described as collaborative sensing practices taking place across multiple subjects and through distinct processes of experience. These modes of sensing could further be referred to as types of "intimate sensing," as Stefan Helmreich has suggested in relation to fieldwork undertaken with oceanographers who employ a complex array of sensing technologies in their research. Sensing, in this account, is comprised of a research "ecosystem," and involves much more than a device focused on an object of study, since bodies enter into a circuit of sensation with instrumentation technologies. As Helmreich writes, "These scientists see themselves as involved not so much in remote sensing as in intimate sensing." Multiple forms of sensing are articulated across different technologies—and so with researchers involved in studying ocean ecologies: "The mediations are multiple and so are the selves."[68]

Influenced by Charles Goodwin's discussion of how forms of "collaborative seeing" are produced within the space of a scientific vessel,[69] Helmreich develops an analysis of the sensing processes that become concretized within these body-environment-technology relationships, where new registers of feeling might sediment through repeated engagement with these devices. The multiple selves that Helmreich discusses most frequently refer back to scientists and crew members on ocean-sensing expeditions, but by extending this approach through a Whitehead-oriented understanding of experience it is possible to include even more expanded collaborative formations of sense. The experiences provided by and through more-than-human processes, as well as the processes that unfold within sense data, in-form a more environmental approach to what might constitute "collaborative" modes of sensing.

Within the area of more-than-human theory, sensation is increasingly understood as distributed in and through more-than-humans in the form of organisms and technologies, together with their environments. At times influenced by Foucault's well-known "death-of-man" statement, media scholars as far-ranging as Friedrich Kittler and Katherine Hayles have in different ways undertaken analyses of media that dispense with an assumed human subject as the principal site of meaning-making in order to recast the relations that concresce in and through media technologies.[70] As Hayles suggests, environmental modes of computation—RFID in her analysis—raise questions about the effects of "creating an animate environment with agential and communicative powers." Such technologies allow us to move toward "a more processual, relational and accurate view of embodied human action in complex environments."[71] Not just sensing but also what counts as the "human" shifts in these scenarios, since computational technologies typically now operate within parallel processes and signal toward a multiplication rather than a centering of subjects.

The subjects that might be discussed as parallel, multiple, or collaborative within environmental sensing extend not just to entities multiplied through more-than-human technologies but also to the incorporation of more-than-human flora and fauna. More-than-human theories of subjects—or ecological approaches to subjects—are becoming increasingly well established not just in media theory but also in philosophy and feminist studies, particularly as articulated in the work of Braidotti, who develops these notions through the work of Deleuze and Guattari (with an emphasis on the notions of ecology developed by Guattari). Braidotti suggests that we begin to work with an "environmentally bound subject" that is also "a collective entity" because "an embodied entity feeds upon, incorporates and transform its (natural, social, human, or technological) environment constantly."[72] In this account, bodies and subjects are even understood as collective information machines of sorts. For Braidotti, "techno-bodies" may be understood as "sensors," or "integrated sites of information networks; vectors of multiple information systems."[73]

Such an ecological approach to subjects resonates with Whitehead's discussion of subjects/superjects, where bodies-as-sensors are expressive and productive of environments. The sensing that takes places is a practice of processing and transforming. If human bodies are sensors, then by extension so too are the multiple more-than-humans that take in, express, and transform environments. As the preceding discussion of the James Reserve suggests, it is relevant to bring these multiple formations of experience to play across human and more-than-human subjects into an examination of the specific distribution of environmental sensor networks in this ecological study site and to consider how sensors are expressive of environments, what new environments and subjects concresce as experiencing entities, and how the sensing experiment might make these experiences possible.

INVENTING EXPERIENCE

From an experimental forest, this analysis of environmental sensing turns back to Turing's countryside—that apparently static backdrop through which sensing was to take place. While Turing imagines a distributed sensing entity processing its bucolic surroundings, in this analysis of test sensors installed in a forest setting it becomes clear that the surroundings to be sensed are in flux and yet formative to establishing conditions and practices of sense. Through this reading, Turing's distributed computer becomes a superject, integrated with and formative of the environments and experiences it would decode. It becomes environmental in that it is an entity that generates the formation for further subject-superject experiences. This approach, as discussed throughout this chapter, provides a way of taking account of the abstractions and entities that lure feeling and settle into forms of environmental engagement.

The environment or milieu as differently understood by writers from Whitehead to von Uexküll, Canguilhem, and Foucault, has been discussed as everything from the conditions of possibility to a zone of transformation and necessary extension within and through which experience is possible.[74] Within the work of von Uexküll, the now well-cited example of the tick that is provoked to act in relation to certain environmental cues is referenced to signal the ways in which sensation is tied to environments and to suggest the species-specific coupling between these.[75] Sensing beyond the human subject can be figured through more-than-human agencies that unfold within environments. But if we take the provocations of Whitehead seriously, then the milieu is not just a site where sensing joins up. Instead, it is also a transformative and immanent process where modes, capacities, and distributions of sense concresce through the experiences of multiple subjects.

Any given milieu or subject/superject is expressive not of scripted coupling as the work of von Uexküll might suggest, but of creativity, as demonstrated in the work of Whitehead and Simondon.[76] If inventiveness is a necessary part of perceptive processes, then the environment-as-agitation necessitates a more ontogenetic, collaborative, and extensive understanding of sensing. In this way, perception might also move beyond the notion of hybridities or even mediations of sense and instead focus on the sensing conditions and entities that concresce, as well as that which environmental perceptive processes make possible, and how inventive processes might further generate new forms of collective potential.

The complex interactions that are the focus of study for environmental sensor systems are transformed through the perceptive processes that these systems generate. The ecological relations that are to be discovered and studied are bound up with the detection of patterns within sense data. Sensor hardware and software do not simply gather sense data in the world, but are part of the process of perceptual possibility, both as more-than-human registers of perception and through making distinct relations sensible as subjects of ecological concern.

The possibility to relate and to make aspects of relations evident is an important aspect of sensor systems, with political and practical consequences. Sensation might be understood as distributed and automated on one level, yet on another level such automation in relation to environmental processes involves not only running scripted functions but also addressing the open and indeterminate aspects of sensors in relation to environmental processes. This is one way of saying that, whatever the computational program, sensors never operate strictly within a "coded" space, but by virtue of drawing together expanded perceptive processes they inevitably make way for a generative technics of environments.

There are political implications to the implementing of sensor processes: relations are not simply discovered in the world, rather they are individuated through these distinct computational sensing processes. These processes further orient environmental practices and politics, where increased data and improved awareness of ecological relationships are expected to translate into an improved ability to manage environments and potentially prevent the spread of environmental damage. These crucial relationships concresce not just through practices of data collection and monitoring, as well as sharing data within larger networks, but also through drawing inferences across data sets that illuminate key ecological relationships that are to become the basis of concern or protection. As Whitehead suggests, that which counts as a form or datum is what endures within a "process of composition," which is expressive of "historic character."[77] What counts as empirical requires acts of "interpretation" but also describes a concrescence that continues to have the force of natural fact. Drawing on Locke, Whitehead notes, "The problem of perception and the problem of power are one and the same, at least so far as perception is reduced to mere prehension of actual entities."[78]

While Whitehead's analysis works across philosophic and cosmological registers, and does not directly address sociopolitical analysis of environments, his work does point toward potential translations to be made across experiencing subjects to political possibilities. As Shaviro suggests, following on Whitehead, experience is a site of potential: "It is only after the subject has constructed or synthesized itself out of its feelings, out of its encounters with the world, that it can go on to understand that world—or to change it."[79] In other words, as Whitehead notes, "How the past perishes is how the future becomes."[80] That which is sustained and that which concresces as a register of novelty are processes whereby experience may give rise to new experiences, interpretative practices, and matters of concern. In a different way, Foucault indicates through his discussions on the milieu that sensory arrangements articulate distributions of power, and involve making ongoing commitments to relations and ways of life.[81] Sensory processes that occur across subjects are expressive of material–political relations and possibilities for participation.

Environmental monitoring through sensor networks is a technoscientific practice that pertains not just to the study of ecological relations but also to newer modes of participatory sensing and citizen-science activity that rely on the use of the sensing capacities on mobile phones and low-cost sensors to track and gather data from environments. While citizen-sensing applications have developed to move these scientific applications into the hands of the general public,[82] even more questions arise as to how or whether sense data makes an effective traversal from data to action. The implications for sensory practices that are articulated within an environmental monitoring context then have relevance for thinking through the processual, relational, and heterogeneous aspects of sensing. Given that the CENS research has moved "out of the woods" to citizen-sensing applications, while at the same time a whole host of participatory applications such as forest monitoring platforms are materializing to protect forests for conservation, how do forests, "citizens," more-than-humans, and sensor technologies converge to invent new forms of politics that are attentive to present matters of concern and those that are yet to come? In the next chapter, I consider this question in relation to a seemingly more prosaic "sensor" and the sensing practices it operationalizes: the webcam.

Figure 2.1. Moss Cam capturing images of a boulder covered with the moss *Tortula princeps* at James Reserve. Photograph by author.

2

From Moss Cam to Spillcam

Techno-Geographies of Experience

Web cameras are now a common technology for remotely viewing ecological processes, including everything from animal activity to freeway traffic. Cameras fix upon phenomena that may be studied over time or in absence of direct human intervention. Any number of organisms, from falcons to badgers and turtles have been "caught" on camera while undertaking migratory or nesting activity that may have been previously unrecognized. These image-based modes of monitoring are also meant to provide important information for protecting organisms.[1] Similarly, web cameras may be used to report on and monitor environmental calamities, as was the case with the British Petroleum (BP) oil spill in the Gulf of Mexico.[2] Concerned onlookers could focus on the ruptured pipe connecting to the underground oil well and watch a steady plume of oil spilling into the Gulf. Image capturer, reporting device, surveillance technology, and visual sensor: the web camera increasingly operates as a generator of environments of attention and concern.

In this chapter, I first address the proliferation of cameras—from web cameras to camera traps, animal-borne cameras and eco-drones—for environmental study and engagement. Rather than discuss cameras in the context of the proliferation of visuality and images, however, I focus on the ways in which cameras, images, and image practices are remade into sensors, sensor data, and sensor practices. As much as they produce images of environmental activity, cameras also operate as sensing and measuring devices, converting physical stimuli into electrical signals that transform into sensor data within extended sensor networks. Cameras-as-sensors concresce as distinct technical objects and relations, and in the process they articulate environments and environmental operations. The becoming environmental of computation, in this case, includes the invention of a "technogeographical milieu," as Simondon calls it,[3] where the webcam is involved in constructing distinct environments that are both technical and geographical.

Web cameras, camera traps, animal-borne cameras, and eco-drones now operate within sensor networks, and the images they produce are often processed as another mode of (image-based) sensor data. These shifts in the practices of processing image data as sensor data are productive of sensor environments that create distinctly different engagements with *imaging,* not necessarily as an a priori fixation of visuality, whether as an epistemological or disembodied register,[4] but instead as a processual data stream that irrupts in moments of eventfulness and relevance across data sets comprised of multiple sensor inputs. Image sensor data, in this case, is part of sensor *systems.*

In this chapter, after first briefly discussing a range of cameras-as-sensors that are in use in environmental study and engagement, I go on to consider the distinct modalities of sensation and environment that webcams articulate. Engaging with long-standing science and technology studies and media theory discussions of visuality, I focus on two renderings of sensation across cameras and sensors in the work of Donna Haraway and Katherine Hayles. In her discussion of the "Crittercam," a camera affixed to organisms, Haraway discusses processes whereby cameras are involved in the "infolding" of sense and how this might reorient the ways in which we consider the "sight" of cameras. Hayles considers how the "distributed" sensation of RFID tags, as a version of a passive sensor, influences the animation of environments and changes the agencies of sense.

I work through these two discussions of sense and technology to consider how sense is embodied and distributed across more-than-human entities (as discussed in chapter 1) and examine how environments are critical to the generation of sense. In this way, I consider how sensor environments shift a focus from *bodies of sense to environments of sense.* As discussed in the introduction, this is another way of engaging not only with the becoming of *subjects of experience* but also with the *becoming of superjects of experience.* Importantly, the way in which I am encountering environments is not simply as a place or site, but as a processual condition that is individuated along with sensing technologies and subjects. Webcams and other cameras-as-sensors operate as technologies that generate new modes of sensor data while also individuating new relations and possibilities for relations within and through environments.

From this perspective, I then consider two rather different web cameras and the ways in which they articulate specific techno-geographies of experience. The first camera, Moss Cam, was briefly discussed in the last chapter as one of several webcams located in the James Reserve ecological study area. On the one hand, this webcam offers distinct ways of thinking about biomonitoring and phenology, where sensors gain access to—or experience—environmental data vis-à-vis a close relationship to monitoring organisms that are expressing changes in environments.

In this sense, the sensor network crosses electrical, physical, chemical, and bio-logical registers of experience. Yet, on the other hand, the Moss Cam operates within a wider array of sensors and sensor data, and is one of several webcams that have been a point of public engagement with ecological research. It is these sensor relationships and sensor environments that I expand upon in the context of the becoming environmental of computation, where webcams become part of sensor systems.

In comparison to the Moss Cam, I consider Spillcam, a webcam that captured a plume of oil leaking into the Gulf of Mexico from the the Deepwater Horizon oil spill that occurred on April 20, 2010. BP made Spillcam available for public viewing as the result of a direct request by U.S. senator Edward Markey through the now-disbanded Select Committee on Energy Independence and Global Warm-ing.[5] Markey argued that the live video feed should be available for the American public to monitor progress being made toward stopping the oil and gas leak and to assist scientists and engineers in estimating the flow rate of the spill.[6] BP made the live video feed available on May 19, 2010, nearly one month after the blowout, and it became an immediate source of public attention. While the Spillcam was made available to attempt to make an image of environmental catastrophe into an actionable and accountable object, the image data captured by Spillcam prolif-erated into multiple other effects and relations.

Figure 2.2. Cornell Herons Pond Cam in Ithaca, New York; a webcam that is part of the Cornell Lab Bird Cams project. Additional birds that may be observed via webcam include barn owls in Texas and Laysan Albatross in Hawaii, among many others. Screen capture.

From Moss Cam to Spillcam, here are two different versions of monitoring and sensing set in motion. Based on a comparative discussion of these two web cameras, I conclude this chapter with an expanded consideration of the ways in which cameras-as-sensors generate techno-geographies of experience. I ask: What are the distributions and individuations of sensing that concretize through Moss Cam and Spillcam? How do these image-sensor-milieus create and express experiences beyond the visual- and human-oriented toward other operations of sense? And how do these cameras-as-sensors generate distinct environments and environmental practices? In other words, what are the ways in which these camera-sensors become environmental, as they make specific techno-geographies of experience?

CAMERAS, CITIZENS AND THE SHIFTING PROCESSES OF IMAGES-AS-SENSORS

Web cameras now make available a vast array of creatures that would typically elude detection. From pine martens and shrews, to bear and deer, salmon and hawks, raccoons and wolves, any number of organisms can be tuned into, whether through cameras at ecological study centers, national parks, zoos, or even individual back yards, as well as Wildlife TV offered up on holiday retreats. Web cameras focus both on the relatively near-at-hand, including squirrels and garden birds, as well as the remote and difficult to access, including whales, polar bears, snow leopards and even (purportedly) Big Foot. It is also possible to take a DIY approach and install your own webcam, for instance, in the form of a bird cam complete with a nest box that connects a camera feed to a television set. One slightly more epic project even proposes to link up animal webcams in a sort of "Interspecies Internet." Similar to the Internet of Things, the Interspecies Internet would allow one to tune in to a worldwide distribution of organisms and interact with them—where for instance, one could sing along with bonobos as they play piano accompaniment from their remotely located home in the zoo.[7]

In addition to web cameras, there are multiple other imaging techniques for sensing and capturing data about organisms and environments. Camera traps are a common and typically more low-tech method for capturing images of rare animals in order to estimate animal populations.[8] This method has most frequently been used to estimate densities of large cats such as tigers, where camera traps will be located in sites with obvious animal activity, such as paw prints or scratch marks. A typical camera trap consists of a thermal or infrared motion sensor, camera and flash, as well as data storage that allows data to be downloaded on a periodic basis.

Animal-borne cameras are now in frequent use as well, where organisms from household cats to marine animals are affixed with cameras, often along with GPS trackers. With these methods, image journeys are created that are meant to provide clues about animals' habits and spatial journeys.[9] Eco-drones are also emerging as yet another camera-based mode of sensing, where wildlife can be

surveyed from an aerial perspective and illegal poaching monitored. These practices, along with all of the above mentioned image-based techniques for monitoring organisms, are often presented as key ways in which to study creatures in
order to protect them, although most frequently the emphasis is on nonintervention and noninteraction in order to study creatures "in the wild." And this is the
point at which public engagement can also come into play, where people may
gain access to organisms without apparent interference or disruption.

From the standpoint of public engagement, the mass viewing of secretive
creatures often constitutes a residual or secondary use for web cameras, where
they are deployed within scientific study in much different ways and for different
purposes. Indeed, what is striking about so many webcam video feeds is that
nothing seems to be happening, and it is difficult to "make sense" of the images
in any usual way. Beyond the frequent error messages, broken links, and notices
that video feeds are offline, typical views captured consist of images empty of any
obvious activity, including nests that have been vacated or forest scenes that are
studies in a minimalist composition.

As I write this text, I am watching a webcam of barn owls on Wildlife TV.[10]
In the frame is a sleeping owl, still as a statue. Flies and midges scatter across
the camera lens, feathers and dung line the floor of the nest box, and a square
light at the back of the frame indicates that beyond this view there is another
world in waiting past the entrance to the box. Some time later, the barn owl
stirs, paces, spreads his wings once, then twice, and launches in a seemingly gran-
diose gesture merely to move to his stoop, another space of waiting. Many images
and video clips made available of sighted animals are selected as "highlights,"
where a notable activity does occur. These highlights are often silent, even black-
and-white night shots, and feature glimpses of animal motion, typically a moment
lasting from thirty seconds to a few minutes, capturing some scurrying and nos-
ing around, a flash of glowing eyes, and a swift exit into some other space.

Within the larger span of inactivity, human viewers of webcam images become
pattern detectors and image analyzers as they wait for signs of activity, which is
detected and reported in forums focused on organisms of interest. Indeed, new
scientific data has emerged from "armchair" citizen scientists watching web cameras, who, for instance, have witnessed a Great Horned Owl attack the nest of a
sleeping Great Blue Heron at 3 a.m. and emailed a report of the activity to scientists at the Cornell Lab of Ornithology who set up the cameras. Here, someone
keeping watch over a bird in the small hours of the night detected new data, since
in this case the heron made an intense screeching sound to ward off the owl, and
this particular heron vocalization had not previously been recorded.[11] Between
eight and ten million people watch the webcams set up by the Cornell Lab,[12] and
these citizen scientists watching webcams are "filling in the dots," since "all these
little data points get put together and then we see a larger picture."[13]

Commenting on another instance of a citizen scientist making a new discovery, this time of an elephant seal eating a hagfish witnessed vis-à-vis an underwater webcam, Steven Mihaly, staff scientist at Ocean Networks Canada, notes, "We have reams and reams of this video data that we could try to mine using data mining techniques, but the best data miners we know out there are people like you—the citizen scientists."[14] Watching, identifying, counting: the work that algorithms, machine learning, or lab techs might otherwise perform becomes the occupation of vigilant webcam watchers, citizen sensors of sorts, who might be as attentive to moments of significant data capture as they are to the attachments they form with distant creatures. Here are images, not framed and fixed, but rather in process and operationalized. Citizen sensing becomes a project of intensive if distant watching, and web cameras become sensors in an extended network that connect up environmental phenomena with actions of detecting, analyzing, and reporting.

Scientific Observation and Practices of In/visibility

Nature is what happens when we are not there, or this seems to be the message with many of these imaging technologies. From the remote to the out-of-hours to the inaccessible or hidden, cameras cross into spaces and temporalities to "sense" what would otherwise be unapproachable. It is by now well established that scientific observation often takes place through visual modalities, which access phenomena beyond the usual registers of human sight.[15] Yet in this case, rather than micro- or macrovision, what unfolds is a mode of vision that attempts to absent humans. Gregg Mitman discusses how the camera and technologies such as biotelemetry and Landsat imaging made possible the active monitoring and protection of "nature" seemingly without human intervention. These sensing devices not only allowed humans to remain relatively remote and distant—even "invisible"—from their sites and objects of study, but the scale of the devices also allowed "the researcher to integrate instruments 'as part of the living system.'"[16] However, human participation and intervention emerged in other ways through the assumed absence of people from wilderness spaces and the perceived need to manage these spaces toward this end.[17]

The objectivity of images, the invisibility of human intervention, the ways in which automation or distributed sensing might provide a greater fidelity to ecological processes: these are currents that have run through science and technology studies and feminist technoscience engagements that examine how images operate within and beyond scientific practice.[18] Stengers suggests that an important move in the study of science and technology is to take seriously the claims made by these fields, since this may open up new or experimental ways of thinking about questions and problems posed and the facts that take hold in particular environments. In another way, Haraway undertakes this project by reworking the "modest witness" of scientific observation to consider how to *refigure the* subjects, objects, and communicative commerce of technoscience into different kinds of knots."[19] Rather than approach vision or scientific observation as presenting a "view from nowhere," Haraway seeks to demonstrate how the embodied and situated encounters we have with visual technologies can provide an entry point for generating other knowledge practices.[20]

With these situated viewing practices in mind, I turn now to consider one particular visual technology—the Crittercam—that Haraway discusses in just this way as an embodied and situated encounter, and compare this to Hayles's discussion of distributed sensing, drawing these together into a consideration and proposal for a particular environmental approach to sensing that reworks how we might describe the operations of cameras-as-sensors. Beyond visualizing the invisible or gaining access to activity that typically goes on in the absence of humans, I suggest that cameras for ecological study are generating new modalities of sense, participation, and environment that exceed the usual considerations of in/visibility and non/intervention.

Crittercam and Infolding Sense

Haraway analyzes the technosensorial implications of webcams through a discussion of Crittercam, a project developed between marine scientists and National Geographic that involves affixing video cameras to various creatures, from whales to sharks to clams (and eventually to land animals), to capture images and details of these organisms' habitats.[21] She focuses on marine deployments of animal-borne cameras, and moves through the different forms of sight that are articulated through the coupling of animal, camera, and often-distant human viewer.

The National Geographic description of Crittercam indicates that this animal-borne camera is "a research tool designed to be worn by wild animals. It combines video and audio recording with collection of environmental data such as depth, temperature, and acceleration."[22] Not simply creating images, the camera is part of a project that produces multiple forms of sensor data, which complement the images gathered. Similarly, the project material notes:

> These compact systems allow scientists to study animal behavior without interference by a human observer. Combining solid data with gripping imagery, Crittercam brings the animal's point of view to the scientific community and a conservation message to worldwide audiences.[23]

The Crittercam produces both scientific data and an appeal for conservation. Haraway addresses these aspects of how the material hardware, the data, the animals, scientists, and more are involved in a complex set of encounters that attempt to provide more information about marine environments and which unfold into a heterogeneous set of relations.

Beyond cameras peering upon animals and their habitats, or humans acting in a simply voyeuristic manner, Haraway suggests that technologies and bodies (of humans and more-than-humans) become articulated as "'infoldings of the flesh.'"[24] The possibilities of perception, in her (post-)phenomenological-inspired account, occur through the meetings of bodies—whether human, more-than-human, or technological. These infoldings are zones of interaction, contact points from which "worldly embodiment" emerges.[25] For Haraway, the emphasis is not on a human sensing-subject who decodes images of animals but rather on the bodily and material meetings (and all that facilitates these meetings) across technologies, humans, and more-than-humans.

These sensory encounters also produce the possibilities for inhabitations, where mutual embodiment is a process of making worlds. From this vantage point, she considers how the Crittercam as a technology is both "always in formation," as well as "always compound."[26] Sensation occurs at the ongoing meetings of *multiple* bodies—to be multiple is to multiply, or to generate shared (compound) experiences and worlds.

Crittercam is a relevant point of departure for this discussion of camera-as-sensor, since with this device Haraway moves from an analysis about the *viewing of* others into a technological relation that involves *becoming with* others. Moving from sensation as a process of perceiving objects to a shared if asymmetrical practice, she articulates how these infoldings occur through the "flesh"—and this flesh, these bodies, and the possibilities of sensation are undertaken together in lived experiences. Haraway's account moves toward unsettling the human as a fixed processor of sensory stimuli and instead considers the collective creation of sensory worlds. Yet is it possible that the worlds that are constituted through these infoldings are more than flesh, and that shifting sensory modalities are not only infolded but also always becoming concrete as *environments* through these processes? Haraway's discussion signals toward but passes over the matter of "worldly embodiment." Yet what worlds not only come into being but also shift to in-form new possibilities of individuals and infoldings through these camera and image encounters? I suggest that the *worldly* aspects of embodiment that are so much a part of this becoming together might also require a more detailed consideration of the role of environments—as processual techno-geographies of experience—in expanding these sensory tales.[27]

RFID and Distributing Sense

Haraway's analysis of the bodily shifts that occur at the meetings of Crittercams, more-than-humans, and humans demonstrates how these technologies are not just reporting devices but rather are generative of experiences and material relations. Indeed, technologies such as cameras and the wider range of sensors and radio-frequency identification (RFID) tags now proliferating in environments do much more than simply allow for new modes of distant observation and data gathering. In addition to shared or remote modalities of sense, cameras and sensors redistribute the locations, agencies, and processes of sensing operations. Hayles suggests that "an animate environment" now surrounds us that is involved in producing sensory information in and through which humans are but another contributor to sensory processing.[28] Sensor technologies not only gather information but also perform distributed sensing processes for which direct human intervention or participation may not even be necessary.

Hayles speculates that humans might even potentially become rather marginalized within sensor environments that form their own sensory exchanges. The implications of such distributed sensing potentially may then involve experiences other than infolding sense. As she writes,

> Combined with embedded sensors, mobile technologies and relational databases, RFID destabilizes traditional ideas about the relation of humans to the built world, precipitating a crisis of interpretation that represents both a threat to human autonomy and an opportunity for re-thinking the highly politicized terrain of meaning-making in information-intensive environments.[29]

With these formations of ubiquitous and embedded computing, Hayles suggests that sense is distributed within and automated throughout environments in ways that challenge human sense-making. Sensory processing is not directed through human sensing-subjects primarily but is instead located throughout automated sensing processes. As it is decoupled from human subjects, sensing as a process of making meaning, and of generating capacities to make sense and act on information, can occur beyond the realm of human intervention. Indeed, many of the cameras trained on environments are not connected up to human bodies or eyes, but rather work through image analytics to detect patterns and send alerts when significant change is detected. In this case, cameras-as-sensors could be seen to operate in ways that Hayles identifies: as seemingly autonomous agents that are recasting practices of sensing and interpretation.

While webcams and related sensor technologies might raise issues of surveillance and monitoring,[30] surveillance is not the only issue to which these technologies give rise. In other cases, early work on web cameras focused especially on new qualities of telepresence, or the ability to know things from a distance.[31] Distance presented a new way of understanding mediation: as sensation understood through a more filtered and remote rather than immediate engagement.[32] Mediation, however, is a term that typically suggests the relative fixity of subjects and objects in an exchange of information—with the filtering process seemingly acting as the location of transformation.

Both surveillance and telepresence are arguably, as Hayles suggests, ways of encountering technology that "are primarily epistemological," or issues about "who knows what about whom."[33] But while monitoring is an important part of the operations of sensor technologies, it may be overemphasized at the expense of also taking into account a range of other sensory processes. Shifting sensor environments present issues that are as much ontological as they are epistemological.[34] Ontologies of sense—and the environmental practices and politics to which they give rise—are then a necessary point of focus for understanding the orientations of these technologies.

Moving from Hayles's observations to those of Simondon, moreover, we could say that an animate environment presents an *ontogenesis of sense,* where such distributions require rethinking sense toward process and practice. Rather than see humans as marginalized entities within animate and sensing environments, we might then consider how human sensing practices are in-formed and individuated differently, how they concresce—even more than infold—with technologies and within environments in different registers. In this way, sense ceases to be an ontologically prior category or relation, where the five human senses are but a rough guide for understanding how sensation becomes possible. Sensor technologies can be understood as in-forming ontological processes that make sensation possible, as well as generating entities and environments that take hold through these sensory processes.

Webcams and Environments of Sense

Distributed sites of sensation, as discussed in chapter 1, include a rather heterogeneous range of participants that contribute to sensory processes, from humans to more-than-humans, technologies, sites, and more. These environments of sensation suggest an alteration to the trajectories through which sensorial encounters are configured, so that sense—or experience—is not exclusively about immediacy or coming into contact but instead also refers to sensory capacities that concresce in relation to *environments* of sense.[35] Experience occurs within a techno-geographic milieu, since it unfolds within technical and geographical conditions that are further generative of worlds, environments, and relations.

Environments do not necessarily refer to a fixed *sense of place,*[36] in this respect, but rather involve the making of distributions and articulations of experience. Environments, furthermore, are not the static ground across which species-sensor encounters are located, nor are they containers for these meetings. Rather, they are a critical contributor to the distributions and possibilities of sense—as in-formed and in-forming milieus.[37] More than the bodies of humans, animals, and technologies infolding to form sensory capacities and meetings, by attending to techno-geographies of experience it might be possible to consider how bodies are not the only zones of sensory formation and processing.

Such an approach, where environment might be understood as a more dynamic condition, further resonates with Simondon's discussion of the techno-geographical milieu. In this sense (not dissimilar to Whitehead's focus on subject-superjects of experience), Simondon attends to the ways in which distinct environments concretize along with the unfoldings of particular technical objects. Simondon notes that a technical object is a unit of becoming, and this becoming extends not just to the evolution of the object itself but also to the environment in and through which it is sustained and interacts.[38] However, this is not simply a project of making environments but rather involves the production of processes that ensure the continuation of those environmental conditions that sustain technical objects. The production of a machine involves the production of an external milieu along with the possibility for these entities and relationships to recur and have a shared effect.[39] This is what Simondon refers to as a techno-geographical milieu.[40]

The technical milieu of the technical object works in relation to a geographical milieu where its operations are translated or transduced. In forming a transductive relation with the geographical and meteorological world in order to perform its operations, the technical object is also acting on that world, just as the geographical world is in-forming the technical object. Neither a dialectical stance nor a simple humanization of nature, Simondon describes this process as one whereby the technical object concretizes the meeting of these worlds, which are not one and the same, and may even be in conflict with one another.[41]

Simondon makes the point that as both an adaptive and concretizing process, the technical object expresses while also creating particular environments that may have had a latent or virtual presence and that spring into life at a certain critical point in the operation of a technical object. This is the point at which a "concretizing invention"[42] generates a techno-geographic environment, which supports the further functioning of the technical object. A viable technical object is the very vehicle through which the possible integration of the technical and geographical is able to occur—it in-forms these shared milieus and it exchanges energy across them so that they become interwoven and creative of new environments and conditions for technical functioning.

Techno-geographies then describe a particular process that resonates with the becoming environmental of computation. Simondon focuses on traction engines and audiometers and electric clocks in the context of his study of milieus and the relative situation of technical objects as elements, individuals, or ensembles. In a different way, I move this discussion of techno-geographies into a consideration of computational sensors to include a focus on the camera-as-sensor that is not just a technical element but also a sort of technical individual, which is expressive of the milieus in which it operates. This discussion also seeks to build on the variations of technologies and sense as discussed by Haraway and Hayles: how infoldings and distributions of sense might create further inquiries into techno-geographies of sense. Environments of sense play an important role in thinking through sensor technologies, including the web camera. I now discuss this techno-geographical orientation in the context of Moss Cam and Spillcam. The Moss Cam and Spillcam might be understood as sensors that are within and constructive of environments, producing visual and other multisensory records of sites, but also making new formations of environments.

MOSS CAM: IMAGING AN ECOLOGICAL OBSERVATORY

Moss Cam—the first web camera I discuss here in the context of techno-geographies of sensing—tracks the processes of *Tortula princeps,* or Star Moss, at the James Reserve. Discussed briefly in the last chapter through the sorts of biosensing and biomonitoring that can be understood to occur through mosses, this particular web camera–sensor is at work in a larger sensor network at the James Reserve. The Moss Cam is a sensor in the form of a web camera, a device that is by no means new or unusual, except in this deployment as part of a sensing lab it begins to raise new questions about the creation and use of environmental images as a form of sensory data within an expanded sensor system.

The James Reserve Moss Cam is one of many web cameras fixed on sites of study, including a Meadow Cam, Creek Cam, and Bird Cam.[43] This pervasive form of remote viewing is commonly found at work in sites of ecological study. Yet web cameras now operate as one of an assortment of sensors for ecological monitoring, which include acoustic, CO_2, light level, and temperature monitoring devices.[44] Although primarily gathered for ecological study, at the time of this fieldwork the moss images were also viewed by interested publics via a website and online repository of images.[45]

The Moss Cam has its gaze fixed on a granite boulder speckled with mounds of Star Moss, which moves in and out of phases of lush green active photosynthesis and patchy brown senescence. Star Moss is a species of moss that is particularly desiccation tolerant and has the ability to dehydrate over the course of months or years and yet regenerate and begin photosynthesis within five minutes of exposure to water.[46] The Moss Cam allows for a long-term continuous observation of the patterns of dehydration and rehydration that characterize the Star Moss lifecycle. The Moss Cam consists of both a video camera and infrared camera, coupled with a weather station that provides data on temperature, moisture, light, and CO_2 levels. Sensor technologies track the chemical, physical, and biological interactions within these ecosystems from a detailed and on-the-ground perspective. Sensation and the environmental formations that concretize here are relational and dynamic processes.

While the web camera is constantly trained to the boulder, allowing any visitor to the James Reserve website to observe the latest state of the moss, it is also set to automatically capture images of the moss at least once per day, and as often as every fifteen minutes, and stores a record of the moss up to 35,040 times per year. In this continual tracking, questions arise regarding how best to manage data, and whether automatic harvesting is the best use of sensors, or whether a more particular or opportunistic use of sensors to gather data at expected times might be devised. Yet in the continual tracking that is made possible, observations potentially also become available not just to scientists but to anyone anywhere who might access particular data sets online. The James Reserve includes a "virtual observatory," with archived data from underwater ecologies to auditory signatures of woodpeckers.[47] Such ecological and virtual observatories are the format through which many sites of environmental interest throughout the world are increasingly made available as remote visual experiences or data sets.

While a web camera, by virtue of being a camera, might be construed as a representational device, in its operation as a sensor, within an extended sensor environment, it operates more as an "imager."[48] As an imager, the web camera activates alternative practices of sense that are connected up with networks of sensory monitoring. This sensory data is crosschecked and compared across data sets and even connected up with other data sets formed from remote sensing images. The images generated in the Moss Cam ecological observatory operate less as pictorial, scenic, or representational registers of environments, and more as continual data grabs. No single Moss Cam image stands alone. Instead, it is plugged into sensory performances and capacities that compare and relate the ecological processes of moss to its own ongoing development and its responses to local conditions.

The Moss Cam documents the daily processes of a patch of moss through an interconnected network of sensors that create a dynamic picture of environmental conditions. Images of moss growth and senescence form a detailed report of this microenvironment at a moss-covered boulder in the James Reserve.[49] The images gathered from the Moss Cam are not typical depictions of organisms, however, since the data are gathered in recurring sequence and understood in relation to other ecological sensing data. In this way, distinct data relations and ecologies concretize through webcam images. These are the techno-geographies that are generated through the operations of cameras-as-sensors.

Characterized less by a comprehensive view and more through amalgamations of sensor data and processes, where one image is a moment articulated within a larger set of relationships, these images as sensor data become dynamic accounts of changing relationships within specific sites, and they fuse together as particular renderings of environments. These images are not fixed representations of environments, but rather are temporal markers of visual sensing data within larger data sets—the environments that can be read through these images must be assembled and compared within and in relation to other data sets and processed through image analytic techniques: environments that concretize are not captured in one image but are formed through shifting relationships across sensing technologies, data, sites, ecological study, and practice, and the more-than-humans that inhabit these sites.

In order to understand the phenology, or seasonal timing of mosses, it is also necessary to compare the time of year, temperature, moisture levels, and light levels, among other data, in order to establish which changes may be occurring—and may be detectable—through this sensory data. These sensory relations become the basis for understanding environmental processes, where a sensory arrangement of moss dehydration, CO_2 levels, and light and temperature levels, as captured through sensors that gauge these phenomena, may be assembled in a distant database through which new sensory practices and questions may arise. The continual capture of images, rather than a single or even sequence-based rendering, in-forms the basis for understanding ecologies—the relationships between organisms and environments. But it also becomes a way of rendering environments as both processual and multisensory arrangements, which are remotely and automatically sensed in the field, and which publics in turn encounter through ecological observatories and websites.

The field of sensory relations multiplies through these studies, as does the need to compare sensed criteria: the time-sequenced imaging of moss is only one aspect of its study, which also involves comparing a wide range of environmental data from many different sensors. These multiple sensors provide a general sense of environmental conditions, and images provide the basis for refining

Figure 2.3. Moss Cam with weather station at James Reserve. Image and infrared data are compared to weather observation data to understand the moss lifecycle. Photograph by author.

observations and estimates in relation to moss growth. This "sensor fusion" provides a more nuanced way to ask ecological questions about the distribution of moss, including why it follows particular distribution patterns, and why it prefers a particular niche. The sensor arrangements and deployments are set up in order to reach toward a better understanding of the ecological processes of moss. Such ecological observatories redistribute sense not necessarily as a single or even infolded zone of contact but rather as a processual relay of relationships, cross-checked and compared, and made possible through the techno-geographies of sensor environments.

Ground Truth

The Moss Cam images and data present the opportunity to "ground truth" other data that is collected through remote sensing. Detailed, on-the-ground observations provide a way to corroborate more distantly gathered data, and so these views from the ground are seen to offer a truth, or "truing," of more abstract observations. Remotely sensed images are typically "cleaned up" based on a number of assumptions and are not necessarily comparable to the fine-grained sensor data captured "on the ground," through microsensing technologies. The concrete environments of sensing then become an important part of how ecologies are understood.

More abstract models and projections might allow for speculation as to the relationship between increased levels of CO_2, increased temperatures, and moss growth, but sensor devices such as the Moss Cam offer a way to study the detailed patterns of growth and responsiveness of a distinct organism in situ. Since traditional methods of studying moss often involve working in a controlled chamber in a laboratory, there is a risk that the study of moss, including the use of sensors, would destructively modify the organisms under study. In order to measure moss CO_2 respiration, it is necessary to have the measurements take place in situ. The environment—and the ecological interactions and processes that occur there—is an important part of the sensory practices for studying moss. Sensing ecological processes is an interaction that requires not necessarily a sensing subject and a sensed object, but rather involves fields of response and resonance that might need to be read through other subject-superjects of experience and practices of detection.

The ground of ground truth is not, however, the final point of resolution in these sensor environments. Instead, it is a reminder of the constant need to draw connections across phenomena. Ground here is connection and concretization. The specificity of observations becomes a way to correct or correlate more remotely sensed data. Ground truth sensor data is then a much different perspective from which to understand environments, in comparison to those more global or aerial views that might guide both ecological study and environmental

imaginations.[50] Sensory practices are bound into forms of environmental understanding, and so are "situated," as Haraway has suggested.[51] Situated sensing and
seeing refer not just to the embeddedness of researchers but also to the concrescence of entities and prehensions that form an expression of any given milieu.
Sensor technologies may be observing a designated ecological site, but sensation
also occurs between humans and more-than-humans, where the status of moss at
any given time may be an indicator of the changes it is sensing. In this way, the
empirical measurements of the environmental conditions of moss express a multispecies sensing of environments, as discussed in chapter 1.

    This change of focus is not to remove researchers or humans from the sen-
sory environments (who are also "embedded" in their own ways), but rather to
consider in greater detail the distributions of sense within the extended environments where the Moss Cam is located. Visualizing and imaging (among other
sensory modalities) are situated and distributed practices, located across a techno-
geographical sensorium of ecological research sites and sensor technologies, website and publics, organisms and sensor data. Such multilocatedness shifts the focus
on sensory modalities from ecology-technology to a field of sensory relations that

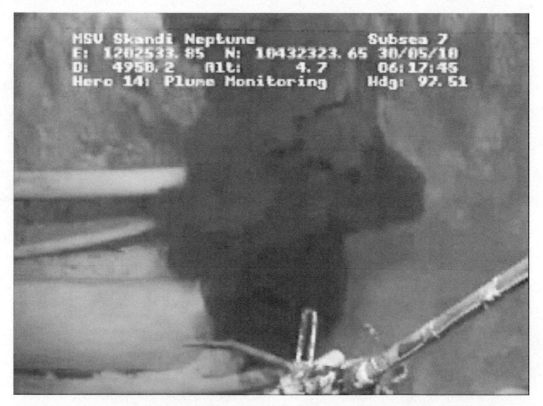

Figure 2.4. Spillcam example footage. BP video from Remotely Operated Vehicle (ROV) monitoring the
plume of oil, gas, and mud escaping from the ruptured BP pipe. Screen capture.

might be seen as a specific techno-geography that is articulated through any given webcam. In this way, environments and ecological processes concresce in relation to technologies. Webcams tune into and take account of particular environmental processes, and the image data they generate are further synthesized with other sensor data, forming subject-superjects of experience. The objects sensed are not just immediate data or encounters but also a non-sensuous field of perception that is part of the "vague" conditions in-forming the possibilities of sense.[52]

SPILLCAM: IMAGING AN ENVIRONMENTAL DISASTER

As mentioned in the introduction to this chapter, Spillcam emerged as a live image of the Horizon Deepwater oil spill, which resulted from an offshore oil rig in the Gulf of Mexico exploding due to a buildup of methane in the drilling riser. The explosion led to the worst oil spill to date, with an estimated 4.9 million barrels of oil leaked—at one point over 88,000 square miles of the Gulf were closed to fishing.[53] Spillcam was the publicly accessible live video made available from remotely operated vehicles (ROVs) that BP used in its attempt to stop the leak, but which became a site of environmental monitoring and call for accountability and transparency.

Markey, who requested the live video be made available, argued the feeds were not the sole property of BP, but rather that the public had "a right to the information that they contain and to be able to see for themselves BP's progress in containing this ongoing environmental disaster."[54] As the offshore oil well involved drilling in the deep ocean, the actual scale and volume of the leak was relatively undetectable. Located five thousand feet underwater, the ruptured pipe and blowout preventer could only be accessed through ROVs, which captured footage of the spill.

Not only would the live video feed provide a tool of public accountability; it could also, Markey suggested, be a scientific research tool, where "our best scientists and engineers" could be provided "with information that could be helpful in developing much needed solutions to the ongoing oil spill, both in terms of subsea operations and surface spill response."[55] When the live video feed was made available on May 19, 2010, it purportedly received upwards of one million views, crashing the Select Committee on Energy Independence and Global Warming's website and temporarily putting the House of Representatives' web system out of order.[56] Spillcam became a popular reference point in news media, and as it received viewers in the millions, the number of sites hosting the live feed also multiplied to three hundred or more, and "Spillcam" entered the English lexicon as a top word for 2010.[57]

As the Spillcam captured and transmitted images of oil leaking from an underwater pipeline in the Gulf, it simultaneously became a way to make BP accountable to publics, while also providing a way to visually monitor the rate of the leak and the likely composition of the material flowing out—whether oil, gas, or mud. At the time of its going online through the committee's website, a message announced, "You are watching a live video feed of the BP Oil Spill from the ocean floor, 5000 feet below the surface," as well as a disclaimer: "Please note that these live streams may freeze or be unavailable at times."[58]

Real Time

The single Spillcam feed multiplied into twelve Spillcams as numerous viewpoints from ROVs were sought to observe and study the plume of oil gushing into the Gulf and the rate and volume at which it was flowing.[59] From Ocean Intervention III, to Viking Poseidon, Boa Deep C, Skandi, Enterprise, and Q4000, the ROVs that provided the video feeds streamed real-time footage of the plume of oil as it drifted through the Gulf. Public fixation upon the video footage, as well as anger and concern about the disaster, was writ large in multiple forum posts on websites that featured the live video. The *Huffington Post* logged a total of 1,038 comments, with proposals to show the live feed in Times Square 24/7 and make BP show the feed on flat screen televisions in every window of every BP gas station across the country. At the same time, there was concern that the video feed provided a false sense of transparency and that the camera views were presented so as to show the least damage, or that since BP was controlling the feed, anyone watching it would be increasing site traffic through BPs websites.[60]

In contrast to the ground truth that the Moss Cam camera-as-sensor is meant to provide, here in the water and spill-filled depths of the Gulf an attempt at providing accountability through a real-time live video feed becomes a site of questioning. The real-time flow of Spillcam images introduces uncertainty, indeterminacy, and even a restless but seemingly helpless fixation. One commentator in the *Washington Post* suggested that watching Spillcam was like viewing a horror film crossed with Andy Warhol's *Empire,* where "there is no sound and nothing happens, except the inexorable, unending flow. You watch a little, and then a little more, and then you can't stop watching as a steady plume of dark brown oil belches upward from the floodlit, rocky ocean floor."[61] Unlike cinema, however, the Spillcam was not a representational narrative or documentary record, but rather a real-time image that captured the ceaseless flow of oil into the Gulf.

Embedded within an urgent flow of events, the Spillcam became a techno-geography of experience, expressive of real-time environmental emergency. Colors of the spill were watched closely as indicators of the progress of the environmental disturbance: the darker the plume the more likely oil was seeping out, the more muddy in hue, the more likely the oil was being stopped and drilling mud and other sediment were the primary effluvia. Drawing on Helmreich, we could say that the "'empiricity' of the spill" played out not just through evidence of oil on the shoreline and models of oil-in-seawater movements,[62] but also through

Spillcam images that tied into these techno-geographical systems of sensing and experience that made the oil apprehensible. Real-time images captured through Spillcam made this technical object into a particular type of camera-as-sensor—working alongside other sensor systems while also organizing attention toward the environmental catastrophe that was unfolding. In the process, observing becomes an experimental action, a real-time generation of a sensory system.

In the confusion of blown-out rigs and leaking oil, toxic dispersant and devastated marine life, loss of livelihood and loss of life, the live video feed provided a shattering if steady stream of images of oil gushing into the deep sea and ROVs at work attempting to plug the well. At the same time, scientists monitoring the site were using any number of methods to assess the volume and rate of the plume, often through sensing techniques that were not visual but rather were acoustic, chemical, geolocated, temperature-based, and focused on producing a molecular "fingerprint" of the oil and gas hydrocarbons in order to trace oil found throughout the Gulf and link it to the BP spill.[63] Attention was directed toward studying the flow of currents with gliders communicating to satellites and with

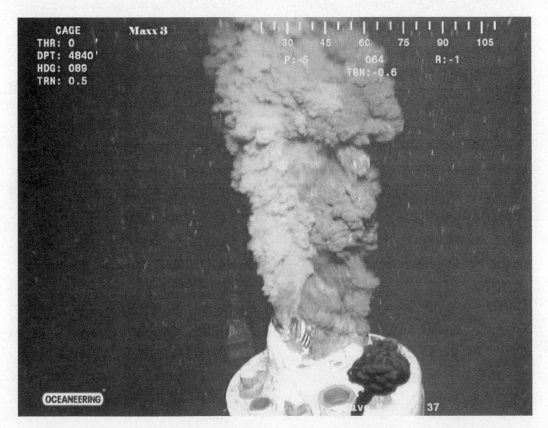

Figure 2.5. Additional Spillcam footage, as captured by the Oceaneering Remotely Operated Vehicle (ROV) monitoring the BP Deepwater Horizon well and broken pipe. Screen capture.

drifter cards retrieved by citizen scientists, as well as to assessing the impact of oil on deep-water coral and benthic organisms.[64]

These various projects that focused on providing more "clarity" within a complex environmental disaster operated alongside and in another register to the Spillcam images. Although the leak was eventually stopped on June 15, 2010, and the well was capped on September 19, 2010, the effects of the oil spill continue to play out in the Gulf region, where human and more-than-human health are affected in ways that are not yet fully established are still being revealed. The "ground-truth" and "real-time" accounts that sensors, whether as images or other forms of marine data, are able to provide is not a clear and easy project, since this is an environment that is shifting and will continue to change as hydrocarbons settle into organisms and ocean sediment, bodies, and instruments, which will continue to individuate in this transforming environment of sense and experience.

TECHNICAL MILIEUS AND AMPLIFYING ACTIONS

Spillcam presents a much different point of view than Moss Cam—the latter, a still, green, and seemingly uneventful moss-boulder slumbering within an ecological study area; the former, a temporary if dynamic instrument of environmental disaster and accountability. But each camera or compound of cameras operates in a more sensor-operationalized milieu and less as an individual and fixed representational image. Each camera-as-sensor articulates different environmental processes. They are bound up with the becoming environmental of computation through image-sense events that produce much different operations and inhabitations of "vision" along with multiple other forms of sensor detection.

In *The Five Senses,* philosopher Michel Serres suggests that we attend to environments and spatial practices as characterized more by "visits" rather than views.[65] With this provocation in mind, it is possible to approach the becoming environmental of computation in a way that does not unify into a singular visual frame, but rather engages with a multiplicity of sensing in and through techno-geographies of experience. With web cameras used for ecological study and citizen sensing, this multiplicity of environments and experiences concresces across ecological study, technological sensing, data collection and processing, public engagement, and environmental action.

Rather than discuss the complementary or alternative contributions that may be made through the expanded array of human senses, from hearing and touch to smell and taste, the distributed sensing of sensor technologies creates an entirely different arrangement of sense. The camera as sensor, imager, and program operates within a sensing environment that is not delineated according to human bodily senses but instead within units of measurability and comparison.[66] Sensing concresces through conditions of fusion, generality, ground-truth, real-time, and transformations across organisms and environments. In this respect, sensing takes place more as a program or operation rather than a process of mediation between subjects and objects. Here, sensing occurs not as a hierarchy of the five senses but as a distinct set of contingent relations that make possible practices ranging from scientific empirical assessment to environmental action, and where processes that may be detected give rise to communication to diverse publics about environmental change.

But these expanded practices of sensing require more elaboration—even within a visual framework. For instance, the Moss Cam, as part of an ecological observatory, is not just a visual technology, it is also a "machinery" that establishes a processual engagement with imaging.[67] "Objects" become expressed as information through processes and infrastructures that enable organisms and environments to be transformed into digital entities. The practices and technologies that facilitate the entry of a so-called "natural" object into a dataspace (such as the lizard discussed in chapter 1) and the sensory or visual processes whereby these transformations are made might even be referred to as a "digitectomy,"[68] where in order to make certain creatures visible or relevant—in this case, sensible as digital objects—they must find expression within distinct practices of scientific seeing.[69]

While an approach that emphasizes the heterogeneous if multiple sites and processes through which organisms are processed and transformed to become observable articulates a compelling map of sense-based transformations, there is still the notion that preprocessed organisms exist, organisms that might be available to literal seeing, but not visible within (constructed) scientific spaces. Arguably, this analysis focuses on scientific processes that render their object visible, whereby natural objects are made into legible items of scientific relevance. Might this view also focus on the ways in which interpreting human subjects make objects relevant for study through observation and instrumentation? If we return to the discussion in chapter 1, however, we might consider how the experiment is always involved with the experience of subject-superjects. Such an approach moves beyond discussions of the real or the constructed to a consideration of the registers of interpretation and experiential arrangements that are put into play with and through sensing practices—which do not, furthermore, hinge on establishing the veracity of a substantialist object for a human mind to decode.

This study on the Moss Cam, Spillcam, and their extended sensor environments then decenters the human subject as the primary locus of sensation and considers how sense data concresce through environments of sensibility and subjects of experience. The points of sensation discussed here are not so much about "nature" being transformed into "data" but rather about how the collecting and processing of data—these programs of sensing—in-form and contribute to the ways in which sensing becomes superjectal, or to how subjects of experience arrive at subject-superjects of experience. This approach also signals the ways in which sensory relations hold together *as processes*. Most importantly, this is a way of considering how facts, organisms, ecological processes, and computationally enabled image-sensors come to have a foothold within particular environments. Organisms are not constructed into digital entities, but rather, organisms concresce and are encountered through computational environments and environmental attachments. In other words, these organisms would not exist without the environments in and through which they are individuated and concresce.

In this sense, new environments come together through sensor technologies—through the altered perspectives, practices, and ontologies set in motion through these sensory processes.[70] To understand these different sensory concretizations of environments, it is useful to consider alternative accounts of sensory operations that occur across bodies, environments, and technologies. Sensing as a techno-geographical project offers up a different material arrangement that displaces bodies as sites of processing and sensing. Bodies constitute distinct if different sites of sensing within a more distributed set of sensor relations. Such sensing operations revise ideas about embodiment or mediation, since not only is there no originary experience to be mediated but also there are many sites through which sensing and experience occurs.

There is a diverse range of sensory studies, briefly signaled at the beginning of this chapter, which move through visual and other sensory modalities to connect up with other possible relational practices. While embodiment has served as a way to rethink visual technologies by challenging the notion of a detached observer, there is still much work to be done in thinking through what new "situated knowledges," as Haraway suggests, might gain a foothold at this juncture. In this respect, María Puig de la Bellacasa suggests, in relation to Haraway's situated knowledges, that by "affirming the embodied and situated character of material and semiotic technologies of vision" it is possible to also "affirmatively transfigure the meanings of objectivity and open possibilities for committed knowledge practices."[71] Haraway contrasts the view from nowhere with situated knowledge, "the view from a body," which challenges the free-floating and disinterested gaze of science.[72] For Bellacasa, this suggests a practice of touch, of "knowledge-as-touch" that would circumvent the possibility of knowing from nowhere.

But this body, which importantly locates sensing and knowing in one type of fleshy-site, may require even further situating. Not only is sensing superjectal and techno-geographical, it is also individuated in distinct ways in relation to the sites, bodies, and technologies that meet there. As the Moss Cam and Spillcam cameras-as-sensors demonstrate, the meetings of technologies, organisms, and sites are situated, individuated, and concresced in environments that are also on the move in these sensory formations and techno-geographical milieus. And it is these very sensory formations that further in-form our ability to respond to environments from a committed inhabitation.

The committed position that opens up through situated knowledges raises a final question about how superjects of experience organize human participation in citizen sensing and even environmental politics. The "armchair" citizen scientists discussed at the beginning of this chapter describe one type of environmental practice that concretizes at the techno-geographical juncture of webcams and more-than-humans, as well as watchful human participants. Sensor-generated ecological data is often gathered with the purpose of articulating more exactly the scale and details of environmental change, but here monitoring extends to include other types of citizen interventions. These watchful humans are seen as vigilant "live eyeballs," as well as caring participants who might even be able to make citizen arrests if they see poaching occurring—where, for instance, drones, webcams, or traffic cameras could capture wildlife trafficking of elephants, rhinos, or large cats.[73]

There is an "ethos of ecological monitoring," as Helmreich has suggested, where sensor technologies act as the "eyes" watching over environments under study.[74] To observe is to watch over, to attempt to mitigate harm, and to act in time to prevent environmental calamity. The fine-grained, pervasive, and constant quality of these observations in-forms our sensory practices in relation to environments: always on, always aware, and constantly gathering information. While Helmreich draws on McLuhan's assertion that media technologies alter our sense ratios and make possible distinct practices within these media environments, the question remains as to which sensory and environmental practices settle in relation to sensor technologies.

It may be that these sensor environments of ecological monitoring comprise an even greater store of information from which to act, where the politics of sensing across these distributed and pervasive sites seem to influence even more extensive possibilities for environmental action. Yet the risk is that monitoring could become an end in itself, where sensory data amasses in excess, but these often-distant sites of ecological study present detailed sensory datasets that do not translate into environmental practice. One way to test this concern may be to return to the sites of sensation and to consider the ways in which infoldings, distributions, and environments of sense may recast the scope of our multiple and diverging environmental engagements.

As I have suggested in this discussion of the Moss Cam and Spillcam, these sensor environments move discussions of sensation—and environments—beyond mediated/immediate, nature/culture, or direct/indirect to cross over into infoldings, distributions, and techno-geographies of sense. These sensory operations do not rely on a framework of mediation, where sensing subjects decode objects of sense, but rather they point toward dynamic formations of sense that concretize through distinct techno-geographical relations. As Combes has written in relation to the milieu of technics:

It is Simondon's virtue to have seen that technics *as network* now constitutes a milieu that conditions human action. Out of that milieu, we need simply to invent new forms of fidelity to the transductive nature of beings, both living and nonliving, with new transindividual modalities for amplifying action. For, in our relation to preindividual nature, multiple strands of relation—to others, to machines, to ourselves—entwine in a loose knot or node, and that is where thought and life come once again into play.[75]

Techno-geographies offer up not just an approach to working through the generative meetings of technical objects and environments but also to the forms of relation that may become sites of invention and amplification. In his discussion of techno-geographical milieus, Simondon is not insisting on a particular relation to environments, but rather is searching for new relationships to technology.[76] His work indicates how, in our techno-geographical milieus, environments and technology might be co-constitutive. The webcam, seemingly an agent of description, representation, and nonintervention, might actually in-form engagements, while environments also in-form webcam processes. From "armchair" citizen science to biomonitoring to petropolitics, the webcam-sensor operates in a modality of "technical culture" and techno-geographical experience, which are further generative of practices that become self-sustaining. In this way, we could ask how cameras-as-sensors concresce with environments, relations, and technical systems, and what possibilities there are for connecting up with worlds of sense that these devices would make present and actionable.

3

Animals as Sensors

Mobile Organisms and the Problem of Milieus

Y͏OUR DAILY WEATHER FORECAST may have been brought to you, in part, by southern elephant seals. In the Southern Ocean, seals tagged with conductivity-temperature-depth satellite relay data loggers (CTD-SRDLs) capture a detailed picture of ocean waters. Sensors travel along with seals to map their dive profiles and to gather a previously inaccessible set of oceanographic data from the ocean surface to depths down to nearly two thousand meters. Based on seals' foraging patterns carried out in response to the Southern Ocean environment, sensor data documenting seal movements, along with temperature and salinity, can lead to further inferences about sea-ice formation, the likely movement of fronts within the Antarctic Circumpolar Current, and global circulation of ocean currents—indicators not just of weather patterns but also of longer-term shifts in climatic patterns.[1]

Relaying across the journeys of tagged elephant seals swimming through circumpolar seas, sensors gather data on location, depth, and temperature that circulate to Argos satellite systems (which consist of three satellites in three orbital planes),[2] then filter into the institutions of environmental science to form seal dive profiles, there to complement remote-sensing and ship-gathered climatology data sets, while informing the nightly weather report. Weather forecasts and longer-term studies of environmental change come together through these multiple and distributed sensing technologies and (more-than-human) journeys.

Sensors are increasingly used in ecological study for the tracking of organisms, often with the direct outfitting of animals with sensor backpacks and radio collars, in order to understand movement and migration. Sensors typically used in tracking include data transmitted through Argos or GPS satellite systems (including environmental temperature and humidity), light levels, acceleration, location, body temperature, heart rate, orientation, altitude, and pressure.[3] The latest wave of computationally enabled sensors for tracking follows from and complements multiple methods that capture movement and migration, including bird ringing or banding, bird observatories, radar and time-lapse film, and isotope and DNA analysis.[4]

Any number of organisms have been outfitted with computationally based tracking technologies, including honeybees with RFID tags,[5] green darner dragonflies with miniature radio transmitters,[6] and Arctic terns with miniature geolocators.[7] Even animals that disappear or die are monitored, as with the well-known case of "Happy Feet," the "hapless emperor penguin" who turned up on a beach in New Zealand. Fitted with an Argos satellite transmitter Sirtrack and sent on his way back to Antarctica, Happy Feet's tracker soon ceased transmitting, which led "to the conclusion that either the satellite transmitter has detached or an unknown event has prevented Happy Feet from resurfacing."[8]

The movements of organisms, from badgers to elephant seals to storks, influence understandings not only of the journeys these animals take but also the environments that they inhabit and rely upon. Tracking devices are often presented as key technologies for studying animals under threat and for informing policy and management decisions so that these organisms may be better protected.[9] Disruptions to habitat, causes of mortality, loss of feeding grounds: these are environmental events that the tracked journeys of organisms may reveal. But these technologies also generate concerns about the changes in activity and performance that animals might experience as a result of wearing the devices and through the stress caused by capture in order to fit the devices.[10] Sensors used for tracking raise questions about the extent to which animal movements and relations to environments shift through the machine-organism milieus that are traversed and inhabited. How are organisms, tracking devices, and milieus transindividuated, and what are the ways in which these technical objects *become environmental* through tracing animals' journeys?

Drawing on fieldwork and informal interviews conducted with environmental and computer scientists, as well as a review of scientific and technical literature and symposia, this chapter examines how the movement and migration of organisms have become key sites of study facilitated through environmental sensors. It asks how understandings of environmental change have shifted through increased levels of monitoring the movements of organisms. It also attends to the ways in which sensors for environmental monitoring undertake distinct journeys and types of attachments in order to travel along with organisms. In this concrescence of machine, organism, and environment, I ask how the milieus of technical and living entities in-form and transform through the tracking process. Rather than see tracking technologies as mirroring devices that invisibly capture hitherto unknown movements and journeys, I consider how these technical objects are involved in individuating organisms and environments as entities in need of further study and protection and as concrescing computational relationships that would activate the practices necessary to achieve these objectives.

In order to investigate these individuations of sensors, organisms, and milieus, I follow three organisms on their tracked journeys. These animals include badgers with RFID collars inhabiting the well-known "ecological laboratory" of Wytham Woods in Oxford; southern elephant seals fitted with data loggers and satellite transmitters, mentioned at the beginning of this chapter, as they dive through the Southern Ocean; and white storks carrying satellite transmitters, both as they are monitored and observed by scientists and citizens in the field and as they move across the platform of an Animal Tracker app set up to engage publics in practices of tracking animals.

As the southern elephant seal that appeared at the beginning of this chapter reminds us, environmental sensor networks involve not just tracking and tracing animal activities but also new forms of computational and collaborative sensing, where (as discussed in chapter 1) sensing is undertaken with and through more-than-human technologies and organisms. In other words, it is not just the organisms that are sensed. Organisms are transformed into sensors that would also communicate registers of animal-based environmental sensation and inhabitation. Much like the moss discussed in the last chapter, these organisms become biosensors of sorts, and through their journeys they provide data about environmental conditions and changes.[11] Through this relay of machine and organism, I then expand upon the sorts of sensing and experience that are articulated in tracked journeys. What are the nexuses, or actual worlds, of sense that occur through these mapped events?[12] And how do they both expand and challenge notions of environmental participation across organisms and machines? By addressing the traversals made across technical and living milieus, I finally consider how organisms and technical objects are not only expressive of living and technical milieus but also indicative of the particular problems they encounter in these milieus.

SENSING MOVEMENT AND MIGRATION

Animals are on the move, and have always been so, but their movements and migrations are emerging as different and more detailed events vis-à-vis data gathered through tracking studies. Why do organisms undertake these journeys? And how do these journeys shift understandings of environments not as fixed territories but as fluctuating zones of inhabitation, sporadic meeting points, feeding, and mating grounds—essential stopovers on some far-flung journey? Studies of animal movement and migration have been undertaken for some time, with tracking and nomadism as much indicators of different practices for studying (and even living with) animal movements as scientific approaches to tracking and tracing organisms through computational sensors and data. The comings and goings of insects, birds, mammals, and fish have indicated seasonal change, habitat disruption, or even impending disaster. Movement, it seems, provides an indication of animal behavior and routine.[13]

Indeed, the emerging field of "movement ecology" is seen to address some of the "unanswered questions in ecology," including reasons for fluctuations in animal populations, where animals are and when they are located in distinct areas, as well as when, where, and why they die.[14] If animals are on the move, so too is ecology adopting more mobile methods in order to understand these questions. As historian of science Etienne Benson discusses in his study of wildlife tracking, *Wired Wilderness,* Denali National Park in Alaska is one such site that at one point went from a practice of reluctant animal tracking to one of perpetual observation, where "new kinds of radio collar" were being deployed that would make it possible "to track wildlife 24 hours per day, 7 days per week."[15] In some cases, arguments are made not just for mapping and tracking animals in relation to distinct research questions but also for "life tracking" animals in order to understand movement, behavior, ontogeny, dispersal, and mortality across lifetimes.[16] In such a scenario, animals would be tagged at birth and spend their entire lifecycle wearing a tracking device that monitors and records their activities.

In a parallel way, tracking animals is a practice that provides more information on the environmental selections and constraints that animals encounter. Movement and migration data gathered from tracking devices can be related to environmental conditions, for instance, as one environmental data platform, MoveBank, makes readily available.[17] In the context of an experimental ecology that translocates tagged birds away from their migration routes and then observes their eventual (correcting) movements, it is possible to draw inferences about the way that birds pick up on the movements in winds, the possible cues made available through shifts in atmospheric conditions, and the flight paths taken, as well as wing beats, heart rate, and energy expended along a chosen path. As ornithologist and advocate of animal tracking Martin Wikelski has indicated in relation to this example, "We can use individual animals as sentinels for the atmosphere if we understand how they use the atmosphere."[18] Through flight path, heart rate, and acceleration data (among other variables), the journeys of tagged animals can provide indirect data on the churnings of air. These data can be further corroborated by comparing environmental data to flight conditions, creating a record of the particular milieus that animals might be navigating through and inhabiting.

Every Animal with a Cell Phone

The technologies used to track animals typically consist of devices that communicate through radio and satellite telemetry, including 3G and 4G mobile phone signals, GPS, Argos satellite systems, RFID tags, and data loggers.[19] While most tracking devices are attached to the external bodies of organisms as collars,

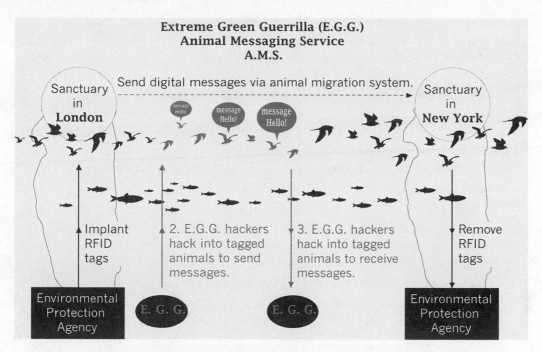

Figure 3.1. "Animal Messaging Service." Speculative system for sending digital messages via migrating animals implanted with RFID tags. Extreme Green Guerillas, illustration courtesy of Michiko Nitta.

backpacks, and epoxied antennae in the case of insects and smaller organisms, some tracking devices are subcutaneously injected (not unlike the RFID tagging of many household pets) and are especially focused on logging heartbeat and temperature. Monitoring technologies are becoming increasingly miniaturized, and the expectation is that a greater number of animals may be monitored through more sensory variables over longer periods of time.[20]

Presenting research at the Symposium on Animal Movement and the Environment 2014, Wikelski recounted a project focused on studying families of geese in Siberia in order to understand their social interaction. Describing the process of tagging the geese, he noted, "So you catch them, you put tags on them and this is really how the world should look like: every animal has a little cell phone, they talk with us."[21] Wikelski indicates, however, "There are still some problems because these things should be much smaller, and they should all be on necklaces." Despite these considerations, all in all, the tracking of geese was "working well,"[22] and was providing new data about their social interactions, as well as the ways in which rising temperatures lead to migration activities. This is seen to be a way, ultimately, to "get closer to the decision-making process of these geese."[23]

Within the proliferating range and type of tracking devices in use, it is possible to combine sensor data with a wider range of data. In addition, Wikelski has noted that there is a further need "to have very miniaturized cameras on these animals," similar to those used on the CritterCam project (and as discussed in chapter 2) in order "to ground truth other sensor data."[24] As Greg Marshall of the CritterCam project has noted, initially the idea to attach cameras and monitoring devices to animals seemed improbable, and the consensus was that animals would not tolerate carrying the devices. However, as he suggests, "most animals seem to care little about the unusual electronic remora appended to their backs. This unexpected finding has increasingly emboldened researchers to consider use of animal-borne imaging tools to study difficult-to-observe animal behavior and ecology."[25] As a result of this and the miniaturization of monitoring technologies, there are now

> more species, gathering richer information, resulting in an expanding body of statistically supported assertions of novel behaviors and ecological relationships. And today, with the ongoing revolution in solid-state imaging systems that integrate video, audio, environmental, geospatial, and perhaps even physiological data streams, we can expect a quantum leap in application of these instruments.[26]

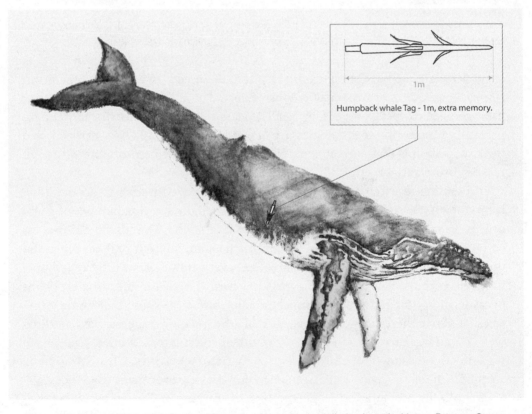

Humpback whale Tag - 1m, extra memory.

Figure 3.2. "Animal Messaging Service." Example of humpback whale implanted with tag. Extreme Green Guerillas, illustration courtesy of Michiko Nitta.

With such extensive plans for instrumentation, across sensor modalities, animals move from being cell-phone equipped to becoming multidimensional sensing nodes that communicate their own bodily conditions, interactions with neighboring animals, movement, and environmental conditions. In this sense, Marshall notes in relation to marine organisms, "animals themselves can now serve as remote ocean observation platforms carrying instruments to characterize habitat over temporal and spatial scales relevant to their basic biology and life histories."[27] Through this ongoing data collection, animals become sensor nodes and platforms, from which some of the "unanswered questions in ecology" are meant to be addressed, while also creating an expansive and even global animal-sensor network that functions as the "pulse of the living planet."

Pulse of the Planet

The notion that animal-sensor networks are providing the "pulse of the living planet" comes from one of the most notable and sizeable initiatives to undertake animal tracking on a global scale: the satellite-based project International Cooperation for Animal Research Using Space, or ICARUS. A global small-animal tracking system led by Wikelski, ICARUS seeks to set up a "remote sensing platform for scientists world-wide that track[s] small organisms globally, enabling observations and experiments over large spatial scales."[28] While the Argos satellite system has primarily been used for tracking and sensing larger animals, as it requires larger transmitters for communication,[29] ICARUS is able to work with smaller tags that communicate locations across shorter distances. The anticipation with this technology is that ICARUS will allow insights into the "dispersal and migration" of smaller organisms, which will "provide a seeing-eye dog for humankind"—in other words, it will enable the use of "the evolved senses of animals for remote sensing."[30]

While ICARUS is set to be launched as an antenna added to the International Space Station (ISS) in 2015, it is also seeking to partner with Russian, Chinese, and European Space Agency (ESA) satellite launches. The ISS antenna is proposed to communicate with small tags weighing less than five grams (and which Wikelski projects will further reduce in size) and which consist of a logger tag with GPS.[31] There are many perceived possible uses of ICARUS that "will enable researchers to answer some 'grand challenges in environmental sciences,'" including understanding the spread of diseases carried by animals, protecting sites key to migration pathways, and establishing relationships between biological diversity and ecosystems, as well as establishing a disaster network based on animal sensing.[32]

These multiple uses of ICARUS for research are based on animals serving as continual generators and collectors of data, which can further be collected in the MoveBank database and compared to corresponding environmental data. All together, these animal-generated and sensor-based data are meant to provide an

ongoing picture of the pulse of the planet. The movement of organisms, the fluctuations of animal populations, the responses to deforestation or other land-use
based events and changes, may all register as flickering patterns of information,
the cadence of a living planetary body: here is another iteration of "program
earth," articulated through the digital monitoring of the movements of innumerable earthly organisms.

More Data = More Engagement

The impetus to collect more data as a way to achieve greater insights is one that
runs through environmental sciences and is a key way to address environmental
change. As discussed in chapter 1, the proliferation of more detailed and more real-
time sensor data is also meant to provide fundamentally new insights into ecological processes. These data-based insights are further meant to bolster conservations
projects, so that more sound and effective decisions can be made. Monitoring, in
this sense, is a practice that is undertaken ultimately to protect organisms.[33] The
"great migrations" are declining,[34] the extinction of species is occurring at an unprecedented rate, and organisms that are vital to food chains are collapsing.

One project, the Tagging of Pacific Predators, or TOPP, deployed 4,306 tags
across 23 species to study events such as the possible collapse of bluefin tuna.[35] In
a video outlining the aims of the project, the point is made that these tagging
initiatives will not only "unlock the mysteries of the deep" but will also provide
new information that can be communicated to publics. As Warner Chabot notes,
"The more that we can give the public the facts they will respond. If you inform
and inspire the public you will empower the public to respond and, frankly, act in
its goodwill and future."[36] More detailed information is a resource that is meant
to motivate conservation efforts, inform policy, and stir publics into action.

The drive to collect more data on organisms as they move around and undertake migrations has in this way influenced numerous citizen-sensing projects.
While animal spotting and bird ringing are long-standing practices within citizen
science, many more projects are emerging that make use of digital devices to
track, record, monitor, and observe moving and migrating organisms. Indeed, in
one example of GPS-based citizen science and environmental monitoring, I can
recall a campaign made by the Wildfowl and Wetlands Trust (WWT) in 2002
for citizens to "adopt a goose," which would be fitted with a GPS tracker to relay
data about a particular goose's journey to the citizen-adopter. As a sequel, perhaps, to the transmitted heartbeats of the Russian dog Laika launched in *Sputnik 2*,
the WWT conservation project monitored light-bellied brent geese and their
migrations from Canada to Ireland. The public could adopt geese with such aus-
picious names as Major Ruttledge and Arnthor, and in exchange receive up-to-the-
minute information on position, speed, and heading, delivered via email or to
mobile phone.[37]

Many citizen-sensing engagements then focus on both gathering data as well as interacting with the greater stores of data now available on animal movements. Gathering information in some cases is the trigger for particular conservation-based actions, including BirdReturns, a project led by the California Nature Conservancy that incorporates citizen-science data from the eBird project out of the Cornell Lab of Ornithology, which maps the migratory patterns of birds as they move through the Central Valley of California. Participants in the project can further submit observations through a web platform or app, BirdLog North America,[38] and all together this citizen-supplied information is mapped onto critical wetland zones. The Nature Conservancy then rents the space from farmers via a "pop-up habitat" scheme, where the farmers allow their land to remain flooded during critical times when birds are migrating through the valley.[39]

The Migratory Connectivity Group notes that multiple citizen-science and citizen-sensing projects actually tend to concentrate on migratory organisms, including birds, invertebrates and fish.[40] Projects in this area working through distinctly digital methods and devices further include the "Tag a Tiny Program," which enrolls the help of recreational fishermen to catch, measure, tag, and release juvenile Atlantic bluefin tuna in order to study their "annual migration paths and habitat use."[41] Many more citizen-sensing projects focus on animals as they move and migrate, from tagged sharks that send tweet alerts when they approach popular beach areas in Australia, to records of migrating eels in the River Thames, as well as projects such as Roadkill Garneau, which asks citizens to record and locate roadkill sightings using EpiCollect, an online platform and mobile app, so that critical sites where movement has gone wrong can be recorded.[42]

There are even videos instructing lay audiences on how to build their own DIY GPS tracking kit by hacking a GPS device to build a bespoke radio collar.[43] And creative-practice projects have created distinct ways of engaging with animal movement by, for instance, being able to engage in a text-message exchange with fish in the Hudson River in New York City. "Amphibious Architecture," a project by Natalie Jeremijenko, David Benjamin, and Soo-In Yang, uses motion sensors, LEDs, and a text-message system to trigger alerts to passersby who may tune into the movements of fish. Sample text messages note, "Underwater, it is now loud. To find out more, text 'HeyHerring' or 'AhoyAnchovie' or 'GreatEast.'" This proof-of-concept project enters into those communicative exchanges with animals that tracking and sensing technologies are meant to enable, albeit with a slightly different approach to the messages that might be shared.[44] In a different way, in her "Extreme Green Guerilla" project, Michiko Nitta has proposed that we might harness the movements and migrations of animals as an alternative and even "green" communication system, where our messages might be more efficiently and ecologically carried by animals crossing oceans and continents.[45]

A common thread across these scientific and creative-practice projects is that communicative exchange unfolds not through speech, but rather through perceptive engagements built up through environmental inhabitations. The prevailing sense with tracking projects seems to be that this is a mode of communication that may be readily accessible to us, where by observing organisms it may be possible to deduce their environmental requirements. Watching, spotting, and reporting journeys; tagging and contributing to scientific monitoring; and amassing collections of migratory data—within and through the interstices of movement ecology projects—multiple projects are contributing to building up more detailed accounts of animals' movement and migration.[46] And in this watching and encountering of organisms, humans, more-than-humans, and organisms are moving through intersecting milieus, forming new nexuses of sense.

ANIMALS AS SENSORS: BADGERS, ELEPHANT SEALS, AND WHITE STORKS

Data sets that are more complete and comprehensive are meant to fill in the blank spaces on our maps of animal movement so that we might "build a global picture of the creatures with which we share this world."[47] With animals serving as sensors and sensor networks, sensor data is meant to function not only as descriptive data but also as material that allows us to infer events from what animals might be sensing and responding to in environments. Animals-as-sensors become subject-superjects in a particular way within tracking projects, where their journeys are meant to communicate the experiences of their environmental encounters. The becoming environmental of computation here occurs through the journeys and tracking that unfold as sensors travel with organisms, as well as through the ways in which *organisms become computational* both as carriers of sensors and through the ways in which their sensory ecologies are meant to provide data and information on environmental conditions. Organisms are thus made to be computational twice over, as they sense and are sensed. I now turn to consider three specific journeys or movements of animals that attend to the ways in which animals-as-sensors concresce as indicators of specific engagements with milieus.

Badgers Socializing in Wytham Woods

WildSensing, an interdisciplinary collaboration between computer scientists and ecologists based at the University of Cambridge and Oxford University that took place between 2007 and 2010, involved a study of badger activity in Wytham Woods near Oxford—a highly instrumented test site known for its ongoing ecological experiments from at least the days of Charles Elton, an ecologist well known for his studies of population ecology and animal invasions in the early to mid-twentieth century.[48] Wytham Woods is a 390-hectare landscape that is "one of the most researched areas of woodland in the world," with numerous monitoring projects underway at any given time.[49] But many of these projects are often set up

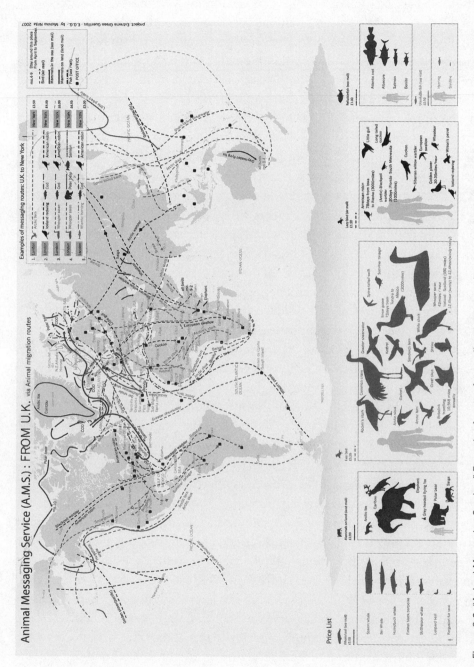

Figure 3.3. "Animal Messaging Service." Example of routes for sending messages via tagged animals as they undertake their migrations. Extreme Green Guerillas, illustration courtesy of Michiko Nitta.

in relation to distinct research questions and concerns and do not join up data sets collected from the site. At the same time, because ecological study and experimentation have taken place over several decades at Wytham Woods, there are extensive data sets and histories of animal observation. With badgers, for instance, data collection extends over the past twenty years, although it may have a larger granularity due to manual observation methods; and badgers have been trapped and released in Wytham Woods for the past thirty years (which has been the usual way of studying animal movement).

The WildSensing project was initiated to establish whether to and to what extent badgers transmitted tuberculosis, for instance, to livestock. Data from these observations were meant to aid in policy and management of badgers at agricultural edges.[50] To undertake this research, the project focused on the social networks of badgers, since as it turns out they have distinct modes of interaction and cooperation. In total, eighty badgers were tagged and caught once every six months over the duration of the project. Animals were tagged with RFID radio collars, which would be released when badgers where thinning. As a result of using RFID for detection, badgers could be sensed underground as well as above ground, but only within the sensor area and not across the entire forest.[51]

In the first iteration of the WildSensing project, badgers were tagged with RFID radio collars that communicated with fixed sensor detection and storage nodes located within a zone of the forest. From these points, field researchers could conduct mobile data collection (which could, theoretically, also be carried out by mobile robots). Within one year, the project collected over twenty-five million records, and so the gathering and transmission of data presented issues for how to structure these networks.[52] Due to the quantities of data collected and transmitted, much of the project focused on ways of duty-cycling data more efficiently in order to save power, which is an ongoing issue within sensor networks.

In the second iteration of the project, an increasing emphasis was placed on working with off-the-shelf sensor equipment. Rather than having fixed sensor nodes in the network, the project instead used the badgers as the mobile sensor network across which data circulated to fixed collection nodes triggered by presence detectors with a fifty-meter radius. The data from these nodes were then either stored on SD cards or transmitted via 3G mobile phone networks several times per day to servers. On the one hand, this approach focused on how sensors learn and adapt to animal behavior. Working with RFID sensors and machine learning in the form of an adaptive algorithm, this approach focused on having sensors operate in response to and at key moments of animal activity. On the other hand, as sensors and animals were paired in this form of environmental monitoring, sending new software over wireless networks to the animal collars also became a way to reprogram sensors without having to catch the animals or adjust the sensor hardware or infrastructure so that the network could be adapted to animal activities.[53]

Emerging within this approach is the use of sensors not just to describe and capture environmental events but also to develop a dynamic evolution of sensors in response to animal behavior such that computation and the distribution of sensation are ontogenetic. While critiques of early tracking devices suggested that they were "'mere descriptions of movement and activity,'"[54] and hence at times considered to be relatively static renderings of environmental processes, increasingly sensor systems are regarded as generating more integrated, adaptive, and actuated approaches to environmental monitoring.

As the WildSensing mobile network developed, it became a system for relating information from animal to animal via radio collars and then on to collection nodes. Animals became sensors and operators in the network, at once collecting data about their activities and location, while also becoming part of the extended computational infrastructure. The network patterns were ad hoc, based on the badger activity, and were not entirely preestablished configurations. The social behavior of the badgers, as well as the microclimate and other environmental conditions at Wytham Woods, contributed to the intersections of technical and living milieus. The sensors and computational network necessary to capture phenomena had to emerge along with ecological events and animal activity, where, for instance, practices of relaying data across organisms and storing sensor data in nodes, then capturing the data through mobile collection, developed as a more effective configuration for sensing the badger activity.

Machine learning here extended not just to parsing environmental data but also to learning animal behavior and reprogramming sensing and collection methods accordingly. In this sense, sensors became organismal *and* environmental. While this was not a completely open process, as sensors are configured to detect certain variables and not others, it was also not a process of complete automation, where sensors might be preprogrammed to detect phenomena according to fixed configurations. If we were to follow Simondon in this regard, how might this contingent approach to sensing shift both technical object and technical milieu in relation to the individuations that occur through encounters with living entities? Rather than approach sensors as "prosthetic" devices, moreover, might we find it more accurate to consider the ways in which these sensor technologies reorganize, in-form, and transform along with the organisms they would track?

Elephant Seals Diving in the Southern Ocean

If the badgers of Wytham Woods presented a quite local and land-based sensor study, then the elephant seals of the Southern Ocean offer up a much different milieu in the form of underwater spaces, relatively obstreperous temperaments in comparison to badgers accustomed to recurrent catch-and-release, and sensor systems that communicate via satellite rather than more proximate RFID nodes. Led through the Natural Environment Research Council Sea Mammal Research Unit at the University of St. Andrews, the elephant seal study set out to research how these animals respond to environmental variability and how this variability might in turn influence population fluctuations.[55]

As a study of telemetry and marine mammals, the project was situated within a larger study of over one hundred mammals and marine species. The elephant seals were tagged with CTD-SRDLs in the first iteration of the study. Data gathered included depth, conduction, temperature, pressure, and acceleration, as well as stereo sound captured at 500khz. The tagged seals not only revealed profiles of their diving habits, where everything from buoyancy to fat reserves can be assessed based on diving details, but they also captured data on the temperature of the Southern Ocean at depths not typically monitored.[56] Translations could then made from elephant seal data to the Argos satellite system to St. Andrews to the Met Office to the BBC weather forecast.[57] Elephant seal data have also been integrated into ocean observing systems, so that the complex conditions of oceans become more thoroughly monitored through the underwater activities of tagged marine animals.[58]

During a 2010 Mammal Society conference focused on techniques for sensing and tracking animals, one of the presenters discussing this elephant seal project considered the possibility for moving from Argos to a GPS/GSM mobile phone system to relay data. Such a system effectively would involve "stick[ing] a mobile phone on seals as a point of connection," and could become a method for communicating with publics.[59] Here, by bringing animal tracking and communication into the realm of mobile phone networks a more immediate contact with the animals would appear to unfold—similar to the geese with cell phones discussed earlier in this chapter. Such a strategy of communicating with publics was in fact implemented when Argos was first used to monitor animals, where emails and updates of tracked animals were regularly sent to schoolchildren and publics.[60] Yet the imaginary of moving from a satellite connection to mobile phone exchanges of data seems to bring the immediacy of animal communication even closer, particularly when animals are sensors of environments.

This is the common thread that arises in public presentations of animal tracking projects, where the ability to communicate with animals is frequently referred to as one of the benefits of these sensor systems. Indeed, the ICARUS promotional material notes, "With the help of new technology, animals will be able to communicate with us, revealing changes, dangers and connections that will help us to have a better understanding of the fabric of life on earth."[61] The ambition is that animals will "talk" to us, and in so doing that they will communicate the distinct sensory inhabitations that they experience. Animals equipped with sensors become, in turn, active sensors able to perform heightened modes of communication. Through this talking, the hope is that we will finally be guided toward making better decisions for preserving the planet.

Tagged elephant seals communicate not only the specific data points of temperature and downward acceleration when they dive but also provide an indication of the multiple milieus that they cross, from the technical milieus of sensor devices, to the lived milieus of the seal, and the transformative or associated milieus across and through which new becomings concretize. This is a becoming environmental of computation and a becoming computational of organisms. Perhaps it is also the becoming organismal of machines. These points will be discussed in the section below "The Problem of Milieus," but I here turn to consider the final animal-sensor journey in the form of white storks that are tracked with satellite transmitters and which also feature on the Animal Tracker app, which is oriented toward engaging citizens in monitoring animals.

White Storks Navigating Aerial Ecologies

The final sensor journey that I discuss in this chapter involves the tracking of white stork migrations across two main flyways through Germany, Greece, Turkey, Tunisia, and South Africa. Based at the Max Planck Institute for Ornithology in collaboration with researchers across multiple countries in the flyways, this study of white storks attempts to gather more complete lifetime tracking data across populations.[62] The storks are fitted with solar-powered GPS transmitters that capture location and body-acceleration data every five minutes. Updates on the locations of the white storks are sent by SMS messages, but the majority of data are downloaded from relatively nearby VHF radio connections, which need to be no more than three hundred meters away from the storks. This means that scientists need to follow—and even "chase"—the white storks from their breeding areas and along their flyways in order to download data.[63]

Part of the interest in studying white storks is in relation to their importance as "sentinels" for environmental events. For example, storks can be found in areas where there are outbreaks of desert locusts, as they will congregate in these areas for feeding. In this way, the storks have been described as "advanced" remote sensors for providing insights into environmental events. Storks are studied for the patterns and energy expenditure of their migrations, how they interact, and where they stop to rest. Body-acceleration data that are gathered from the storks' journeys allow researchers to estimate energy expenditure and behavior, which can further indicate environmental events.[64] In addition, attention is given to where and why animals are dying, as this could influence land-use decisions about which habitats to conserve.

As many storks die in remote places in Africa, there is a need to retrieve transmitters and the high-resolution data they contain. In this way, another sort of

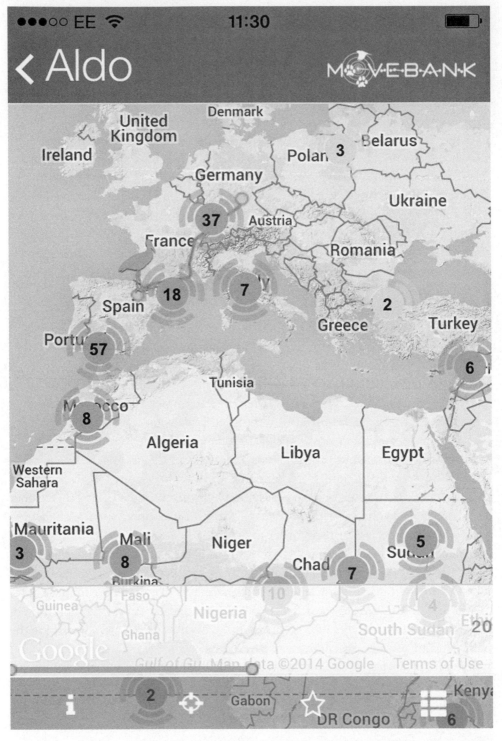

Figure 3.4. Aldo. Animal Tracker app showing locations of tagged white stork in relation to movements of eighty other tagged white storks as they undertake migrations. App developed through MoveBank and the Max Planck Institute for Ornithology. Screen capture.

citizen sensing emerges, although in this case participants are referred to as "collaborators," who help to find and return the transmitters from the dead birds. From Malawi to Sudan, people return the birds and transmitters, often accumulating stories of how and where the storks may have died, including becoming entangled in debris from rubbish dumps.[65]

It is these practices of attending to where storks are, and retrieving them when they die, that Wikelski has suggested is the "future" of citizen science, when people are specifically engaged with observing animal movement. Wikelski has further suggested that animals may be better protected if people know where they are, since they are less likely to be hunted if they are watched over. To this end, the Max Planck Institute for Ornithology has developed a wildlife tracking app, Animal Tracker, which is a tool for citizen engagement that provides relatively real-time data about animal locations and allows us to "observe animals . . . virtually in our cell phones every day."[66]

Some of the first tagged and tracked animals to feature on the Animal Tracker app are the eighty white storks that are under study.[67] Users of the app can see the migratory routes of the white storks and put them on a "watch list" so that notifications are sent "when they do something"—and news of the storks can also be shared on social media while records of the data are kept in the MoveBank archive.[68] As I click through the Animal Tracker app, I find a white stork located just outside of Nuremburg, Germany, in the small town of Forcheim. Named "Aldo," this white stork last registered activity the previous evening, and is tagged with a DER AT881 (eobs 3946) sensor. I learn through searching outside the app that the digital telemetry company e-obs makes "high-end digital tags for the study of animal behavior," with a focus on "lightweight GPS tags."[69] This is the device that allows the white stork's movements to be transmitted and eventually displayed through the interface of the app.

Aldo was born in 2014 in Vorra, Oberfranken, Bavaria, and was one of three chicks in the nest. His siblings are Amos and Resl. I am able to click on Aldo's two-week and one-year movement data, and I see from the two-week record of his movements that he has hovered around the town of Hochstadt for a while, then quickly moved over to Forcheim, dipping down to Erlangen, and then back up to Forcheim, where he currently rests. If I look at the record of his one-year movement data, I can see that he has not traveled far from the place of his birth. I can also mark Aldo as a favorite stork so that I can receive updates. Several weeks later, I see that Aldo and many other storks are on the move from Germany to the south of France, on to Spain and Morocco. While Aldo is near the Spanish Pyrenees I can zoom out even further to see stork movement down the second flyway, across Greece and Turkey in to Sudan and down to Tanzania.

If I happen to meet Aldo in the field, I can also add my own observations of his activities. The "add observation" section of the app asks:

What is the animal doing? Is it alone, together with conspecifics, or with individuals of other species? Is it feeding? Can you identify its food? There may be a
multitude of other things that are interesting for you. Please do not hesitate to
report them. Your photos and observations will instantaneously be published in
the "Animal information" section in Animal Tracker and will contribute to the
data file of this individual. Your observation is a direct and very important contribution to our science and helps us to understand the life of "our" animals much
better. Thank you very much for your contribution![70]

I mark Aldo as a favorite stork, and begin to click through the map looking for
other birds: they are everywhere, in small towns and mountain valleys, some even
appear to be waiting at cafes, pinpointed on top of restaurant signs and in town
centers. I learn about Wanderer, the stork born in a birch tree, and Isolde, who
was born in southwest Germany but is now located in Parc Natural Régional de
Camargue in the south of France.

Commenting on the possibilities of tagging and so sensing with and through
animals, Wikelski notes,

What's also interesting is that we can use this kind of information as a sensor
network. . . . Animals *are* the most intelligent sensors that we have. If we have an
intelligent sensor network that is linked together around the globe then we can
gather amazing information about the environment.[71]

Indeed, in this project of tracking animals and thereby "decoding the intelligence
of animal behavior," even further applications have been proposed for using animals as sensors. If white storks are sentinels for certain types of population fluctuations of organisms they feed on, for instance, other animals could be understood for the clues they provide about possible disasters that may be imminent.
Wikelski has thus tested the anecdote of whether animals are able to predict
disaster by tagging goats near Mount Etna and testing their movement patterns
in relation to volcanic eruptions As the goats are sensitive to the eruptions, they
demonstrate "strange and erratic behavior" that can be captured through tagging. The data gathered from mapping the goats' behavior in real time could
be used to trigger an alert for an imminent volcanic eruption up to 4 to 6 hours
before the event occurs.

A patent is pending on this "Disaster Alert Mediation Using Nature," which is
an invention for "a method forecasting an environmental event" that involves collecting behavioral or physiological data from a population of animals, comparing
it to a baseline data set, and establishing an alert for moments when a threshold is
crossed in the comparison between baseline and current data. The patent includes

a software program that is able to execute the steps necessary to analyze data and trigger an alert.[72] Wikelski sees this alert system as "really useable" as a "technical device," and is seeking investment in order to create "a global animal observation system" especially useful for "areas where people don't have much money."[73]

From sentinel white storks to citizen-sensing apps to global animal observation systems, animals are increasingly made into sensor nodes and networks that would inform us about critical environmental conditions and their responses. Yet what are the implications of these burgeoning animal-sensor networks? And what sorts of animal-human-milieu interactions might unfold through the more pervasive project of tagging numerous organisms? I take up these questions for the remainder of this chapter, specifically attending to the traversals made across organisms, sensing, data, and milieus.

THE PROBLEM OF MILIEUS

The ways in which animals are becoming both sensor nodes and parts of extended sensor networks raise questions about how these tagged and tracked individuals traverse and inhabit milieus. In this discussion of milieus, both technical and living, I am drawing on the work of Simondon and Canguilhem, who in varying but shared ways were interested to account for the ways in which individuals (per Simondon) and organisms (per Canguilhem) are formed and in-formed by encountering "problems" in their milieus. As Canguilhem has suggested in his analysis of milieus, how organisms encounter the problem of their milieu is how they become. Yet these problems are different for different organisms.[74] As Simondon similarly articulates, the problem of the milieu is a condition for inventive responses, which is also a condition for individuation.[75] As milieus are sites of inventive encounters and responses to problems, moreover, it is not possible to limit the relations and capabilities that individuals might draw on and express in addressing the problems of their milieus.[76]

This approach to organisms/individuals and milieus has several points of resonance for thinking about the implications of tracking animals and using their movement patterns as extended sensor networks. Humans in the form of scientists and citizen scientists have largely formed the problem of milieus as one of gathering more data in order to address environmental change. In this sense, understanding how to respond to the problem of our shifting milieus has become a project of ensuring there are no "blank spots" on our maps of environmental change. This problem-logic is influenced by the notion that when data sets are the most complete we will assumedly have the most advanced ability to manage environments. In turn, the problem of our milieus has also become one of monitoring all manner of environmental phenomena, including tracking organisms for the clues they provide about the worlds that they inhabit and how their worlds may be changing.

There are a curious series of translations that take place across animal-sensed milieus, tagged organisms, and generated data, since we could ask whether organisms are having to inhabit our encounters with our problem-milieus by living with tags and tracking devices, potentially for their entire lifetimes. Yet how do these intersections of encounters with milieus transform animals as they encounter their milieus and the problems of their milieus: Does the situation of wearing tags and tracking devices change the ways in which organisms encounter their milieus, while also in-forming their problems? It has been recognized in scientific literature on movement ecology that tagging can and does change the activities of organisms.[77] Questions have also arisen as to whether it is always instructive to tag organisms that are under threat, as the process of capturing, tagging, releasing, and monitoring may contribute to the stress of animals.[78]

But tagging and tracking are not just issues of intervention in order to gain a more accurate picture of organismal activity. There are also points of consideration about how monitoring devices and practices in-form the milieus and perceptive exchanges of organisms with those milieus, since this is also the very thing that would be mobilized, whether for conservation and policy or for disaster networks. Canguilhem has critically noted that a danger with some forms of science, such as physics, is that they can be based upon a universal milieu that speaks neither to the perceptive experiences of organisms nor humans. If science is in the world, however, as Canguilhem suggests, it must admit to a diversity of milieus.

Perception is the way in which organisms go about encountering and fashioning their milieus. Sensing is then a key practice for working through problems of milieus.[79] As Canguilhem writes, "In fact, as a proper milieu for comportment and life, the milieu of man's sensory and technical values does not in itself have more reality than the milieus proper to the woodlouse or the gray mouse."[80] No milieu or experience of a milieu is more real than any other, unless we adhere to the universal milieu of science, which establishes a version of the real that disqualifies all others.[81] Following Whitehead, to account for the experience of the woodlouse and the scientist, we would have to make room for the "pluralistic realism" of environments and inhabitations.[82] Yet this is not a description of an absolute relativism, but rather of accounting for the a/effects that different inhabitations within distinct milieus express.

Indeed, proposals to use animals as sensor networks on one level seems to take on the approach of diversifying the sensing-milieu exchanges that occur across individuals. The encounters of organisms with their milieus provide another empirical basis for understanding environments and make room for the experiences of other organisms. And yet, in attending to the diversity of exchanges within milieus, a consistent if universal mode of capture is employed in the form of sensing and tagging devices. Here, one might ask whether it is perception (rather than a milieu) that has been transformed into a universal reality, whereby sensing devices, the variables they would measure, and the unfolding of sensing processes are made generalizable across organisms as an exchange of information. These generalized modes of information-based perception, furthermore, might be described as distinctly cybernetic operations, where sensing of milieus produces information that is the basis for actuating and producing further effects in milieus. Rather than the physics of a universal milieu, sensors might have given us the cybernetics of generalizable perception and experience.

Working across Canguilhem and Simondon, one could then ask: How do milieus and perception shift, both for organisms and devices, when sensing is primarily undertaken and filtered through tracking and tagging technologies? Working laterally with Simondon's discussion of the associated milieu, we could say that technical objects concretize technical milieus in a way that could be compared to Canguilhem's articulation of how organisms at once encounter and concretize their milieus. The difference, following Simondon, between technical and living milieus would be the way in which living milieus can be self-reproducing, whereas technical milieus are self-reproducing only in distinct circumstances where they operate as natural objects, and even then they imply the contribution and intermediation of the living entities that made them—in other words, humans.

In traversing these different milieus, we could say that it is the living milieus of tracked organisms that begin to resemble the operations of the technical milieus of technical objects, since animal sensing becomes equated with computational sensors. By virtue of being equipped with sensors, animals' perceptive encounters with their milieus are transformed into informational exchanges through computational sensor networks. A response to an environmental event is a sensor-actuator exchange of information. An adaptation to an environmental event is a calculative decision, arrived at through an analysis of energy expenditure and environmental cues. Organisms' perceptual engagements with their milieus become informational not simply in the way in which they are in-formed but also as digital operations generative of computational data. Such an approach in part fits with the more recent notion that all of "nature" is composed of information and so is inherently computable.[83] But it also coincides with the longer histories of cybernetics where informational exchanges have been put to work to explain everything from ecosystems to population collapse.

Sensing of environments is then generally understood with tagging and tracking studies to fit within an informational logic of sensing stimuli, transferring signals, and actuating responses. Yet in what ways might this informational approach to perception preconstitute the possible modalities and relations of individuals as they interact with their worlds? A flight path chosen becomes a matter of a response to wind direction and speed and an organism's internal calculation about energy to be expended to reach a particular destination. Rather than this being a question of what is captured and what is not—a usual way of attempting to make room for all that is in "excess" of scientific endeavor—one might suggest this is a way of making particular worlds and milieus in which the problems of organisms are articulated and acted upon. Environments and environmental change become informational problems. These are the informational-environmental-organismal processes, in other words, whereby we are working through the problem of our milieus, which are increasingly sites of environmental concern, as well as pre-supposing the perceptual-milieus of other organisms. We might then ask how such an approach to working out the problems of our milieus might also in-form our possible becomings in relation to how to "protect" organisms and their milieus. The becoming environmental of computation and the becoming compu-tational of environments are processes that concretize these extended political and ecological effects.

Machine and Organism

While we could discuss the animal-sensor networks that come together in move-ment ecology as hybridities or infoldings of sense, as discussed in chapter 2 (and throughout this study), I am interested to maintain a focus on the *environmental* operations of perception (rather than attend to different conjugations of subjects and objects, nature and culture). At the same time, it is useful here to turn to a particular discussion that Canguilhem raised in relation to machines and organ-isms that provides insights into the ways in which perceptive capacities may be understood, potentially through machinic, and later cybernetic, forces.

Organisms have circulated through computational and cybernetic imaginar-ies for some time now, from dolphins studied for sonar sensing and later taken up as a topic of interest by Gregory Bateson, to Nicholas Negroponte's gerbil-based interests as displayed in the "Software" exhibition, and many more besides.[84] Automata studies have looped, continuously it seems, across organismal and tech-nological modalities of sensing: linking, comparing, and fusing these to arrive at a more perfect union.

In his chapter "Machine and Organism," Canguilhem works through "the mechanical theory of the organism" to consider how philosophers and scientists alike often "have taken the machine to be a given," not only as though it is the concretization of scientific theory but also as though it provides an originary template for explaining the functions of organisms. But he sets out to demon-strate how "biological organization" is anterior to machines, so that life cannot simply be described through reference or analogy to machines. Across Descartes to Taylor there unfolds a certain mechanistic analysis of organisms that accounts for some outputs and not others. From Canguilhem's perspective, there is a need to "inscribe the mechanical within the organic."[85] He writes:

> We must admit that, in the organism, a plurality of functions can adapt to the singularity of an organ. An organism thus has greater latitude of action than a machine. It has less purpose and more potentialities. The living organism acts in accordance with empiricism, whereas the machine, which is the product of calculation, verifies the norms of calculation, that is, the rational norms of identity, consistency, and predictability. Life, by contrast, is experience, that is to say, improvisation, the utilization of occurrences; it is an attempt in all directions.[86]

I read this assertion less as insisting on an essential organic foundation against and with which mechanical operations unfold and more as claiming that life is not reducible to any one version of mechanical rationalization (such as automatism, about which Simondon had much to say) because this approach would delimit what is otherwise an opening into experience and potential. The "function" of organisms cannot be definitively narrowed down to a singular process, since this would be to further obviate the potential for inventive encounters with the problem of milieus.

Canguilhem's grappling with machine and organism extends to Simondon's consideration of technical objects and speaks in particular to his critique of automatism (and Wiener's version of cybernetics) as reducing technical engagements to limited functions without the potential for becoming or invention.[87] Simondon was critical of both the conflating of animals and machines as basic units of responsiveness within Wiener's cybernetic theory (which seems a distinct continuation of the Cartesian legacy of automata),[88] as well as the ways in which—as automata—organisms could be "capable only of adaptive behavior."[89]

With sensors and tracking devices for studying animal movement, what emerges in part is just this cyberneticization of environments and organisms, where sensing becomes a way to home in on responsiveness, where movement and migration are assumed to be largely indicative of a series of adjustments and programs of responses that organisms make in relation to environments, and where sensing is an exchange of information. In this way, computation becomes the unquestioned originary machine that would in-form how we understand organismal and environmental processes. The "behavior" of organisms then becomes a series of information-based calculations and adaptations.[90] For both Canguilhem and Simondon, however, living is an inventive perceptual response to milieus. The problem of milieus gives rise to becomings, and not mere adaptations. Technical objects and technical milieus, moreover, can be understood as particularly human-oriented ways of working through problems of milieus in ways that *might* be inventive, but which are inevitably expressions of value: of which problems matter, and how they are to be addressed.[91] The potentialities of organisms (and machines) are thus individuated in these shared but differently articulated problems, milieus, and relations.

SHARING PERCEPTUAL WORLDS, ENVIRONMENT AS PARTICIPATION

Where does this discussion bring us in considering the implications of computational sensing practices for tracking migrating and moving organisms? I have not sought to articulate a position for or against these practices. Instead, I have attempted to find a way to address how the activities of animals are parsed, what this means for our understandings of their perceptual worlds and milieus, and how we might return to consider that which our informational-based ambitions and methods tune us into. If these data gathered through tracking techniques are meant to influence citizen sensing, policy, and conservation, then how might we also make room for attending to the very particular worlds and inhabitations that are accounted for, and the milieus, inventions, and becomings that might remain off the computational map, no matter how many more data points we add? The point this raises in relation to monitoring the movements of organisms is that tracking technologies are expressive of computational and cybernetic logics that parse organismal "behavior" as a particular response to milieus. But this articulation of relations is not definitive. Instead, it makes particular worlds within which we understand the goings-on of animals. It is another versioning of a programmable and programmed earth.

Part of this project might then involve attending to the nexuses of sense that are formed through distinct subject-superjects, as Whitehead puts it in his "philosophy of organism," since it is at these points where "actual worlds" form that are specific to perceiving and experiencing subjects.[92] This is also why, as I articulated in the introduction to *Program Earth,* we might consider the multiple earths that concresce through the project of programming the planet, since not only is the planet "wired up" in numerous ways but also there are many organisms that differently express their experiences and attendant worlds as part of their inhabitations.

In this study that works across Simondon and Whitehead, among others, it becomes ever more evident that what we take for a subject or individual is not a pregiven matter. Instead, individual entities are articulated through these processes of individuating and concrescing. Why is it important to return, continually, to this point? Because it is through addressing the entities, experiences, and worlds that are engaged in processes of becoming that we might attend to the ways in which feelings traverse worlds, organisms are affected, and individuals feel themselves to be more-than-one. As Combes writes in relation to Simondon:

> The "perceptive problematic" is that of the existence of a multiplicity of perceptual worlds wherein it is always a matter of inventing a form inaugurating a compatibility between the milieu in which perception operates and the being that perceives; and this problematic concerns the individual as such. Why insist

here that we are speaking of the individual *as such?* This is because the affective problematic is, inversely, the experience wherein a being will feel that it is not only individual. To put it more precisely, affectivity, the relational layer constituting the center of individuality, arises in us a liaison between the relation of the individual to itself and its relation to the world.[93]

We might say that sensor technologies—across uses that involve tracking and much more—assume that the perceptive problematic is settled, and the task that remains is to document the facts of movement in order to account for behavior. In this respect, we might say that a certain cybernetic mechanization of organisms and their milieus has settled into practice, whereby informationally based modes of sensing eliminate both perceptive problematics and the possibility for inventive experience. As Simondon reminds us, the perceptive problematic is also a matter of participation, where perceptual encounters with milieus are articulations of a feeling for more-than-one, and a feeling for worlds.

Simondon has then suggested that the ongoing resolution of problems in a milieu is the basis by which an organism continues to individuate itself and be individuated. Perception, in this case, is not a matter of an organism decoding an external form for calculative gain but of articulating relations within their milieus.[94] All of the entities involved in individuating, perceiving, encountering, resolving, relating, and worlding can shift in these processes, which concretize through participation in milieus. As Simondon writes:

> A relation does not spring up between two terms that are already separate individuals, rather, it is an aspect of the *internal resonance of a system of individuation.* It forms a part of a wider system. The living being, which is simultaneously more and less than a unity, possesses an internal problematic and is capable of being an element in a problematic that has a wider scope than itself. As far as the individual is concerned, participation here means *being an element in a much larger process of individuation* by means of the inheritance of *preindividual reality that the individual contains*—that is, due to the potentials that it has retained.[95]

From this perspective of encountering individuals through processes of individuation, Simondon further notes, "It now becomes feasible to think of both the internal and external relationship as one of participation, without having to adduce new substances by way of explanation."[96] Participation is a way of working through the problems of individual entities, milieus, and relations. These are conditions and entities that form through the very processes and experiences of participation. Participation is in-formative and inventive, rather than a register of programmed behavior or responsiveness.

For this reason, organisms might also disrupt the projects and studies in which they find themselves, from destroying cameras to removing tags to disappearing from the radar.[97] As Stengers has noted, "The construction of an experimental device" in no way "ensures that the being we wish to mobilize will agree to show up."[98] Rather than seeing these as instances of bad data or outliers in a research study, we might instead tune into these events as instances where participation in actual and perceptual worlds occurs in different registers, often in relation to different problematics, where one milieu or perceptual problematic is no more real than another. The worlds and milieus to which we attend are specific expressions of problems and commitments. The fact that these are more-than-one makes for what Stengers terms "cosmopolitics," which admits the coexistence of different and even contrasting practices and worlds.[99]

In order to move beyond the delimitations of machines and organisms, technical and living milieu, we might finally consider how, following Simondon, it might be possible to adopt an approach to the *technicity* of sensors and tracking devices in order to consider the sorts of relations that they put into play and how these relations are not the only possible modalities for being and becoming.[100] By addressing the technicity of sensors, we might find it is possible to articulate different perceptual problematics as well as different potentialities of sense, while considering how these potentialities are bound up with making milieus and actual worlds.

Even more than studying the computational *object* of sensors, by accounting for the *technicity* of sensors we might consider the participatory relations of sense, as well as the becoming environmental of sensor devices across particular modalities and with specific organisms and milieus. While tracking devices provide one approach to encountering the perceptual world of organisms through measuring variables as indicators of behavior, there also are ways in which we might consider how, following Anna Tsing, "our observations of non-humans present continual challenges to our cultural agendas that require new inflections and transpositions of our cultural 'sense.'"[101]

Organisms are tuned to particular problems in their milieus. They are affected in differing ways according to their interests as well as what the milieu proposes—and they have effects on their milieus. This is another way of coming around to the discussion of affect and being affected, a topic traversed by Spinoza and Simondon, as well as Latour, Stengers, and even writers on environmental topics, where the capacity to be affected might further in-form our conservation and environmental practices. Affect has to do with participation, and it may further spark an ethos in relation to environments. Encounters are critical to this process, since as Didier Debaise has written, "The relation between the ability 'to be affected' (passive potentiality) and 'to affect' (active potentiality) is complex, as the living can neither be explained by its environment nor by its components.

Everything happens in the encounter."[102] We might, from this perspective, approach animal tracking as well as citizen sensing as practices that unfold as necessarily inventive encounters. At the same time, at these nexuses of sense and in this becoming actual of worlds, we might consider how our information-based environmental problems affect and in-form the problems and milieus of other organisms, both in their lived actuality and in the ways in which we generate problems to be acted upon, of and for these organisms.

II

Pollution Sensing

Figure 4.1. Kilpisjärvi Biological Field Station and site of arts-and-sciences residency. Photograph by author.

4

Sensing Climate Change and Expressing Environmental Citizenship

We are now in the mountains and they are in us.
—John Muir, *My First Summer in the Sierra*

We are in the world and the world is in us.
—Alfred North Whitehead, *Modes of Thought*

IN LATE SEPTEMBER 2011, I traveled as far north as I had previously ever been to spend a week in an arts-and-sciences laboratory at a biological field station in Lapland. The coordinates, 69°03′N, 20°50′E, might locate this site on established maps. But during the time I spent here, I found this northern location began to multiply and fluctuate as a concatenation of milieus, processes, and subjects. While the participants on the residency program were here as a group of artists, writers, and scientists engaged with developing experimental modes of fieldwork, we were also located in varying proximity to exurbanites and lifelong natives, tourists and seasonal workers; fishers and farmers; Sámi, Finns, Norwegians, and Swedes. Also in this region were reindeer and dueling lemmings, crowberry juice outlets and imagined cloudberry sightings, forests dense with mushrooms, moss-covered granite boulders, drifts of mountain birch and Arctic scrub, grazed-over lichens, and fjords with rivers emptying deliveries of trout, as well as northern lights, chainsaw art, gift shops piled high with sauna kits, and mythic mountain giants once engaged in a wedding brawl.

Fennoscandia, as this area is also known geologically and geographically, might then be referred to as multiple milieus—sedimented, in process, or yet to come. While I was participating in this residency, I spent my time in and around a biological field station set up to monitor ecological processes in the Arctic environment. Kilpisjärvi Biological Station is a site of long-standing environmental monitoring, and it has become a place where computational environmental sensing also now occurs. The monitoring that takes place at the station ranges from weather stations and webcams to fieldwork for studying Arctic lake ecology and lab analysis of biological samples.

The working group in which I participated, Environmental Computing, focused on the ways in which data are gathered and worked across different fields of artistic and scientific inquiry.[1] My particular interest was in the way in which environmental monitoring articulates distinct practices and politics of environmental citizenship and how traversals might be made across environmental data and action. By engaging in this experimental fieldwork residency, I hoped to consider how environmental data come into formation, what technologies and practices are mobilized to detect and gather data, and how technical and aesthetic capacities for sensing concretize ontological commitments to make distinct environmental processes evident or relevant.

Kilpisjärvi Biological Station is located in an area where the effects of climate change are more acutely experienced, since warming in the Arctic is occurring at a much faster rate than in many parts of the globe. Climate change becomes a recurring factor that in-forms how and why environmental monitoring takes place and the environmental data that might be generated. Sensing of temperature in air, water, and soil; inventories of organisms and pollutants; and samples of pH in lakes and streams are examples of monitoring practices that can accumulatively demonstrate how environments are changing in relation to a warming planet. But alongside scientific practices for documenting environmental change, there are also the lived experiences of humans and nonhumans who differently express the effects of climate change as a planetary event.

In this chapter, I take up practices of climate change monitoring in the Arctic to ask: How do we *tune into* climate change through sensing and monitoring practices? What are the particular entities that are in-formed and sensed? How do the differing monitoring practices of arts and sciences provide distinct engagements with the *experiences* of measurement and data? And what role do more-than-humans have in expressing and registering the ongoing and often indirect effects of climate change, such that categories and practices of "citizenship" and citizen sensing might even be reconstituted?

To consider these questions, I walk through fieldwork and observations gathered from my time spent at the Kilpisjärvi Biological Station, and I compare computational and environmental sensing practices across science and creative practice that attend to environmental and climate change. Based on this material, I consider how the scientific measurements of climate change often take place through the gathering of data on *individual variables* such as temperature or concentrations of CO_2 (as a sort of pollution sensing) and how creative and community-based monitoring projects might differently attend to expressions of climate change as they occur through *connections* made across entities and milieus.

While climate change is an environmental event that may create a particular conception of the planet as a planet in crisis,[2] the Arctic concretizes and materializes the ways in which the global effects of climate change register in much different ways within complex systems. Here is another version of a program earth—not *one world* in crisis, but a multiplicity of effects expressed through climate change as a comprehensive yet differently articulated environmental event. Following Whitehead, this is another way of saying, in relation to the earth and climate change, "any description of the unity will require the many actualities; and any description of the many will require the notion of the unity from which importance and purpose is derived."[3] Climate change provides relevance for monitoring environments as they add up to a global register of planetary transformation, yet the many actualities are the ways in which this event is studied, which can materialize climate change as a differently distributed and connected-up event.

I further discuss how these new arrangements of environmental monitoring and distributed sensing might shift the spaces and practices of environmental participation, both within environmental citizenship actions and through creative-practice projects that take up citizen sensing as a tactic for engaging with sites of environmental concern. How do these modes of monitoring, within the context of environmental change, influence practices of sensing and expressions of citizenship? By attending to the effects of climate change on the multiple experiences of organisms, moreover, how might citizen sensing rework the *citizen* in citizen sensing to direct attention toward other extended expressions of environmental citizenship? The becoming environmental of computation and the becoming computational of environments in the context of climate change monitoring then describe processes where different techniques for creating evidence of climate change can make the multiple entities and relations that are affected by this planetary event more or less pronounced.

ECOLOGICAL OBSERVATORIES AND MONITORING ENVIRONMENTAL CHANGE

Kilpisjärvi is at once a specific site for field study, as well as an environment connected to ongoing changes in the Arctic and beyond. From greater concentrations of persistent organic pollutants (POPs) to increasing temperatures and shifts in land use, the Arctic is a region undergoing considerable changes. As both field station and laboratory, Kilpisjärvi operates as a kind of ecological observatory. It is a site for monitoring environmental changes and recording those observations, often over extended periods of time.

Observatories emerged along with nineteenth-century "observatory sciences" for the study of astronomy, which began to focus on "precision measurement, numerical data processing, and the representation of scientific information on a global or cosmic scale."[4] Observatories, especially contemporary ecological observatories, then become distinct types of technical objects that function in and through their relationship to the milieus in which they are situated. While environmental monitoring at observatories may not initially have been established to study climate change, the decades-long stores of data that observatories now hold have often provided useful records for understanding how environments have changed over time. Environmental monitoring with sensors is then situated within longer histories of measurement practices and the sites where measurement would take place.

Kilpisjärvi Biological Station is a participating field station within multiple ecological "observing" networks, from Sustaining Arctic Observing Networks (SAON) to SCANNET: A Circumarctic Network of Terrestrial Field Bases, as well as the International Network for Terrestrial Research and Monitoring in the Arctic (INTERACT).[5] Kilpisjärvi is also a station within the Arctic Monitoring and Assessment Programme (AMAP), the international governmental body working under the Arctic Council that gathers and reports scientific findings to influence policy and environmental practice. Numerous monitoring initiatives connect up in the Arctic. AMAP links to these initiatives to improve Arctic observation in relation to environmental change.[6]

Climate change monitoring is a key activity in the Arctic that demonstrates how this region fluctuates and is subject to the migrations of other milieus as they travel toward and accumulate in the North. Planetary warming is taking place in much greater intensity in the Arctic regions due to the circulation of atmospheric and ocean currents toward the northern regions.[7] Indeed, the changes in the Arctic are even expressed in terms resonant with information theory, where the intensity of change in the Arctic can be read as a much clearer "signal" in the "noise" of environmental data. As one report on the Arctic in the Anthropocene notes, "Because changes in the Arctic are happening fast and the signal emerges clearly from the noise, in many ways the science of change is currently easier to study in the Arctic than in most places."[8]

More pronounced environmental changes are detectable across organisms and ecosystems, as well as cultural practices in the Arctic. At the same time that a more legible "signal" of climate change may be detectable in the Arctic, there is a relative scarcity of monitoring data from these regions over time, since they are much more difficult to observe year-round. Although climate change monitoring occurs across a planetary realm, not all locations are monitored to the same degree or extent. Historic records may not exist in all locations, or observing systems may not be monitoring essential climate variables as systematically as required by IPCC standards, which leads to perceived gaps in datasets that can also skew models and forecasts. In this respect, questions arise as to how accurate modeling of this region may be, and in some cases observed changes are even more pronounced than climate models may have initially forecasted. Despite these gaps, with the increasing drive to instrument the earth, planetary data on

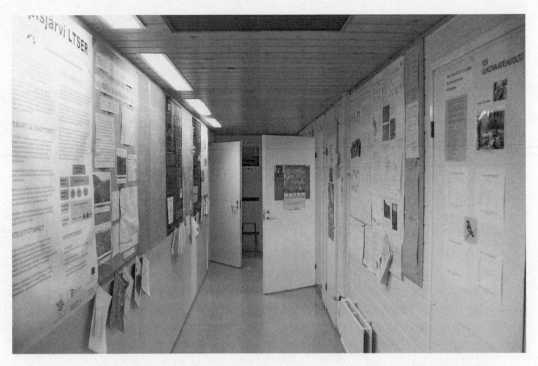

Figure 4.2. Kilpisjärvi Biological Field Station laboratory and posters of monitoring studies conducted in the region. Photograph by author.

the whole are increasing, as are abilities to process data gathered.[9] Climate change monitoring at once take places through contemporary observatories that produce ongoing records of environmental change, as well as through historic and paleoclimate records that extend across deep time. Environmental monitoring records are also assembled from research that may have been initiated for purposes other than studying climate change.

Fifty Essential Variables

Science-based practices for monitoring climate change typically take place in multiple locations across air, oceans, and land, from the tropics to the poles. Measurements of observed change are the key data that are fed into climate models and that project future scenarios for climate conditions. In order to gather the data that are the basis for observed change, direct and ongoing measurements as well as historic and proxy measurements are gathered in relation to fifty "essential climate variables." These variables include everything from air and sea surface temperatures to carbon dioxide levels, ocean acidity, soil moisture, and albedo levels (or the ability of surfaces to reflect solar radiation). Environmental monitoring of these variables provides data that supports the International Panel on Climate Change (IPCC), and they form the basis for networks of observation that take place around the world, as well as historical observations that are mined to provide further climate change data.[10]

Measurements gathered in relation to more contemporary events are collected through airborne instruments, satellites, ocean vessels, and buoys, as well as terrestrial monitoring stations such as carbon flux towers that can be found dotted around the globe. While fifty essential climate variables are routinely sensed and measured for evidence of climate change, most discussions focus on the rising concentrations of CO_2, which correlate to increasing global average temperatures. The current concentration of CO_2 currently hovers around 400 parts per million (ppm), a level that was last reached in the mid-Pliocene, two to four million years ago, when sea levels were up to twenty meters higher than present-day levels.[11] If the additional greenhouse gases currently monitored are added to this measurement, then the current CO_2 equivalent is even higher, at 478 ppm.[12] The Keeling Curve register of CO_2 concentrations is regularly updated through measurements that have been taken at Mauna Loa, Hawaii, and the South Pole since the late 1950s; and carbon-observing satellites capture detail about global concentrations of CO_2.[13] The systematic measurement of increasing parts per million (ppm) of CO_2 in the atmosphere has become an important part of an expanded climate monitoring infrastructure, where observatories fold into observatories.[14] And ice cores provide data on CO_2 levels that stretch as far back as eight hundred thousand years ago. As the latest IPCC Assessment Report 5 has noted, from these studies it has become clear that "the main contributors to increasing atmospheric CO_2 abundance are fossil fuel combustion and land use change."[15]

Climate change monitoring produces pronounced and startling encounters that unfold across environmental datasets. Rates of greenhouse gas rises in the atmosphere are referred to as "unprecedented"[16] and connected to increases in air temperature in the troposphere, marine air temperature, sea surface temperature, ocean heat content, temperature over land, water vapor, and sea levels, as well as decreases in glacier volume, snow cover, and sea ice.[17]

While systematic observations of climate change continue to take place in observatories such as Kilpisjärvi, at the same time many of the effects of climate change are incurred not simply through shifts in datasets or increases in temperature and GHGs but also through "indirect effects" that occur through complex changes in environments and organisms. Monitoring of CO_2 might also include observing and documenting the rapid rates of deforestation, which as carbon sinks can contribute to the overall increase in planetary carbon budgets. Ocean acidification occurs through increased CO_2 absorbed in these sinks, which in turn can lead to a reduction in biodiversity of organisms that require a more alkaline ocean environment. While CO_2 has not typically been subject to regulation as an air pollutant, as it seemingly has no direct health effects (as many other pollutants for which concentrations are regulated do), it becomes a polluting compound

Figure 4.3. Kilpisjärvi Biological Field Station monitoring literature, *The Moth Monitoring Scheme*. Photograph by author.

through its indirect transformations of entities, environments, and ecological processes. The "pollution sensing" that climate change monitoring involves, in this case, follows different trajectories where CO_2 becomes a transformative and polluting entity in relation to complex systems.[18]

SENSE DATA, SENSING DATA

As discussed throughout *Program Earth,* environmental sensors have become a common device within ecological study. The extensive monitoring and observation networks in place and proposed are a way to assess changes in the Arctic environment due to climate change, long-range transport of pollutants, and other environmental events. The collection of sense data through computational sensor technologies can establish how environments are changing and how planetary events register at different locations and through different organisms and ecologies throughout the Arctic.

While environmental sensors are instruments used by scientists studying phenomena such as climate change, sensors have also migrated into citizen-sensing projects, where grassroots science conducted through computational sensing kit is seen as a way to encourage environmental engagement and improve possibilities for addressing environmental change. This is done, in part, by making citizens into scientists, but then the practices and abstractions of what constitutes scientific activity also change. There are numerous citizen-sensing projects for undertaking carbon reporting and carbon accounting in multiple forms, where the monitoring of climate change transforms into a political activity of sorts.[19] At the same time, multiple projects allow citizens to monitor and take account of deforestation and even report instances of illegal logging, since deforestation is a considerable contributor to climate change. Rather than discuss these types of citizen-sensing projects in relation to climate change monitoring, however, I attend to the ways in which climate-related sensing practices occur across sciences and arts. The indirect effects of climate change are materialized across organisms and their milieus, but these effects are differently registered within sensing and monitoring practices. How might a consideration of how climate change monitoring is undertaken open up the ways in which we formulate this environmental event as a project of gathering data in order to act on that data?

Just as scientists are increasingly deploying sensors in order to take ongoing rather than discrete measurements of ecological processes, creative practitioners are also developing new practices in relation to computational sensors in order to gather and repurpose distinct sense data about environmental phenomena. These monitoring initiatives include artists as well as indigenous peoples in different types of monitoring projects that might, on the one hand, rework science- or social science–based approaches to sensing environments and, on the other hand, articulate environmental engagements through different relations and registers of perception.[20]

Our working group on Environmental Computing was interested in these particular uses of sensors across arts and sciences and how these practices generated distinct if not new ways of understanding environments. At the same time, it was clear this was a crosscutting area of interest, since numerous participants within other working groups of the residency also had their own mobile sensors for undertaking field investigations, including geophones and hydrophones, YSI water sensors, light sensors, and more. The station where we were based also prominently held an array of meteorological sensors on its rooftop. A webcam regularly produced images of the site, and these were also streamed online. In addition to the many mobile devices used in the field, environmental sensors in use at the Kilpisjärvi Biological Station were connected up to a sensor data platform, where relatively continuous data streams provide indications of ecological processes.[21]

As part of our experimental field laboratory, questions arose as to what the particular objectives of environmental monitoring are. Monitoring, as a practice of sensing, raises questions about who or what is undertaking sensing practices, how this in-forms what counts as "sense," and what types of milieus concretize in the process. Does monitoring in some way already presuppose a certain set of

Figure 4.4. Mount Saana. Field and study site adjacent to Kilpisjärvi Biological Field Station. Photograph by author.

practices that assume distinct ways of accessing and studying environmental phenomena? Perhaps processes of sensing environments with computational sensor technologies demonstrate how these devices do not so much detect data "out there," as discussed in chapter 1, but instead give rise to distinct ways of articulating environmental sensing across multiple organisms and processes. Given that the aim of this residency was to create experimental fieldwork engagements, we took a walk to the nearby Mount Saana in order to consider these different approaches to monitoring environments and what new arrangements of sensing and milieus came together.

Walking to Mount Saana

Mount Saana has been an ongoing site of Arctic mountain lake research, as included in the Arctic section of the IPCC fourth assessment.[22] Poster presentations and scientific reports available in the library of the Kilpisjärvi station captured research on studies of how warming temperatures in the Arctic and at Lake Saanajärvi have led to increased levels of biota. As Lake Saanajärvi's average temperature has hovered around -2.7°C, it has historically had an absence of biota such as algae. But through the collecting and recording of sense data including temperature, water samples, sediment samples, oxygen measurements, and analysis of diatoms as bioindicators, evidence of increasing levels of biota has emerged. The warming of Arctic lakes, in other words, is in part expressed through the increasing numbers of organisms populating these waters.

One of the most pronounced changes in the Arctic is in relation to the loss of snow and ice cover, or the decline of the cryosphere, where not only rising temperatures but also disappearing snow and ice impacts the cold-weather organisms that might inhabit these environments.[23] Lake Saanajärvi has been a monitoring site since 1996, and based on observed and projected changes the warming of freshwater lakes in the Arctic is expected to lead to numerous shifts in these ecosystems, from shifting temperature and water chemistry to increase in vegetation, decrease in water levels due to higher transpiration levels, and shifts in organisms that may reach a temperature "threshold" for survival in these conditions.[24] This one lake that we are traversing around is a site of ongoing monitoring and measurement, generating data about the complex if often indirect effects experienced across organisms and milieus. As a field location that is also a site of sample gathering and analysis, the lake demonstrates the complex practices that are undertaken to sense environmental change and how these are brought into scientific research. At the same time, the lake raises questions about how experiences of environmental change register and what modalities of data are generated to document these changes.

Measurement as Experience

The scientific study of Lake Saanajärvi involves measuring set variables and systematically gathering data in order to establish conditions of environmental change over time. These observed changes provide the quantitative data points

Figure 4.5. Map of Mount Saana within larger context of Finnish and Norwegian Lapland. Photograph by author.

that are both evidence for historic and present shifts in environments as well as reference points from which climate models may be produced to project toward future scenarios. Measurement is at once inherited, present, and speculative, in this sense, since it accumulates through ongoing records, sustains practices in the present, and also projects toward future milieus in and through which measurement according to established variables will continue to have relevance.

While I was based in Kilpisjärvi, I spent some rainy afternoons looking over scientific papers, AMAP reports, and reference books in the field station library. One of these texts, *Measuring the Natural Environment,* outlines the central role of measurement to studying environments. Under the heading "The Need for Measurements," the opening line of the first chapter notes,

> Whether it be for meteorological, hydrological, oceanographic or climatological studies or for any other activity relating to the natural environment, measurements are vital. Knowledge of what has happened in the past and of the present situation can only be arrived at if measurements are made. Such knowledge is also a prerequisite of any attempt to predict what might happen in the future and subsequently to check whether the predictions are correct.[25]

Measurement, according to this text, allows observations of environments to become relatively systematic and continuous. Instruments that monitor environments become assessed for the degrees of accuracy they allow in obtaining "hard"

Figure 4.6. Kilpisjärvi Biological Field Station library with literature on monitoring and measuring the environment. Photograph by author.

facts about environments. Sensors working through computational modalities are part of the progression of developing more accurate instrumentation, which also allows for greater automation of measurement without the need for human intervention.

The variables according to which environments are measured and sensed, from radiation and temperature to precipitation and wind speed, become the individual data points according to which systematicity and continuity are achieved. Yet just as many questions have arisen, often within the context of science and technology studies, which ask: Why are these variables assessed and not others? What commitments are made when some forms of data become the benchmark for evidence gathering?[26]

At the same time, other questions arise in relation to how individual variables as entities are drawn from the complex environments in which they are in play in order to be measured. In his comparison of laboratories and field stations, Robert Kohler has suggested that the identification of single variables for measurement is largely influenced by laboratory methods that do not always translate well to field-based study. Alternative methods that ecologists working in the field have adopted have included developing instruments such as an atmometer, which rather than singling out variables for measurement attempts to study the complex interaction of variables by emulating the experiences and processes of transpiration within a plant, for instance.[27]

Measurement becomes a way both to individuate entities and to express and concresce possible relations across variables, entities, and milieus. Technologies for undertaking measurement are not simply techniques for gathering facts, as it were, but are also ways of identifying particular registers of data as important and relevant. The basis for understanding environmental change as *change* is understood through modes of systematicity that focus on particular variables over time, so that shifts within variables establish the evidence for degrees and intensity of change. These registers of change then further influence how we come to understand and operationalize environmental problems such as climate change through shifting variables: 2°C as a threshold for the increase in the mean temperature of the earth, 400 ppm or 350 ppm as notional limits for CO_2, and policy measures oriented toward achieving these targets.

How then does the decades-long practice of gathering measurements of set variables compare to the more itinerant sense-gathering of a walk to Mount Saana? In what ways does a possibly more random or momentary recording of field phenomena with sensors compare to these practices for detecting change? Do sensory investigations need to be guided by more than technical "probing," or a documentary tracing of audio, video, tracing, and indexical capture of momentary phenomena? Scientists typically collect data to research particular questions about environmental change—for instance, asking how the long-term temperature of an Arctic lake has shifted—while many artists' experiments might focus more on the phenomenal or sensory aspects of data gathering, such as capturing the sound and light of that same lake within discrete moments, in order to visualize or sonify experiences in the field. What counts as data in scientific and creative practice differs, as do the motivations for the collection and use of data. Calibration, protocols, and measurement techniques in response to the variables studied influence the quality of data gathered. But these "variables" might be approached and evidenced much differently across creative and scientific practice.

Whitehead has suggested that measurement is a way of sensing and experiencing. I take this to mean not simply that bodies are tuned to or mirroring the capacities of measurement but that the practices of *parsing experience as measurement* give rise to perceptive entities and occasions: they are ways of making measurement possible as a distinct experience and mode of individuation.[28] If, rather than see the relationship of monitoring instruments to environments as a transparent rendering into data (or, alternatively, as a mere construction) we instead regard it as a process of selecting, experiencing, and concrescing *worlds,* then an expanded repertoire of measurement practices might also emerge. Experience as it transforms into measurement might occur through identifying single or complex variables, or it might also occur through measurements that capture the connectivity of events, or the comparative registers through which one environmental occasion might be a suitable proxy for another occasion or practice.

Science-based climate change monitoring often focuses on measuring single variables that are correlated across data sets, so that part of the difficulty has been to establish the connection between increased CO_2 levels and temperature (not to mention anthropogenic influence). Variables that are stabilized for one type of measurement then require the difficult work of being knit together—historically, at present, and in the future—through supercomputers that can "crunch" the data and establish recurring connections in relationships between variables that correspond to larger patterns of environmental change.

Measurements that focus on experience, relation, and the complex interplay of variables have been referred to in other contexts as a sort of citizen science, or folk measurement.[29] In this respect, we might consider how "community monitoring" and "community data" projects that are underway in the Arctic are not simply expanding the number of participants who would be involved in citizen-sensing projects. Instead, these projects incorporate parallel and distinct ways of experiencing environments through different engagements with measurement. Not necessarily tuned to a project of single variables, these monitoring projects might instead engage with felt experiences of alterations in climate, new timings of seasonal events, and organismal activity that fluctuate together—or suggestions for new observations that might be recorded in relation to emerging phenomena such as melting permafrost.

Many projects are now working in this space, from the Exchange for Local Observations and Knowledge of the Arctic (ELOKA), which was launched during the 2007–2009 International Polar Year (IPY), to the Atlas of Community-Based Monitoring in a Changing Arctic, which documents traditional knowledge and community monitoring activities. Also included is the Arctic Perspective Initiative, an artists' project that develops a DIY environmental sensor network for studying flora and fauna through computational techniques, and which focuses on installing sensors for community-oriented scientific research.[30] Neither a simple project of asking Arctic "natives" to do science within a usual scientific register nor an easy claim to a more "holistic" ecological understanding, these projects instead provide different angles for encountering environments through experiences that can also be differently individuated and operationalized through measurement.[31]

If we apprehend measurement as experience and individuation and as necessarily giving rise to the entities and relations that together articulate distinct types of collective potential, then measurement might proliferate in such a way so as not to demand that only one version of measurement may stand in as "fact." Nor is this a simple project of relativizing all measurement through experience. Instead, data that are generated as measurements of experience are expressions of the subject-superjects that are undergoing situated and felt changes in relation to a warming planet. Measurements are articulations of relevance: for science, of which variables matter as key indicators of change in environments; for creative practice, of which articulations of experience might reveal different ways of taking account of environments and environmental change; for community monitoring, of which connections and lived experiences might be crucial for sustaining or enabling particular environmental relations.

Besides the different ways of monitoring environments across arts and science practices, this perspective also shifts when we consider the ways in which the multiple other inhabitants of milieus, including more-than-humans, sense environments. In this way, during our environmental computing group we also found ourselves engaged in discussions of indicator species, of lichens and mosses and other organisms that can be studied as expressions of environmental processes, whether for atmospheric pollutant levels, radioactivity, or different types of mineral depositions in soil.[32] Organisms may be studied as proxies of environmental processes, yet they are not parsing single variables as much as living with changes in complex environments over time. Organisms inherit those changes, work through the new collectives that form, and attempt to gain a foothold to ensure their ongoing existence.[33]

Organisms that experience climate change are in one way measuring and monitoring shifts in their environments. As registers and expressions of environmental change, organisms further become environmental media of sorts. At the same time that the planet is experiencing increasing levels of greenhouse gases and temperatures, Arctic environments are also sites where organisms and their ecological relations are shifting, as lived experiences of their environments. If we consider the expanded registers of measurement and experience that occur through community monitoring, then we might also attend to the ways in which citizen sensing shifts if the experiences of more-than-human organisms were more fully incorporated into accounts of climate change. Who or what is a citizen? How is environmental citizenship articulated in relation to climate change? And how does this matter for how we sense and monitor climate change?

TEXTING FISH AND TALKING WITH DOLPHINS

While environmental sensing technologies may have developed through applications at ecological observatories such as Kilpisjärvi, among many other milieus, these devices have been taken up in creative-practice projects that explore how sensors can be constitutive of new relationships while raising questions about how sensing as measurement occurs across sciences, arts, citizen engagements, and more-than-human organisms. These differing encounters with environmental monitoring demonstrate how measurement, particularly in relation to climate change monitoring, can be a practice whereby new processes of sensing, expressions of milieus, and practices and formations of citizenship emerge.

Creative-practice projects that deploy environmental sensors often focus on ways of monitoring pollution. Such pollution-sensing projects often attend to

Figure 4.7. Reindeer near Mount Saana. Photograph by author.

urban air pollution, as in projects such as Area's Immediate Reading (AIR) by Preemptive Media or Feral Robotic Dogs by Nathalie Jeremijenko, which will be discussed further in chapter 6.[34] In another way, projects such as "Amphibious Architecture," as discussed in the last chapter, raise questions about how monitoring and sensing take place through extended environmental relationships, where the transmission of text messages becomes a sort of "spectacle" for connecting up usually disparate human and more-than-human urban dwellers.[35] How, through the use of environmental sensor technologies, might we begin to understand urban environmental health as shared across fish, humans, and river ecologies? Who or what counts as a citizen if citizenship is articulated through cross-species sensing practices? Can fish become citizens? Or does citizenship involve a different set of practices through these inhabitations?

During my time in Kilpisjärvi, we organised an evening salon to consider this set of questions in relation to environmental monitoring. Within our specific discussion of environmental sensing and computation issues, our group considered the topic of how to understand the *"citizen" in citizen sensing*. We began our conversation by asking who or what is a citizen, and how different notions of "citizen" might influence the type of sensing that might take place. We also asked how citizen sensing might shift when we trouble assumptions about who or what is a citizen in these projects.

We discussed additional examples of citizen-sensing projects from Beatriz da Costa's Pigeon Blog, to Safecast, a project for detecting radiation after the Fukushima nuclear fallout in 2011, to the dontflush.me project, which uses proximity sensors to inform New Yorkers when to avoid flushing the toilet when the sewer system may be at capacity and in danger of dispersing waste into the harbour.[36] Other projects, such as Vatnajökull (the sound of), allow listeners to phone up a melting glacier in Iceland, while Pika Alarm puts mountain rodents to work as sentinel species for climate change.[37]

In many citizen-sensing projects, environmental monitoring and data-gathering practices activate or enroll a certain hypothetical "citizen" that is already built into mobile devices and social media. By using social media, citizens are seen to be empowered to undertake newly informed, connective, and collaborative projects. While we had initially hoped to develop speculative practices around what other possible forms of citizen-sensing practices might look like if new formations of citizens were introduced, many discussants were concerned about the use of the term "citizen" to describe more-than-humans. Don't citizens have free will and rights? Aren't animals simply the props for human experiments into sensing? Are these sensing practices perhaps even exploitative? How could a tagged reindeer possibly be counted as a citizen? In this way, one discussant asked, "Is this about trying to talk with dolphins? I know of an artist who tried to do that and he went a bit mad, actually."

Other examples of citizen sensing emerged in our discussion at this point, which began to test the idea of new arrangements of citizenship. One project reference, the Million Trees NYC project in New York, was cited as an example of a practice where crowdsourcing was used to identify where trees might be planted in the city.[38] Once planted, the trees could be monitored in order to ensure their longevity. Such a practice of urban tree stewardship implies a relationship with the trees, and environmental citizenship might be practiced through sensing— with or without computational devices—trees and their local environment.

While the extension of citizen practices to more-than-human entities might press at the limits of *common sense*, in many ways expanding the scope of citizenship through sensing may be one way to develop strategies for finding new politics of subjects, as Braidotti suggests, which are environmentally connected.[39] In another way, and working laterally from the subject-superject discussions developed by Whitehead, generating a new politics of subjects also entails generating a new politics of milieus. Milieus in this way might be understood not as populated by humans sensing and acting on environments, but rather as entities involved in generative sensing arrangements that might produce new milieus and possibilities for engaging with milieus.

In other words, rather than see environmental citizenship as something that inheres within a preformed and exclusively human subject position, I am instead interested to consider how practices unfolding across human and more-than-human registers might enable distinct articulations of environmental citizenship. The notion of "environmental citizenship in the making,"[40]—in other words, a processual and relational approach to environmental citizenship (here working with a Whitehead-inspired approach to citizenship)—potentially opens up ways of thinking through the sensing practices and entities that concresce through monitoring and experiencing environments. This has relevance in relation to climate change monitoring, since multiple entities might be involved in expressing or individuating experiences of climate change, which has further implications for how we encounter the measuring that citizen-sensing projects might undertake.

Expressing Citizenship: Reworking Collective Experience

These different sensory engagements could be seen as ways to open up milieus through different encounters or distributed ways of expressing environmental processes. Milieus are expressed through effects and experiences of human and nonhuman inhabitations. This approach suggests that fluctuations and expressions of milieus run through and are differently carried by the multiple inhabitants of milieus. In this sense, while rooted in place, fieldwork milieus also travel and change across the subjects and communities they affect. There are multiple modes of sensing that are activated in relation to climate change, so that we might ask: Who is undertaking the observation of what in this planetary and ecological observatory? Or rather, how are experiences constitutive of worlds, and how does this world-making in-form possibilities for citizenship, collective engagement, and participation?

On one level, the Scottish-American environmental writer John Muir captures this sense of experience and world-making when he writes of his travels in the High Sierra Mountains of California, "We are now in the mountains and they are in us." Included in the epigraph to this chapter, Muir's statement seems to be a recognition of the ways in which milieus and subjects commingle. Yet on another level, when Whitehead writes, "We are in the world and the world is in us," he is signaling toward one of his key concepts about the ways in which subjects are always part of specific and concrete occasions that are constitutive of worlds. The earth of program earth is then a distinctly experienced and formative entity, proceeding from an objective datum to become a felt and experienced entity for a subject.

Perception, moreover, is distributed in and through these worlds through multiple subjects and processes. As discussed in chapter 1, in Whitehead's approach, all entities are in some way "taking account" of their environments. In this way, subjects are always what he calls "superjects," which are bound up with and concresce through actual occasions.[41] A subject-superject is not only a human figure but also necessarily includes rocks, animals, and plants. At the same time, these entities and relations are not fixed, nor are they singular or necessarily always overlapping, but gain a foothold through the distinct types of "interpretation" or *expressive* experience that each organism undertakes.

In-Forming Environments

Computational sensors, in addition to stones and animals, are also expressions of environments. Whitehead's approach suggests that these multiple modes of planet sensing might expand here not just to encompass milieus as in process but also to gather together multiple modes of sensing that are in-formed through the expressive activities of multiple subjects. The becoming environmental of computation extends to the ways in which sensors are expressions of environments and to the ways in which organisms register as environmental media working through and expressing changes in climate. Sensor networks are not just formed by bits of circuitry and code but also in-formed through exchanges of energy, materializations, and relations that concresce across organisms and that are brought into practices of measurement with climate-change monitoring.

While the project of collecting as much data as possible is then seen to be critical to the study of environmental change—where there are no blank spots left on the map, as discussed in the last chapter—what counts as "information" may not be as self-evident as first assumed. Data in the form of essential variables may be gathered in one way, and yet in-formation is proliferating in the ways in which organisms are experiencing environments and recasting how we might think about measurement and the conditions of citizen sensing that might register those experiences. Yet more information does not automatically lead to a more effective set of political actions for addressing climate change. This other way of encountering the "informating" of environmentalism, as Kim Fortun has discussed, often attends to the ways in which environmental change is an epistemological project, where forms of knowing might shift in relation to information technology toward "comparison, extrapolation and cumulative effect," thereby "displacing the dominance of linear constructs of causality in sense-making practices."[42]

Yet as I am suggesting here, by attending to the experience of measurement and the incorporation of climate change into organisms as environmental media, a different set of encounters concretize that might be described as the in-forming of environments and environmentalism through sensor technologies. As discussed throughout *Program Earth,* Simondon's notion of *information as a process of in-forming* presents a way to move beyond a substantialist tradition of form and matter to address how individuals, milieus, and relations *take form* through exchanges of energy, resonance within systems, and changes in intensity.[43] Simondon describes the processes whereby entities and milieus are in-formed as *transduction,* which signals toward the ways in which "'an activity propagates from point to point within a domain, while grounding this propagation in the structuration of the domain.'"[44]

Such a trajectory might further be understood as a reworking of collective potential from what Simondon has called the "preindividual reserve" that is a nonessential nature prior to individuation and which is shared across "objects" that are natural and technical. Any subject is at once individual and preindividual, which makes it a "more-than-individual being" that is continually worked out within collectives.[45] In this way, part of this working-out that characterizes "'the problem of the subject'" takes place through "the heterogeneity between perceptual worlds and the affective world."[46]

The individual is always bound up with a preindividual reserve, and the human is always more-than-human (where what we understand to be human is not a site of exclusion or exception, moreover). Any individual entity is formed by and working through a perceptual and affective connection to collective worlds and is constituted through "'the activity of relation'"[47] that does not precede individuals or collectives but is the "transductive reality" that emerges along with them.[48] If we further bring Whitehead and James to bear on this understanding of relation, we might say that relations can also consist of negative prehensions, or disjunctures. Relations, whether conjunctive or disjunctive, are always as real as things. This approach matters in relation to sensing climate change, since it in-forms how the effects of climate change are experienced and how notions of what constitutes

Figure 4.8. Mushrooms near Mount Saana. Photograph by author.

an expression of citizenship in citizen-sensing practices might in-form monitoring practices.

COMPLICATING CITIZENS AND SENSING PRACTICES

"Citizen" is an ambiguous term and attractor that travels across environmental discourses and practices. What does this term mobilize in concrete occasions, and as discussed in chapter 1, how does it act as a "lure for feeling?"[49] The fluctuating milieus and sensing subjects that are in-formed and expressed through the differently distributed experiences of climate change, and which are the topic of this chapter, suggest that new arrangements of citizen sensing—and environmental practice and politics—might concretize here.

As this discussion of sensing climate change across scientific, creative, and citizen-sensing practices has suggested, citizens might no longer be conceived of as exclusively human subjects endowed with rights, but rather through relationships that at turns make us responsive to changes in our environments or otherwise generate alternative ways of engaging with the multiple modes of sensing that take place in milieus. In this way, Stengers writes, "what Whitehead calls a subject is the very process of the becoming together, of becoming one and being enjoyed as one, of a many that are initially given as stemming from elsewhere."[50]

Subject-superjects are diversely distributed, continually in formation, and also generative of and generated through practices such as citizen sensing and environmental monitoring.

Citizens, in this case, might then be defined not only (or even) through those more traditional inheritances of a subject bound to a nation-state but also as subjects that are expressed through *environmental practices that are constitutive of citizenship.* These practices within environmental citizen-sensing projects often consist of monitoring, gathering, and reporting. The relationship between digital technologies, practices of environmental sensing, and citizen engagement becomes an important point of focus, since environmental monitoring activities involve not just gathering data but also performing particular types of citizenship through sensing technologies.

Environmental-computing monitoring projects raise questions about who or what sense data are for, which interpretive practices are productive of citizenship, and what new collectives sense data might mobilize. Such an approach to the multiple if divergent and differently captured expressions of milieus may be a way to open up speculative citizen-sensing scenarios to consider new arrangements of citizenship, expanded entities and processes of sensing, and new ways of articulating milieus within practices of environmental monitoring that attempt to respond to the ongoing event of environmental change.

How might we account for the varying "ecology of practices,"[51] as Stengers terms it, which unfold through these differing ways of engaging with and experiencing climate change? The field-science approach to environmental problems is to observe, monitor, and gather as much as evidence as possible to make a case from empirical data for certain analytical theories about environments. In another way, this scientific approach to ecology might then be used to mobilize a political ecology of acting on this information. But what is the relationship between scientific and political ecology within this trajectory? Might there in fact be a failure to make clear connections between scientific ecology and political ecological projects in such a framing? At the same time, environmental monitoring by citizen sensors or scientists often assumes the automatic effectiveness of citizens gathering data en masse in order to influence environmental politics. The sensing operation is instantaneously made into a citizenly operation, without any clear indication of how this progression, mobilization, or translation is to unfold.

Consider how climate change is configured through these practices of citizen participation and sensing. Also consider how these formations of citizen sensing operate as practices of individuation and attachment. Environmental citizenship is expressed through digital technologies, and different forms and practices of sense are ways of making operational distinct affective and political capacities. In keeping with the discussion of folk measurement above, the possible limitation with citizen sensing is that science, in the form of measurement, becomes the primary mode of experiencing environmental change, which does not auto-matically translate from data into political change.

So we might ask, following Whitehead, what are the concrescences of sens-ing, environment, citizens, and environmental matters of concern? If citizen sens-ing and sensing climate change are not simply projects of gathering data in order to influence policy makers and so contribute to climate change discourse and practice, then what is it that concresces here? It may be that the motivations for using sensors stem from the perception that effectiveness is bundled into these technologies. Or it may be that a community of citizens engaged in sensing is in itself a mobilizing force to which the sensors play a secondary role.

Beyond a mode of environmental citizenship that might consist of emulating the scientific (in this sense, positivist) gathering of data, typically in order to make visible the invisible aspects of climate change, an expanded ecology of practices and ecologies of citizenship might attend to how different entities are experienc-ing and prehending, or taking up, the effects of climate change, and how these prehensions work across entities and milieus to give rise to shared but differently articulated experiences. Participation in environments and environmental change is not exclusively a volitional exercise of democratic agency, but rather participa-tion concresces across multiple entities involved in working through the shared occasion of climate change.

In Stengers's discussion of citizens and citizenship in the relation to an ecol-ogy of practices, politics is a condition of invention—and struggle—so that any ecology of practices, any formation of relationships, cannot be assumed to be neutral or given. This is a situation where all subjects are involved in "collectively inventing the world we all have in common."[52] But collective invention cannot proceed simply through an assumed starting point of what is a citizen, what is human, what is a relation, or what is political. As Stengers writes, these and many other "categories must be *complicated*"[53] if political invention vis-à-vis an ecology of practices is to occur.

In this zone, where complication might also be understood as a condition of experimentation, we might ask how a different approach to citizens and subjects could in-form new engagements with climate change. An inventive politics of subjects that is attentive to the realness of relations (and their disjunctures) might generate new understandings of citizenship as involving becoming and belonging with extended more-than-human communities and milieus.[54] And from this pro-liferation of subjects a proliferation of program earths might also occur, as milieus formed through diverse experiences for expressing climate change. Folding back into scientific practice and citizen sensing, we could bring Haraway's suggestion for "another science" into consideration here, which, in accounting for multiple subjects, also accounts for "the sciences and politics of interpretation, translation, stuttering, and the partly understood."[55]

From Radical Empiricism to Planetary Propositions

But this is not, finally, a proposal for an obfuscation of a knowledge project such as climate change monitoring. Instead, by considering how to expand the subjects involved in "citizen" sensing, as well as accounting for experiences as generative of different forms of measurement, it might be possible to develop a radical empiricist approach to climate change monitoring. Such an approach would take seriously James's suggestion that relations are as real as things. From such a perspective, different ways of taking account of climate change might be registered that could in-form the invention of politics that occurs in relation to this differently materialized and materializing planetary event.[56]

As Kim TallBear has discussed, people (and organisms) can be decimated when their relationships to environments become untenable, or when those milieus disappear or change unalterably.[57] This is not merely about the destruction of people as bodies or human subjects but also about the destruction of people as they are in-formed and individuated through and along with the milieus they inhabit and the organisms with which they cohabit. Here, indigenous and feminist technoscience meet radical empiricism, where relations are always as real as things. And yet, a call to limit global warming to a 2°C threshold might not capture these relations, saddled as it is with the problem of translating a single if consistently measured variable into a political modus operandi.

What would a radical (and speculative) empiricism of environmental monitoring and data sensing involve?[58] Would the experiences of multiple other sensing subjects become part of how we engage with environmental change not simply as an epistemic or informational project but as something that in-forms individuals and actual worlds? How might this approach further in-form and rework citizen sensing as well as practices that are constitutive of citizenship? In-forming environmental citizenship, in this sense, is as much a speculative as descriptive undertaking, since the expanded subject-superjects of experience give rise to relations, collectives, events, and accounts of environmental change that are in the making.

While climate change monitoring is based on data gathered from observed changes it is also propositional, since it requires models and forecasts to anticipate what the further effects of environmental change might bring. As Paul Edwards has suggested, these are forms of "provisional" knowledge that are always being reworked and rewritten, both as climate pasts and climate futures.[59] As much as *provisional* forms of knowledge, however, we might also say that climate change monitoring is *propositional* in the ways in which it articulates the inherited and experienced effects of climate change and proposes future scenarios for action. These propositions have effects both in establishing particular realities and in inviting us to encounter problems in ways that will continue to provide relevance to those facts.[60]

While some dominant approaches to measuring and documenting climate change seem to easily translate from an observation of any single variable to a *control* of any single variable, such as proposals for geo-engineering or ecosystem services where environmental processes are one more data point to be engineered, other propositions and approaches might materialize through attending to possibilities for reworking the subjects and practices of climate-change monitoring. What a complicated and complicating approach to citizen sensing suggests is that we not simply consider what monitoring data makes evident but also experiment with the new subjects, experiences, relationships, and milieus that monitoring practices might set in motion. With such an approach, we might also develop ways to invent new collectives and politics relevant to the concerns of climate change.

Figure 5.1. Microplastics. The majority of plastics in oceans, particularly in the "garbage patches," consist of small-scale plastics that are pellets from plastics manufacturing or have abraded to smaller sizes from microplastics. Photograph courtesy of U.S. National Oceanic and Atmospheric Administration (NOAA).

5

Sensing Oceans and Geo-Speculating with a Garbage Patch

Located across the world's oceans are several sizeable concentrations of plastic debris that have variously earned the title of "garbage patches." The Great Pacific Garbage Patch in particular has become an object of popular and scientific interest. It is an environmental anecdote to confirm our worst fears about overconsumption; and it is an imagined indicator of what may even outlive us, given the lengths of time that plastics require to degrade. The garbage patch is in many ways an amorphous object, drifting through oceanic and media spaces as an ominous form that focuses attention toward the ways in which oceans have become planetary-sized landfills.

"Discovery" of the garbage patches is often attributed to Charles Moore, a captain-turned-scientist who deployed and publicized the term to describe his observations of a high concentration of suspended plastics in the clockwise currents of the North Pacific Gyre. In so doing, he brought the phenomenon of plastics in the Pacific to greater public attention.[1] However, oceanographer Curtis Ebbesmeyer originally coined "garbage patch" as a term to describe the tendency for flotsam to collect in sub-orbiting gyres.[2] Although scientific observations of the circulation patterns of gyres and the accumulation of debris had taken place previously,[3] Ebbesmeyer and Moore both suggest that it was the naming of the ocean debris as "patches" that eventually galvanized attention for this issue.[4] Anecdotally, the garbage patches have become one of the most potent figures for environmental concern, where the imagining of vast stretches of oceans choked by plastics is at once a media device for expressing the worst of the destructive impacts of humans on the planet and also an attractor for stimulating scientific study into ocean plastics, since it is a topic about which citizens frequently make inquiries to environmental agencies.

Popular imaginings of the Pacific Garbage Patch have included comparisons of its size to the state of Texas, or suggestions that it is an island that might be named an eighth continent, formed of anthropogenic debris. Upon hearing of the concentration of plastic wastes in the Pacific, many people search for visual evidence of this environmental contamination on Google Earth. Surely a human-induced geological formation of this magnitude must be visible even from a satellite or aerial view? However, because the plastic wastes are largely present as microplastics in the form of photo-degraded and weathered particles, the debris exists more as a suspended soup of microscopic particles that is mostly undetectable at the surface of the ocean.

While Google Earth may be a platform for visualizing and locating ocean data,[5] this visualization technique presents a much different approach to "sensing" than seeing the patch as a photographic object. The inability to locate the garbage patches on Google Earth, a tool for scanning the seas through a conjunction of remote sensing, aerial photography, and online interfaces, even gives rise to popular controversy about how to locate the patch and whether the plastic conglomerations are actually present in the oceans, and if so, how to address the issue. The relative invisibility and inaccessibility of the patches render them as looming imaginative figures of environmental decline and yet relatively amorphous and unlocatable and so seemingly resistant to environmental action. All of which raises the question: To what extent do environmental problems need to be *visible* in order to be actionable?

The difficulty of visually locating the patch as an identifiable object reveals how the garbage patch is on one level a "myth" about how plastics accumulate in the oceanic gyres. While plastic exists in considerable quantities in these areas where currents converge into still expanses of oceans, the form that the plastic takes is often in varying stages of decomposition, suspended within water columns, sedimented on sea beds, and even filtered through various organisms that ingest these particles. Several scientific entities such as the U.S. National Oceanic and Atmospheric Administration (NOAA) have gone to lengths to dispel the "myth" of the garbage patch by clarifying that the patch is not literally a surface coating of plastics, but more of a zone with higher concentrations of suspended plastics and especially microplastics.[6]

Yet the term "patch" persists in use, not least because it brings increased public attention to an environmental issue somewhat removed from everyday experience. Scientific agencies such as NOAA explain that the patches do not assemble as islands of plastics, but they continue to use the term as shorthand to describe plastics concentrations; while the media use images of accumulated concentrations of plastics in urban harbors, for instance, to stand in for the more distant and difficult-to-visualize garbage patches; and artists focus on sites such as the Midway Atoll to capture the effects of macroplastics that wash up on islands proximate to oceanic gyres, and which are often taken to be representative of the general constitution of the garbage patches.[7] The patch is a concept that accumulates uses, images, and imaginaries, where the more complex and amorphous garbage patches resist easy identification.

What sort of "myth" might the garbage patch embody? What sort of object or occasion of pollution is this? And how might it be monitored? Rather than seek to "dispel the myth," how might the garbage patch constitute a sort of geo-mythology, or a tale of uncertain earth events and forces? Geo-mythology is a term coined to describe the ways in which distinct earth formations are often explained by myths that capture how they came to be.[8] From this perspective, the ways in which environments and earth features form is not just a matter of geologic process but is also a social, cultural, and narrative process that conveys imagined or actual accounts of how earth formations come to be recognizable objects. From volcanoes to floods, these formations and events are generative of geo-mythological narratives. Google Earth has even been used as a tool to identify these formations and to provide legitimacy for these stories as attached to actual earth objects.

Based on accounts of monitoring plastics in the oceans and locating debris concentration zones, I adopt and adapt the term geo-mythology, which might typically exist as a narrative form explaining earth events and formations of indeterminate origins, and move toward what I develop as a more *geo-speculative* approach to consider how the uncertain and indeterminate aspects of the garbage patches give rise to environmental monitoring practices for bringing these newer and more fluid geological objects into "view." This geo-speculative inquiry is then less focused on explaining the *origins* of the garbage patch. Instead, the geo-speculation developed here considers how much of the uncertainty around the gyres involves exploring what kind of earth or ocean object the garbage patch is, and even more, what *potential* events and effects may unfold through this shifting formation.

Plastics have inevitably been present in the oceans for many decades, but at some undefined moment the concentration of plastics in oceans accumulated to a concentration that constituted an at-times indeterminate and speculative object of study. I am interested to take up the ways in which the indeterminate and changeable qualities of the garbage patch focus practices for exploring what this concentration of plastics in the ocean consists of, including how to monitor these plastics and how to reduce the problem of plastics by involving citizens in sensing oceans and tracking marine debris. In this chapter, I consider two primary aspects of the garbage patch that pertain to its materiality and ongoing circulation. In the first instance, I look at how garbage patches are identified and studied and situate these longer-standing practices within current monitoring practices where oceans have become sensor spaces. Oceans are not only increasingly full of plastic debris, they are also highly instrumented spaces where a vast range of monitoring activity is underway. I then consider how attempts to track or locate trash, whether through Google Earth or citizen-sensing apps such as the Marine Debris Tracker app, work in and around the fact that microplastics—small-scale and invisible plastic particles—are the common material components of the garbage patch. I also look at techniques for mapping the circulation of ocean debris and focus on the Global Drifter Program as one ocean observation project among many that has deployed buoys equipped with sensors that communicate with satellites, and which are used to study the drift of plastics and other debris in the oceans.

This chapter considers how environmental monitoring techniques that are often developed for purposes other than sensing plastics are subsequently tuned in to the drift of oceanic debris. How do the shifting and concretizing materialities of the garbage patch in-form the technologies that come to be used to monitor them? How are littered oceanic spaces entangled in the becoming environmental of these computational monitoring technologies? With these questions in mind, I explore how environmental monitoring techniques "sense" an object such as the garbage patch that is relatively invisible and continually in process. Imperceptible and removed from immediate experience, this pollution event and object raises questions about what citizen-sensing approaches to environmental monitoring are able to evidence, and act upon, when environmental pollution unfolds through registers of non-sensuous perception.[9]

This chapter finally attempts to craft a geo-speculative account of garbage patches in order to consider how the potential effects of this amorphous and changeable object in-form practices for monitoring and remediating this environmental phenomenon. Such an approach works with a generative understanding of the garbage patch as a processual technoscientific object. The garbage patch is not a singular object but is constituted through multiple objects—or "societies of objects"—that concresce within these ocean ecologies and which are of indeterminate and ongoing duration, since plastics persist and transform in environments for indefinite periods of time. Drawing on Whitehead's discussion of societies of objects, I suggest that the forms of relation across the actual entities of the garbage patches become self-sustaining within the particular environment they form and inhabit.[10] The sensors and sensing practices that would track and monitor these processual objects then become environmental along with marine debris in very particular ways, as drifting and circulating objects within enfolding gyres.

LOCATING A GARBAGE PATCH

In more current scientific literature, the Pacific Garbage Patch is often referred to as the Eastern Garbage Patch. This patch area is located between Hawaii and California within the North Pacific Subtropical High, a shifting zone of high pressure and relatively calm water. The Eastern or Great Pacific Garbage Patch is not the only location where marine debris collects in the Pacific, however. A Western Garbage Patch has since been identified near Japan; and the Subtropical Convergence Zone at the transition zone between the Subpolar Gyre and the Subtropical Gyre is also noted for its tendency to collect large amounts of marine debris.[11]

These debris collection zones are also connected to five identified oceanic gyres, including two in the Pacific, two in the Atlantic, and one in the Indian Ocean.[12] As Howell et al. write, the Pacific subtropical gyre is the "largest circulation feature on our planet, and the earth's largest continuous biome."[13] Gyres tend to spiral or converge inward, and the North Pacific Subtropical Gyre—the larger system of which the Eastern Garbage Patch is but a small part, or a "gyre within a gyre"[14]—has been estimated to be roughly between seven to nine million square miles.[15] Of the many forms of marine debris floating through oceans and seas, plastics are the primary form of waste moving through and collecting in oceans. Sixty to eighty percent of all marine debris in oceans is composed of plastics.[16] Plastic fragments sifting through ocean waters most often travel from land-based sources, typically migrating from urbanized areas, wastewater, landfills, and plastics manufacturing sites into oceans. A smaller proportion of plastics derive from marine-based sources, including offshore shipping and fishing activities.[17] Yet with all of these forms of primarily plastics-based marine debris, plastics circulate from manufacturing, use, and disposal to become wayward and often unidentifiable objects congealing in the shifting spaces of oceanic gyres. And marine debris caught in gyres may circle around in cycles that last from six to twenty years or more.

The accumulation of plastics in oceans and seas is increasingly remaking oceanic materialities and environmental processes.[18] However, plastic in gyres assembles less as an identifiable mass of plastic and more as a suspended soup of finer plastic fragments and microplastics. The 2001 Marine Pollution Bulletin article in which Moore and his collaborators describe their findings of plastic to plankton comparisons in the North Pacific Gyre indicates that up to 98 percent of the plastic material gathered through trawls of the Pacific gyre were composed of finer plastic particles. Of these finer particles, "thin films and polypropylene/monofilament line" were present as identifiable plastic, while "unidentified plastic" in the form of plastic fragments were the main types of plastic sampled.[19]

If Google Earth or a satellite view of the garbage patch proves to be an impossible undertaking, it is because the plastics suspended in oceans are not a thick choking layer of identifiable objects but more of a confetti-type array of suspended plastic bits. Practices of sampling plastics in areas of high concentration of marine debris involve working with fine-mesh trawls. These trawls are able to collect microplastics across a range of visible and relatively invisible sizes. Establishing a universal standard for microplastics as smaller than 5 mm has been an important step in regularizing the study of microplastics in seas, since plastics break up into such a wide range of forms and sizes.[20] Microplastics were settled at a certain size in order to study the effects of this distinctive and pervasive category of plastics. Some plastic fragments and objects are large enough to pose ingestion hazards to marine organisms—these are often termed macroplastics.[21] Other plastic fragments are so small as to be invisible and undetectable, or to be readily ingested by many marine organisms without immediately obvious effect. Size is an important indicator in assessing ocean plastics since, on the one hand, there is the risk of entanglement and ingestion hazards and, on the other hand, there are more unknown issues as to how smaller particles of microplastics may transform ocean ecosystems. The garbage patch in this way generates additional objects that concresce through the complex processes of plastics drifting through oceans.

The impacts of microplastics are in many ways still somewhat uncertain, and may have a web of effects that may range from adsorption, chemical transfer of persistent organic pollutants (POPs) (among other substances), endocrine disruption, alterations in plankton feeding habits, decreasing biodiversity, and shifts in climate change.[22] Numerous "data gaps" exist in relation to microplastics.[23] Plastics are known to adsorb and concentrate chemicals such as POPs from seawater and transport these substances to other locations.[24] But how do chemical substances migrate into and across organisms, and what effects do these substances have on organisms over time? How might they cause endocrine disruption within marine organisms and humans?[25] What effect do microplastics have on plankton and insect populations, and how might this also affect food webs, biodiversity, and climate change by altering the composition and source-sink dynamics of oceans?[26]

Attempting to establish the "matters of fact" related to garbage patches in the oceans is an experimental process that is less about how to demystify the garbage patches and more about how the ongoing attempts to make sense of the garbage patch and the effects of plastic are bound up with these complex constellations of objects that also pose pressing "matters of concern."[27] Here, attempting to establish verifiable circulations of plastics in oceans is not about dispelling fictions but about experimental modes of narrating, testing, and sensing that bring the garbage patch as an object of concern toward a space of workable interpretations and engagement. Such an approach to experimenting is rather different than the experimentation that might take place within environmental sciences, for instance, which is typically understood as a process of testing interventions in order to form new hypotheses.[28] Such experimentation with matters of fact and concern constitutes an intervention of a different sort, which questions how technoscientific objects and environmental pollution are made evident and how they may generate other objects as part of the processes of identifying, locating, and monitoring their presence.

"Data gaps" about garbage patches are also an important part of how matters of fact in relation to plastics are experimented through matters of concern. These gaps serve to mobilize experiments on how plastics transform and rematerialize in oceanic spaces. Gyres with higher concentrations of marine debris, and in particular the North Pacific convergence zone, are primary sites where questions related to the effects of plastics unfold. Yet from this perspective, the garbage patch is one of several objects that concresce in the oceans. Locating the garbage patch as an object is not a simple delineation in space but a processual unfolding of ongoing potential effects. The garbage patch is thus not one object, but, following Whitehead, a *"society" of objects* that in their interaction give rise to new and ongoing relations, formations, and actual occasions.[29] Locating the Pacific Garbage Patch is not a matter of demarcating a stable continent of plastics on a satellite map. Instead, the garbage patch requires locating objects within objects— objects that are "intra-acting"[30]—and giving rise to new potential effects and environmental conditions. A geo-speculative approach to a garbage patch grapples with the very ways in which the object-ness—and the *potential* object-ness—of a garbage patch is never stable or given but in process and giving rise to new engagements. So too do monitoring practices and technologies become entangled within these processual societies of objects.

Plastics in oceans indicate how the materiality and interaction of these multiple intersecting objects are continually generating new conditions, objects, and societies of objects. Adsorption of chemicals may alter the habits of some marine organisms; degradation of plastics may shift the composition of source-sink dynamics; new microbes may emerge in the new plastic niches that form.[31] The garbage patch is a site where objects proliferate. As plastics fall apart, they generate new effects, occasions of becoming, and processes of materialization.[32] Plastics as they persist in environments are characterized less as a condition of inert objects taking up space and more as a condition of material persistence and transformation across environments and organisms. With these changeable conditions and objects, sensing practices also are established in relation to shifting pollution events to be monitored. In the next section, I discuss how some of these practices have emerged in relation to marine debris and further discuss how oceans have become instrumented spaces that have given rise to a vast range of sensing practices across computational and other monitoring modalities.

MONITORING A GARBAGE PATCH

Locating the garbage patch is on one level bound up with determining what types of plastic objects collect within it and what effects they have. Yet on another level, locating the garbage patch involves monitoring its shifting distribution and extent in the ocean. As has been discussed so far, the garbage patch is not a fixed or singular object, but a society of objects in process. The composition of the garbage patch consists of plastics interacting across organisms and environments. But it also moves and collects in distinct and changing ways due to ocean currents, which are influenced by weather and climate change, as well as the turning of the earth (in the form of the Coriolos effect) and the wind-influenced direction of waves (in the form of Ekman transport). As an oceanic gyre, the garbage patch moves as a sort of weather system, shifting during El Niño events, and changing with storms and other disturbances.[33]

How does the garbage patch become detectable while it is also in process? Techniques for studying marine debris typically coincide with techniques for studying ocean circulation. In some cases, flotsam is directly observed and modeled as a way of gauging likely movements of debris across ocean currents. A well-known study by Ebbesmeyer focused on the movement of bath toys (ranging from ducks, frogs, beavers, and turtles), which spilled from a container ship that was washed overboard during a storm in 1992.[34] Based on beachcomber efforts, as well as by identifying the bath toys by serial number and mapping and inputting coordinates into the Ocean Surface Current Simulator (OSCURS) computer program, Ebbesmeyer developed a circulation model that gave the locations of gyres (which corresponded with related gyre studies) and the likely time that objects spent in gyres.

His "flotsametrics" technique drew on decades of studies that have attempted to discern patterns in ocean circulation by mapping the pathways of flotsam. Here, the drifting message in a bottle, or MIB, is a classic reference point for studying and experimenting with objects as they travel in oceans. One MIB was recently found in Scotland, which had a release date of 1914.[35] But numerous other experiments have been developed alongside these circulation studies, including a 1976–1980 experiment by NOAA that set loose "tens of thousands of plastic cards in response to significant oil spills along the East Coast from Florida to Massachusetts."[36] Working on behalf of NOAA, Ebbesmeyer collected these plastic spill cards over time, some of which drifted in oceans for over twenty-five years. Given the length of these drifts, Ebbesmeyer estimates the plastic spill cards may have circled between seven to nine times around the North Atlantic Subtropical Gyre—the Atlantic version of the Pacific Garbage Patch.[37]

"Traceable Drifter Unit," or TDU, is the term that Ebbesmeyer uses to describe flotsam that is released en masse (with releases exceeding one hundred thousand drifters) in the ocean and that yields data relevant to ocean surface currents. These TDUs have ranged from Guinness beer bottles to MIBs with biblical or governmental messages, as well as material from known container spills. As Ebbesmeyer writes of flotsam drift studies, "By their endurance for as long as a century, flotsam provides a tool for tracing long planetary drifts. Drifters riding the global conveyor belts, for example, require twenty years to circle the earth."[38] The ways in which flotsam travels, drifts, and collects in oceans may be studied over long periods of time, and the different exit points for flotsam to head toward coasts, or extended times in which it takes to reach coasts, may indicate just how long

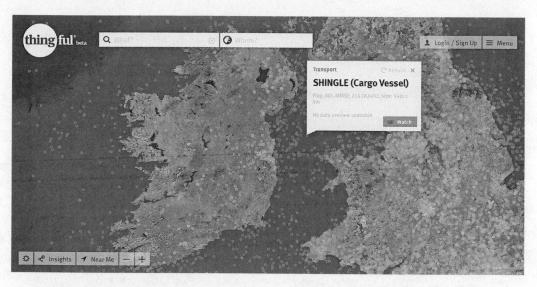

Figure 5.2. Thingful. A beta-phase Internet of Things platform for mapping and viewing sensors worldwide, which includes multiple examples of sensors in marine environments such as the cargo vessel shown here. Screen capture.

marine debris remains within oceans and in particular how many times debris circulates around ocean gyres. The convergence zones are not just collections of primarily plastic stuff but also metamorphosing oceanic repositories that include items from the early boom years of plastics, sporadic spills from container ships, and passing fashions in consumer goods. The packaging, films, fragments, and assorted objects that cycle around gyres may remain there for many decades to come, eventually forming new oceanic environments and influencing organisms and food webs.

Oceans as Sensor Spaces

Beyond following the movement of objects and TDUs set loose in oceans, many more monitoring practices have developed to observe ocean circulation and the likely movement of marine debris, including airborne sensors, coastal webcams, drifter buoys and tracers that communicate to satellites, remote sensing via satellites, and even apps that citizens can use to document marine debris sightings.[39] Oceans have become sensor spaces with an extensive array of sensing nodes and drifting sensor points that can be found on buoys and hulls of boats, underwater gliders, and Argo floats (or instrument platforms for observing oceanic temperature, salinity, and currents). While many sensors are in place to take temperature observations, as well as feed into climate change monitoring and modeling, other sensors are used to survey noise underwater in order to prevent damage to marine organisms' ability to navigate these spaces. Marine traffic tracking sites also document the movement of container ships and other large vessels; and some new platforms and maps such as Thingful focus on capturing objects within the Internet of Things and reveal just how densely populated oceans and seas are with sensing devices.[40]

Ocean-observing platforms span across ships, buoys, Argo floats and subsurface drifters, remotely operated vehicles (ROVs), autonomous underwater vehicles (AUVs), satellites for ocean research, aircraft, unmanned aerial vehicles (UAVs) or drones, HF radar, and drilling platforms.[41] All of these are sites and instruments where ocean monitoring takes place. The importance of monitoring oceans has increased considerably, since oceans are the primary sink that absorbs both CO_2 and heat, and the dynamics of these sink-based processes are less well understood in relation to climate change.[42] On the one hand, there has been a lack of monitoring in the oceans, which current practices are attempting to mitigate. On the other hand, the current spread of instrumentation is leading some researchers to propose remote access to the ocean from any number of sensor networks. As Helmreich writes in one instance about the proposed establishment of a "distributed ocean observatory," this project would involve "a network of remote sensing buoys that can provide continual Web access to data from the sea" and "would allow scientists to sit in their living rooms gathering oceanographic data."[43]

The becoming environmental of computational sensors in oceanic spaces involves the instrumentation of oceans with extensive sensing networks as well as the reworking of the environments in which sensing takes place (from underwater to living rooms). Yet computational sensors become environmental in yet another way, where sensors themselves might be adapted to ocean environments and processes, with drifting buoys, Argo floats, and sensors on vessels circulating through oceans across surfaces, subsurfaces, and at depths now down to six thousand meters.[44] And as sensors fill these spaces and provide monitoring data, they also generate other sensor tales, including observations about the likely drift of marine debris through ocean currents. Oceans might then be seen as an environmental medium with medial effects. I now turn to consider two projects that express much different types of sensing practices in relation to marine debris to consider how apps and oceans, drifting buoys and marine debris, unfold through distinct monitoring practices that attempt to prehend oceanic plastics.

Citizens Tracking Marine Debris

Numerous citizen-science and citizen-sensing initiatives now exist that study ocean environments. While beachcombers of all sorts have been involved in collecting debris, identifying organisms through ocean sampling days, and even mailing in plastic-pellet samples to scientists to aid in research projects,[45] newer citizen-sensing projects are adding to the numbers of data points collected from oceanic

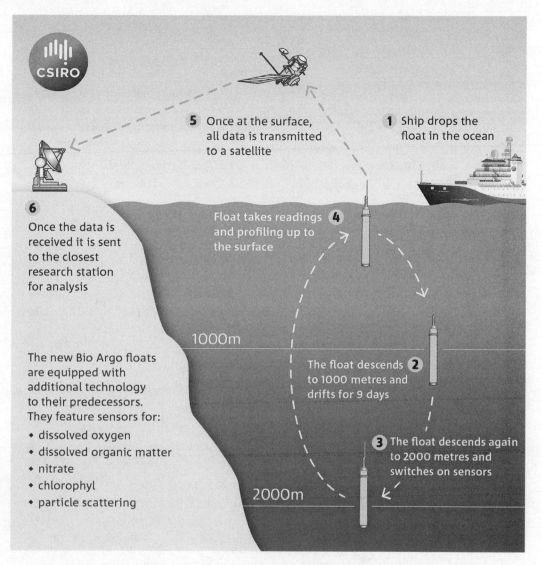

CSIRO

5 Once at the surface, all data is transmitted to a satellite

1 Ship drops the float in the ocean

6 Once the data is received it is sent to the closest research station for analysis

Float takes readings and profiling up to the surface **4**

The new Bio Argo floats are equipped with additional technology to their predecessors. They feature sensors for:
• dissolved oxygen
• dissolved organic matter
• nitrate
• chlorophyl
• particle scattering

1000m

The float descends **2** to 1000 metres and drifts for 9 days

3 The float descends again to 2000 metres and switches on sensors

2000m

Figure 5.3. Argo Floats. Diagram of the deployment of Argo floats with sensors, as well as the communication system for sensing the marine environment and sending data to satellites. Screen capture.

spaces by providing sampling data on marine microbes, plankton densities, and seabed temperatures.[46] There are citizen-sensing initiatives focused on the use of ocean robots, and citizen-science initiatives engaged with monitoring and identifying marine debris.[47] And to return to the introduction of this chapter, there are Google Earth applications for monitoring the distribution and drift of Argo floats, including projects where schools can "adopt an Argo float" to observe monitoring data gathered in the Southern Oceans or in European seas.[48]

One more well-known project situated within these citizen-sensing initiatives is the Marine Debris Tracker, which allows users to identify, document, and map sightings of marine debris.[49] Developed by Jenna Jambeck along with Kyle Johnsen in the Southeast Atlantic Marine Debris Initiative within the College of Engineering at the University of Georgia in collaboration with the NOAA Marine Debris Program, the app was featured as an "App We Can't Live without" in the 2014 Apple Worldwide Developer Conference video, alongside apps such as Tumblr and Pinterest.[50] The app was first released in 2011 and has since been downloaded 10,000 times. Despite these numerous downloads, there are "700 registered users," and of these users 15 to 20 participants "report debris on a near-daily basis." While the number of participants for such a lauded app might seem rather low, the intensive gathering efforts of these participants has nevertheless resulted in "32,899 data points—entries which total about 345,000 individual pieces of trash," including "15,500 cigarette butts found by users on St. Simons Island to plastic jugs floating in the ocean off the coast of Costa Rica and plastic bags near the coast of Brunei."[51]

Proximity to a littoral zone is not a prerequisite for using the app—in fact, some marine debris sightings have been submitted from places as far-flung as North Dakota, presumably with the logic that all things lead to the ocean. With this in mind, I download the app to test the process of submitting entries. The opening app screen notes, "Leave only waves and footprints behind," and allows me to either check the terms of use or go on to "Track Debris." The terms of use ask me to ensure that the information I supply is accurate and also warn that I may be barred from using the app if it becomes apparent that the information I supply is inaccurate. With the "Track Debris" section, I have the option to "Log Items," and from a pull-down menu can identify the item I have found, whether aerosol cans, balloons and/or string, building materials, buoys and floats, cigarettes and cigarette lighters, fishing lures and lines, flip-flops (which have their own separate category, distinct from shoes), gloves, jars, jugs, and plastic or Styrofoam fragments, as well as rope, silverware, six-pack rings, straws, tires, plastic toys, batteries, and fireworks, among many others.

I trawl around through my local gutter, since in fact the Thames River is nearby and this in turn flushes out to the sea. Here, I find numerous bits of cigarette butts and food wrappers. I log the food wrappers, noting a quantity of two and a description of "hamburger wrapper." I need to turn my location services on in order to allow the app to automatically log my location. Once logged, the item shows up on a map, and becomes one more of the many data points of tracked trash. While the app does not necessarily seek to create a "global picture of debris since data entry relies on volunteers," it does hope to provide detailed views of specific locations where users regularly log debris items.[52] If I look over the map and data for where marine debris has been logged, I can see entries that span from Iran to Omaha, Nebraska.

Ocean environments are currently under stress, with increased acidity due to rising CO_2 levels, depletion of fishing stocks, and even the collapse of organisms not well understood. Here, an app that focuses on marine debris orients attention toward logging and mapping the extensive array of litter that flows seaward. Observations made on land and at littoral zones are facilitated and shared through an app that allows citizens to sense and track trash that will largely consist of macroplastics—and have yet to disintegrate into microplastics. If we were to plumb the depths of media theory, we might find this is a rather different engagement with media content and technology than many, even computational, forms to date, since the app functions as an almost understated naturalist's notebook that can be shared and pooled across participants. While the focus in this app is on identifiable debris, it begins with an initial tracing of the journeys that debris might make to oceans and seas. Once in the water, debris can take on yet another journey of indeterminate material transformations, splintering into microplastics and moving in and through organisms that ingest this debris. And yet these app-based citizen-sensed "sightings" of plastics open into a geo-speculative set of encounters: How will these debris and debris mappings generate new modes of environmental engagement? What effect, if any, will they have on the orbiting garbage patches?

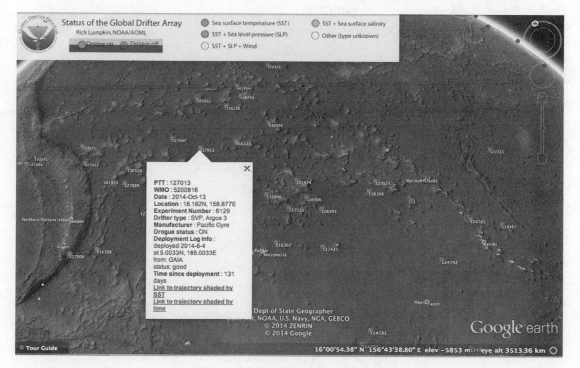

Figure 5.4. Global Drifter Program. Map of drifters in operation as they sense the marine environment near the "Great Pacific Garbage Patch." Global Drifter Program and Google Earth. Screen capture.

Global Drifter Program

Another ocean-sensing project working in a more scientific register across techniques of drifter tracing and sensor communications, the Global Drifter Program, has deployed tracking buoys that communicate with satellites to establish circulation patterns in ocean currents. Along the way, the drifters have also become devices for establishing the likely movements of marine debris, since where the drifters collect is likely to indicate the same locations in which other flotsam collects.[53] The Global Drifter Program consists of a platform of more than 1250 drifting buoys that have been deployed over several decades spanning from initial development in 1979 to current annual mass deployments to monitor the oceans.[54] The buoys monitor the upper water column and provide information on ocean surface and atmospheric conditions, as well as fluxes between air and sea. Run through the Atlantic Oceanographic and Meteorological Laboratory (AOML) in Miami, Florida, the drifters are deployed at study sites and then circulate across oceans. Detecting and sensing sea surface temperature, barometric pressure, wind velocity, ocean color, salinity, and subsurface temperatures, the buoys monitor ocean conditions primarily to determine weather and climate patterns. As they circulate, the buoys send 140-character messages on location and ocean conditions—what physical oceanographer Erik van Sabille has referred to as "Twitter from the ocean."[55] Part of the Global Earth Observation System of Systems (GEOSS) of monitoring technologies, the Global Drifter buoys also link up with earth models to provide forecasting data.

In addition to functioning as weather, climate, and circulation observation devices, the drifters have provided detailed and longer-term data on the likely movement of debris in oceans. A high proportion of drifters has gravitated toward the five gyres, and in this sense the drifters have provided further data for establishing where gyres are located and how long drifters or debris may converge in these areas.[56] Through studies that use Global Drifter data, the formation of a sixth Arctic gyre has been identified, as well as observations about the ways in which patches are "leaky" and circulate debris across regions, potentially over a timespan of centuries.[57] The drifters are in many ways proxies for demonstrating how debris travels over time in oceans, how debris converges in gyres, and the length of time it may take debris to exit convergence zones (if at all) and wash up in coastal regions. The drifters were not originally developed as monitoring devices to study the accumulation of debris directly, since they focused on ocean circulation patterns. But the drifters became an imported technique for studying how debris circulates and settles in ocean spaces in relation to the study of ocean circulation. The drifters also eventually become debris, as they have a limited (five-year) battery life, and cease to function due to mechanical error, environmental stress, and more.[58]

The Global Drifter Program potentially not only corroborates or qualifies prior and differing studies on ocean circulation but also provides a more real-time observation platform for understanding how gyres may shift—and debris concentrations along with them. In many ways, the ongoing deployments, shifting oceanic trajectories, and real-time communication of the drifters are practices that emerge in relation to and through a fidelity to the shifting technoscientific objects under study. The sensing and satellite-linked drifters enable sensing practices that are able to more continually monitor these shifting object conditions and processes. Debris concentrations—whether differently termed and identified as the garbage patch or gyre or convection zone—exist as objects within objects, and these objects change the other objects with which they are intra-acting. New object conditions are continually unfolding here, from changes in chemical and biological conditions to alterations in habitats, shifting locations of the garbage patch due to ocean and atmospheric circulation, and even changes in climate and environments.[59] Within these societies of objects, sensing buoys also concresce along with the circulation patterns and debris under study, thereby materializing a distinctly environmental and oceanic form of computational sensors.

SOCIETIES OF OBJECTS

With the oceanic sensor spaces and two monitoring projects discussed above, plastic marine debris concresces as a society of objects with and through which sensing practices and technologies emerge. "The character of an organism," Whitehead suggests, "depends on that of its environment."[60] The organisms and medial forms drifting through ocean spaces are in-formed by this oceanic environment that is the "datum" and that provides conditions for concrescence. Yet at the same time, "The character of an environment is the sum of the characters of the various societies of actual entities."[61] This is not to say that environments entirely consist of societies of entities, but rather that no society of objects exists without its environment (or datum), and no environment is without its entities or societies of objects.

Environments infuse the characters of societies and entities, but societies are not all there is. At the same time, environments are conditions in which facts and entities take hold, have relevance, and endure. Societies in-form milieus; but milieus also in-form societies, forming conditions for their endurance as well as creative advance. Societies might be seen to be different than "the social," in this sense, since neither are societies a simple assemblage nor are they a relation articulated in advance of the individuating and coming together of entities into collectives. If the last chapter considered the ways in which monitoring climate change might provoke reworkings of how we approach the "citizen" in citizen sensing, this chapter suggests that monitoring plastic pollution in oceans might give rise to reworkings of societies and collectives that are variously understood to be the object of monitoring, as well as the environment in and through which facts and concerns take hold as establishing the relevance of this datum.

As Whitehead, and later Stengers, have suggested, the ways in which organisms take hold and take account of their environments are the key to understanding how relevance is established. "As far as the way in which the living organism 'holds' *qua* enduring is concerned," Stengers writes, "it certainly exhibits a selective character, as is indicated by the relevance of such technical terms as 'detect,' 'react specifically to,' 'activate,' and so on. Yet it is a selection that endures."[62] Endurance is not an individual condition but a process that unfolds through a "dynamics of infection"—where infection is indicative of a "value . . . on what is prehended" so that "when a being endures, what has succeeded is a co-production between this being and 'its' environment."[63] We might refer these notions of infection, endurance, and value also to Whitehead and James's conception of "solidarity," where "the one and the many" are involved in processes of detecting, reacting to, and activating.[64]

For Whitehead, societies are not entirely composed of the "living," since such a nexus does not account for the complex prehensions of environments that occur across organic and inorganic entities. As Whitehead writes, "All societies require interplay with their environment; and in the case of living societies this interplay takes the form of robbery."[65] Here is a condition where a living nexus makes use of material bodies that together can form various types of "structured societies" that span from "crystals, rocks, planets, and suns," and, if Whitehead were writing now, might be a list to which "plastics" could be added. As an inorganic foodstuff of sorts, a scaffolding, infrastructure, collection zone, and supporting environment, plastics concresce as particularly contemporary societies of objects along with the living entities that would inhabit and ingest these materials.

Why bother with this account of societies of objects? What work does it do? I would suggest that plastics and garbage patches as societies of objects point us not just toward the environments and entities—as well as sensing practices—that form with and through these materials, but also to the speculative aspects of garbage patches as a form of oceanic pollution that might be monitored. My point, then, is to bring this discussion closer to the geo-speculative beginning of this chapter, where I suggested that what matters is less the origin of the garbage patch as a geo-mythological figure but more the geo-speculative unfoldings of the multiple and indeterminate garbage patches as they generate new environmental effects and societies of objects, and thereby become sites for establishing ongoing matters of fact and matters of concern.

FROM GEO-MYTHOLOGIZING TO GEO-SPECULATING WITH A GARBAGE PATCH

Such a shifting society of objects, which is in process and so oriented toward further potentialities,[66] gives rise to distinct sensing practices for engaging with these generative—and often non-sensuous—material and technical milieus. The garbage patch is an entity where plastics are in process and circulating across oceanic systems. As discussed here, the plastics that drift through oceans and debris patches are indeterminate objects of study that are often approached obliquely through their nearly imperceptible materialities and unknown potentialities. Given its material constitution, the garbage patch is also not external to that which inhabits it, but occupies the many different organisms that live amid it, where many organisms filter plastics through their bodies. As an ocean-in-the-making,[67] the geo-speculative force of these object-events unfolds as a space of potential objects to come and of indeterminate environmental events to *make sense* of. In this oceanic zone, there is a distinct becoming environmental of computational media that emerges through the wayward drift of plastic and organismal societies of objects.

Oceans and objects are sites for sensing practices in the making. Drifters and sensors, together with studies of particle movement and ocean currents, are *both* abstract approaches to understanding the garbage patch, as well as concrete things in the world that mobilize matters of concern.[68] This is a way of understanding concern less as a cognitive habit of mind and more as an "affective tone."[69] In other words, it engenders a condition or environment of relevance, endurance, and infection. In this way, concern is closely aligned to propositions and speculation, because concern constitutes a proposition about what matters. Shaviro suggests that, in many ways, "concern for the world, and for entities in the world" is also an aesthetic engagement, a provocation (or infection) that involves resonant registers of experience.[70] The garbage patch on one level could be seen as a particular "lure for feeling" that as a proposition even suggests what counts as matters of fact.[71] Such abstraction, as a lure, is not separate from concrete events but instead is an attractor for identifying that which matters and how to make sense of that experience.[72]

In this geo-mythology transformed into a geo-speculation oriented toward garbage patches and environmental monitoring technologies, these societies of objects turn out to be always in the middle of things. Experimental practices and compound objects converge in these oceanic gyres. Being alert to the garbage patch and debris concentrations in the oceans might then require developing an attention to the generative and potential materialities that may continue to unfold through these objects—and not simply making sightings of fixed macroplastics in singular or amassed form. In many ways, this geo-mythology finally shares a sideways correspondence with that earlier plastic *mythology* rendered by Roland Barthes. In his concise postwar account of plastics, he describes plastics as "the stuff of alchemy," through which the "transmutation of matter" takes place.[73] His description charts the "transit" of plastics from raw material to any number of objects. This study of the garbage patches has, in a related but different way, dealt with the transit of plastics from discarded object to environmental and oceanic entity. As plastics break down in oceans, they concresce in oceanic gyres, filter through marine organisms, alter environmental conditions, and turn up as objects of concern for monitoring techniques and citizen-based engagements. The same plastic changeability—or plasticity—that Barthes expounded upon as giving rise to an infinite array of consumer goods redirects toward a material changeability that influences and transforms environments on a planetary scale. The "transmutation" that takes place here is equally subject to speculation: What potential events and objects will concresce through these plasticized oceans, marine organisms, and technical objects?

The garbage patch emerges not as a single fixed object but as a processual and speculative society of objects. On at least one level, it is present as the product of technoscientific advances in materials, where plastics give rise to new environmental and technoscientific problems as a result of the solutions they initially presented. On another level, monitoring the plastic waste requires new technologies of observation, from remote sensing to distributed sensor buoys, to bring plastic as marine debris to attention. Such technoscientific observation techniques focused on marine debris in the gyres inevitably also mobilize responses for remediating and managing the issue of plastics in the seas. In this sense, the garbage patch in its intractable plasticity gives rise to technoscientific practices not just to monitor but also to repair, control, or otherwise manage this object of study and concern.[74]

How does the relationship between monitoring and intervening in the garbage patch influence this society of objects and the practices employed to study and respond to it? Intervening within and developing strategies for addressing the garbage patches may, on one level, appear to require a beach-cleaning effort or antilitter campaign. Yet on another level, designing practices for engaging with the indeterminate and ongoing interactions and societies of objects may be one way to encounter the garbage patch as a space in which to experiment with the matter-of-factness and concernedness of plastic objects as they transform in oceans. Within this space, new understandings of "response-ability" may also proliferate[75] in terms of how the relations between objects are articulated abstractly and unfold concretely, how societies of objects attract and mobilize distinct types of technoscientific and environmental practices, and how the material occasions of oceans are not a remote object of study, but rather are an actual occasion in which we are now participating and through which we will continue to be affected.

A key question arises from this study of the garbage patch as a generative technoscientific and computational object, which is what other forms of technoscientific engagement and sensing practices might be necessary not just to articulate a project of environmental awareness (which is what the project to identify the patch as an explicit and *visual* aberration perhaps demonstrates) but also to generate an object that provokes new forms of environmental participation and attention to the *eventual* effects of our material lives. What experimental forms of politics and environmental practices might we develop that are able to attend to these indeterminate and emergent matters of concern? A repurposed geo-speculative account of the garbage patches might then attend to the indeterminate edges of technoscientific objects and sensing practices and to the modes of engagement yet to be experimented and generated at these sites. Perhaps this geo-speculative account gives rise to the need for a cosmopolitical approach to technoscientific objects such as the garbage patches,[76] which do not concretize as much through their performative or instrumental capacities as they do through the debris of once-useful applications such as plastic that have acquired other capacities and material effects beyond what was anticipated. Here, new societies of objects emerge from the remains of technoscientific pursuits and in turn give rise to new monitoring practices for studying these residual and yet generative objects with unknown and indeterminate effects.

Figure 6.1. London Air Quality Network (LAQN) station. There are over one hundred air quality stations in the LAQN, most of which are managed and run by the King's College Environment Research Group (ERG). Photograph by Citizen Sense.

6

Sensing Air and Creaturing Data

We find ourselves in a buzzing world, amid a democracy of fellow creatures; whereas, under some disguise or other, orthodox philosophy can only introduce us to solitary substances, each enjoying an illusory experience.

—ALFRED NORTH WHITEHEAD, *Process and Reality*

IF YOU SHOULD FIND YOURSELF standing outside the Hobgoblin Pub on New Cross Road in the Borough of Lewisham, London, you might notice a grayish-white box approximately two-and-a-half meters high scrawled with a faded and cascading line of graffiti. Wedged in the space between buildings and facing outward toward the road, the air vent and monitoring equipment at the top may be one of the few details that betray the purpose of this structure, which is to measure air quality at this fixed spot in London. One of the stations in the London Air Quality Network (LAQN) that covers thirty-three boroughs, this monitoring station contributes to the hourly indexes of air quality and news of pollution "episodes" in London. Detecting sulfur dioxide (SO_2), particulate matter 10 and 2.5 (PM 10, PM 2.5), as well as nitrogen oxide (NO) and nitrogen dioxide (NO_2), the station generates data that indicate whether the UK is meeting EU air quality objectives for both short- and long-term emissions of pollutants.[1] The data also contribute to environmental science research and are managed and made available by the Environmental Research Group (ERG) at King's College London, where this network is managed and run.[2]

Passersby may experience, in a potentially fleeting way, the connection between this station, the local air quality, and the data it generates, which typically circulate in spaces of environmental science and policy. The air quality data that are generated at this fixed site are black-boxed and located in spaces somewhat remote from experiences of air quality on the street. Air quality data are not typically present at the point of encounter with this station, but instead are located in more distant spaces of laboratories and servers, where data are gathered and processed to influence the management of environments and air quality.

In order to make air pollution data gathered by this station and the approximately one hundred other stations in the LAQN more accessible, King's ERG has designed a London Air app to allow people to observe emissions levels at key monitoring sites and to make inferences about their own personal exposure when passing through these sites. While this strategy moves toward making the data of fixed sites more accessible through an air quality app, the pollution that individuals experience in their everyday trajectories may be quite different than the types of pollution that are captured through fixed monitoring sites generating data that are averaged over set monitoring periods. The New Cross Road station, for instance, typically records an annual exceedance of NO_2 at this fixed point—a pollutant formed through combustion of fuel that is largely the result of high levels of automobile use in the city.[3] Yet all along New Cross Road individual moments and locations of exposure may give rise to a far different set of pollution "episodes," with much different consequences for urban dwellers in these areas.

Inevitably, the question arises as to how individuals may map their own mobile exposure to air pollution, which is likely to differ from the fixed sites of the official monitoring stations. As discussed throughout this study, environmental monitoring is proliferating from a project undertaken by environmental scientists and governmental agencies to a practice in which DIY groups and citizen sensors are now engaged. One attempt to sense air quality beyond fixed and official monitoring sites has included community deployments of diffusion tubes, a low-cost analogue method for gauging air pollution but which requires weeks-long deployments of tubes that are then sent off to labs for analysis and data production. Here, the process of gathering air pollution data may be democratized, but the generation and analysis of meaningful data take place in remote laboratory settings.

More recent citizen-sensing projects that deploy lower-cost digital sensors and smartphones have focused on monitoring air quality levels in ways that attempt to make environmental data more immediate and connected to experienced conditions. One of the primary ways in which such citizen-sensing projects have sprung up is through direct engagement with monitoring environmental pollution. While some citizen-sensing projects use the itinerant aspects of individual exposure to environmental pollution as a way to experiment with mobile-monitoring practices with which fixed sites of detection cannot compare, including Preemptive Media's "Area's Immediate Reading" (or AIR), which consists of a mobile and individual air monitoring device for gauging individual exposure to air pollutants;[4] other projects, including Safecast, suggest in relation to environmental disturbances such as the Fukushima nuclear fallout that official or government data may not always be available or trusted, so that alternative data sources may be necessary in order to gauge exposure to pollutants of immediate concern, such as radiation levels.[5]

Whether displaying pollution levels or developing platforms to make pollution information more readily available, many citizen-sensing pollution projects attempt to make the details of environmental pollution more instantaneous and actionable. An even more extensive range of pollution-sensing projects have turned up in this area, from Common Sense's work with fitting air quality sensors to street sweepers in the Bay Area, to any number of citizen-sensing kits and devices that use low-cost electronics, including Speck (for PM 2.5 sensing) and AirCasting (for NOx sensing).[6] Citizen sensing is a strategy that often attempts to translate practices of monitoring pollution from the spaces of "expert" scientific and government oversight into practices and technologies that are available to a wider array of participants. As the EPA has noted in its work on surveying and assessing the rise in citizen-sensing practices and low-cost monitoring equipment, air pollution monitoring is no longer confined just to official networks and the professional practices of scientists and technicians, but is proliferating into new types of uses that might, they anticipate, even begin to "supplement" regulatory approaches to air pollution. "New breakthroughs in sensor technology and inexpensive, portable methods," one EPA report notes, "are now making it possible for anyone in the general public to measure air pollution and are expanding the reasons for measuring air pollution."[7] With these citizen-sensing practices, data shift from having to meet a regulatory standard to ensure policy compliance to indicating change, and in the process instigate different citizen-led actions.

In citizen-sensing projects, more extensively and democratically gathered data are typically presented as "the reasons for measuring air pollution," since it is through collecting data that everything from enhanced participation in environmental issues to changes in policy are hoped to be achieved. The impetus to monitor and gather data is bound up with established (and emerging) processes of understanding environments as information-based problems. Within citizen-sensing projects, data are intended to be collected in ways that complement, reroute, or even circumvent and challenge the usual institutions and practices that monitor environments and manage environmental data. Data are seen to enable modes of action that are meant to offer effective ways to respond to those problems. With more data, potentially more accurate data, and more extensively distributed data, environmental problems such as air pollution are intended to be more readily and effectively addressed. Data are intertwined with practices, responses to perceived problems, modes of materializing and evidencing problems, and anticipations of political engagement. But how are air quality data constituted, through expert or citizen practices? How do differing practices of environmental monitoring in-form the character and quality of data gathered, as well as the possible trajectories and effects of those data? What are the instruments, relations, and experiences of air quality data generated through these distinctive engagements with environments and technology? And in what ways do environments become computational through the use of low-cost air pollution monitoring technologies?

In this chapter, I consider how citizen-sensing practices that monitor air pollution experiment with the tactics and arrangements of environmental data. These monitoring experiments, however, are not just a matter of enabling "citizens" to use technology to collect data that might allow them to augment scientific studies or to act on their environments. Rather, as I suggest throughout *Program Earth,* computational-sensing technologies are bound up with the generation of new milieus, relations, entities, occasions, and interpretive registers of sensing. The becoming environmental of computation describes this process. Sensor-based engagements with environments do not simply detect external phenomena to be reported; rather, they bring together and give rise to experiencing entities and thereby actualize new arrangements of environmental sensing and data. The production of air quality data through environmental monitoring generates distinct subject-superject entities and occasions for generating and making sense of that data—as scientific facts, matters of concern, or even as inchoate patterns produced through unstable technologies or sporadic monitoring practices.

As a central point of focus, this chapter then crucially asks in what ways environmental sense data emerge not through universal categories or forms but as concrete entities—or *creatures*—that concresce through processes of subjects participating in environments and environmental events. "The actual world is a process," Whitehead writes, and this "process is the becoming of actual entities. Thus actual entities are creatures; they are also termed 'actual occasions.'"[8] Actual entities are creatures, or lively meetings of entities that form routes of experience. In this sense, the process of gathering air pollution data might be identified as more than documenting static facts of air quality at any given time or place and instead be approached as a practice that gives rise to entities and modes of participation that transmit data in particular ways and along distinct vectors of environmental participation.

Working with this Whitehead-inspired analysis of how concrete entities of environmental data materialize through pollution sensing, I then consider how environmental-sensing projects are processes of what I call *creaturing data,* where the actual environmental entities that come together are creations that materialize through distinct ways of perceiving and participating in environments. These creatures may have scientific legitimacy. Or they may form as alternative modes of evidence presented in contestation of scientific fact. But in either or both capacities, they are *creaturely* rather than universal arrangements of data.

The point of attending to the creaturing of data is to at once draw attention to the concrete actual entities of data—even the "accidents" of data, as Whitehead would have it—and to take into account the "conditions" that give rise to and sustain these creatures of environmental data.[9] Created data are not an abstract store of information or something to be coherently visualized, but rather are actual entities involved in the making of actual occasions and material processes. Data may typically appear to be the primary objective of environmental sensing projects, which focus on obtaining data to influence environmental policy and practices, but along the way the relations and material arrangements that data gathering sets in place begin to creature new entities that concresce through monitoring practices. By turning to the creaturing of data, I consider in the projects that follow how data mobilize or underwrite environmental practices. At the same time, data might fail to materialize as anticipated, and through this process activate participatory arrangements that might be quite different from those intended. The failure of environmental sense data to translate into an easy spur to environmental action can even be an important way in which the creaturely aspects of data concresce. Data unfold not simply through instrumental or even epistemic registers but also as attractors and attachments. It is through these attractors and attachments that experiments with environmental citizenship—and not just sensor technologies—also develop.

The general ethos of many DIY- and citizen-sensing projects has been that by enabling and democratizing the monitoring of local environments, it may also be possible to achieve increased engagement with environmental concerns. These projects test, experiment with, and mobilize alternative modes of environmental citizenship. Yet in what ways do practices of environmental monitoring with sensing devices give rise not just to experimental modes of participation and civic engagement but also to different modalities for experiencing environmental pollution through monitoring practices that generate air quality data? Within these projects, how does the experience and experiment of air pollution and air quality data become a site of political, as well as potentially affective, engagement? How do the creatures of environmental data become points of attachment for influencing and in-forming environmental concern and politics?

Through this discussion of citizen-sensing projects that develop experimental and creative approaches to monitoring air quality and generating environmental data, including Feral Robotic Dogs, the Pigeon Blog, and Air Quality Egg, I further consider how these technological modes of sensing generate distinct practices of environmental citizenship in and through engagements with data. The projects I discuss below involve the creaturing of data in a double sense, since they also deploy more-than-human participants, including robotic dogs, homing pigeons, and plastic eggs as concrete entities for drawing together citizen-sensing practices. While these environmental sensing projects are, on one level, focused on creating opportunities for citizen sensors (of sorts) to generate their own data, on another level these projects also create additional data in and around the practices they set in motion. This is not just the data of environmental phenomena observed and monitored but also the data of not obtaining what was expected, of shadowing events in different ways than a run of quantitative data might evidence, of generating residual and qualitative data from the eventfulness of environmental monitoring, of creating different patterns of data rather than adhering to accuracy as the sole criterion for data legitimacy, and of mobilizing alternative

Figure 6.2. London Air Quality Network app. This app gives a general sense of air quality, from low to medium and high levels of pollution, across London boroughs. Photograph by Citizen Sense.

creatures of data, such as dogs, pigeons, and eggs (and their extended milieus), within the distributed digital infrastructures of environmental monitoring.

How might we describe the processual and creaturely entities that concresce through different practices of monitoring the air? What does the air (and "the environment") become through monitoring devices, and what are the ways in which it concresces and becomes involved with experience, particularly if we consider experience as "constructive functioning"?[10] What are the relationships, political engagements, and ways of mobilizing data that make for the most a/effective environmental practices? And if monitoring and citizen sensing are emerging as new modes of environmental participation, in what ways do these experiments further enable practices for engaging with and addressing air pollution, and for speculating with environmental politics? These are some of the questions that arise when considering how a creaturely approach might shift the ways in which data are seen to materialize and gain perceptive power.

CREATURING DATA I: MONITORING AND MATERIALIZING AIR

Where does my body end and the external world begin? . . . The breath as it passes in and out of my lungs from my mouth and throat fluctuates in its bodily relationship. Undoubtedly the body is very vaguely distinguishable from external nature. It is in fact merely one among other natural objects.

Breath is an example of the difficulty involved in delineating where the body ends and the world begins, as captured Whitehead.[11] It may be somewhat common-place to note that breathing is a process in which we are all involved, necessarily, to sustain our bodies. Whitehead's articulation of the ways in which—through breathing—bodies cannot be conceived of as discrete from environments and other entities on which they rely further indicates the ways in which this process concresces into environmental, political, social, and more-than-human occasions. Breathing articulates distinct subject-superject relationships, since it involves more than the simple fact of needing to breathe and extends to the sites, entities, and conditions involved in exchanging air, which can be polluted and irritating to spe-cific organisms, as well as given to remaking the bodily capacities of organisms as they live and endure within particular ecologies of air.

The ways in which bodies, environments, and the multiple substances that per-colate and stir through any given patch of air come together can be distinctly in-fluenced by what is actually in the air. Beyond the usual list of atmospheric gases, including nitrogen (78 percent), oxygen (21 percent), argon (1 percent), and vari-ous trace minerals, pollutants may be present in quantities that register as parts per million (ppm) or parts per billion (ppb) molecules of air. CO_2, as discussed in chapter 4, has been measured at Mauna Loa, Hawaii, to now be measurable at 400 ppm and rising.[12] Pollutants, in other words, are often present in seemingly miniscule quantities and yet are able to disrupt and remake environments, bodies, and ecological processes on local and global scales. Beyond trace gases, however, a whole range of coarse to ultrafine particles also chuck through the air, from dust and skin flakes to diesel particles, to the airborne residue of grilled ham-burgers and more. The air further exchanges materials with the soil and oceans in a complex cycling that influences weather and climate. Airborne specks and remainders simultaneously issue from and are exchanged to reshape the bodily and atmospheric inhabitations underway in any given environment.

Gases and particles that are actually monitored in relation to managing air quality then constitute a select portion of all the substances mixing within the air. While there is a wide range of pollutants circulating through the air, the EPA has designated common or criteria pollutants that are regularly monitored and are notable for their effects on human and environmental health. These pollut-ants include CO, nitrogen oxides (NOx), SO_2, PM 10 and PM 2.5, lead (Pb), and ozone (O_3).[13]

Smokestacks and chimneys have served as the industrial icons of air pollu-tion, and sulfur-saturated skies were the historic events that led to the formation of clean air legislation in many Western countries, from the Clean Air Act in the UK (1956) to the Clean Air Act in the United States (1963), as well as evolving air quality objectives in the EU (2008).[14] Inevitably, the "cleaning up" of Western skies often raises questions about how coal-fired manufacturing may have been displaced rather than remedied, as the exceedingly high levels of pollution in China and other Asian manufacturing hubs demonstrate.[15] Beyond coal use, pollution now emanates from a range of sources and is not always present primarily as SO_2 that forms from the burning of coal. Instead, NOx and PM are the relatively colorless and odorless pollutants of increasing concern and that are primarily generated in urban areas from automobile traffic as well as the heating of buildings. Unlike SO_2, which is a far more visible and palpable pollutant both in its immediate presence and eventual effects in the form of acid rain, defoliation, and more, NOx and PM tend to be less immediately evident as key pollutants, but they have considerable effects on human and environmental health.

Defining what counts as air pollution is far from a straightforward matter when the evidence of harm potentially becomes more difficult to establish.[16] Institutional and governmental monitoring networks typically identify pollutants of concern in response to health research that provides evidence for levels of harm caused by particular pollutants. As part of the Global Burden of Disease 2010 study, outdoor air pollution was identified as a leading cause of death, contributing to heart, lung, and cardiopulmonary disease, which are now particularly linked to PM 2.5 exposure, which are also less evident as pollutants.[17] In many ways, health research influences environmental policy, which sets targets in relation to which monitoring networks set criteria for monitoring, as well as providing air quality forecasts, management, and mitigation.

While the impacts of air pollution on human health are one of the key motivators for establishing air quality standards, often the means of monitoring and enforcing these standards can miss the localized pollution experienced by individuals. Environmental and individual health are bound up with articulations of what does and does not count as a pollution episode and what may constitute an excessive level of pollutant exposure. Emissions of a certain pollutant at a given site in a city may be within an acceptable range, but individual exposure may vary considerably. Air, noise, and water pollution are local if distributed environmental disturbances that many urban dwellers experience on a regular basis, although for some more than others since sites of pollution are often concentrated in lower-income urban areas.[18] Emissions and exposure mitigation have then been identified as two different ways in which to monitor and manage air quality: one addresses fixed sites and reductions of air pollutants; the other attends to how individuals may manage their individual experience to lessen air pollution exposure, such as monitoring and taking alternative routes through cities, although not necessarily attending to overall reductions of air pollutants.

There is an extensive literature that discusses citizen engagement in monitoring air pollution, although often at the level of how citizens respond to or aid scientific findings, or how they collect evidence themselves in order to contest or augment official air quality readings.[19] In this second approach, analyses of participatory- and environmental-justice-focused engagements with air pollution have discussed the many and even noncomputational ways in which air samples may be collected in order to influence environmental science and politics. Global Community Monitor, for instance, is an environmental activism group engaged in a DIY-bucket collection method to monitor air quality in places such as neighborhoods adjacent to oil and gas refineries located in regions known as "Cancer Alley."[20] While the bucket becomes a device for collecting air samples in a more democratic and local way, the analysis of the air samples must still take place in laboratories (similar to the diffusion-tube air analysis) that are not sites of citizen engagement. Such projects present a low-tech way of conducting a version of citizen science, which are largely focused on environmental activism and justice, but they also raise questions about how air quality data might be experienced in more real-time situations and how data might become admissible as evidence for making environmental claims. This is where many citizen-sensing initiatives attempt to make a contribution by providing more immediate and accessible access to air quality data. But this approach is also not without its problems, as will be discussed below.

Data Becoming Relevant

Articulations of personal, urban, and environmental health shift across these different strategies for addressing air pollution. Practices of monitoring pollution at the citizen or individual level provide a way to counter or redress the possible gaps in data, but there is more to these projects than this, since in mobilizing sensors to bring environmental monitoring into a more democratic, if often individual, set of engagements, new material-political actors, engagements, and experiments concresce—along with new political (im)possibilities. With many of these projects, the question arises as to how data become relevant. Air pollution data might become relevant through health research that establishes high levels of morbidity due to particular air pollutants, or through scientific monitoring networks that identify pollutants exceeding accountable limits, or through concerns for certain environmental effects, from acid rain to eutrophication, which unfold with excessive levels of pollutants.[21]

Relevance is a term that Whitehead uses to address the ways in which facts have purchase, and the "social environments" that are set in place in order for facts to mobilize distinct effects.[22] Relevance is a critical part of the process of creaturing, since creaturing involves the ways in which creativity is conditioned or brought into specific events and entities. The ways in which creatures gain a foothold, in other words, are expressions of relevance. Social environments are integral to the immanent processes that condition and give rise to creatures— they do not exist without the formation of creatures, and they continue to co-evolve as the situations in which creatures make "sense" and have effect.

Environments, as understood here and throughout this study, are then at once an "object" of study as well as a mutually in-formed and coproduced relation through which monitoring practices and gathered data take hold and gain relevance. The relevance of air quality data is not determined through absolute criteria, since these criteria shift depending upon modes of governance, location, and more. If data are understood instead as perceptive entities, it then becomes possible to attend to how data are differently mobilized and concresce within and through practices. Data in one context might have the status of facts, and in another context might galvanize a much different set of a / effects. As the EPA has expressed in its analysis of new modes of environmental monitoring, "types of data" and "types of uses" are interlinked.[23] Data typically only become admissible for legal claims when gathered through specified scientific procedures and with quite precise (as well as expensive) instrumentation. There may also be situations in which data are "just good enough" for establishing that a pollution event is happening, for instance.[24] Yet it remains a relatively open question as to what the uses and effects of data gathered through citizen-sensing technologies might be, since these creatures have arguably not yet settled into entities for which relevance is expressible. In other words, how do citizen sensors undertake actions with and through air pollution sensing practices and data? Could it be that the environments of relevance for this data are still in formation?

At this point, it might be easy enough to make a statement about the ways in which environmental monitoring technologies "construct" the air and the problem of air quality. While this inquiry works in a way parallel to constructivism, it also attempts, following Stengers, to think of constructivism not as a process of making *fictions,* but rather of making realities concresce and take hold—or gain a "foothold," as Stengers has discussed elsewhere.[25] As discussed in chapter 1 of this study, sensors are part of generative processes for making interpretative acts of sensation possible and for attending to environmental matters of concern. The environments, arrangements, and practices that are bound up with how facts take hold and even potentially circulate with effect are then a critical part of any study into how expanded and differently constituted air pollution data and data-gathering practices might have relevance and be able to effect change.

This approach to constructivism is different from a poststructuralist rendering, since ideas and language do not *mediate* things, but rather things concresce as propositional effects.[26] As Whitehead notes, every fact must "propose the general character of the universe required for that fact."[27] Here is another aspect of *tuning,* which is not just a process of making particular modalities of sensing possible across subjects, environments, and experiences, but also involves the tuning of facts and the conditions in which those facts have relevance. If facts require particular social environments in order to have relevance,[28] this does not make them illusory. Rather, it draws attention to the conditions needed for facts to have effect. In this way, facts are creatures, since, as Whitehead elaborates:

> Each fact is more than its forms, and each form 'participates' throughout the world of facts. The definiteness of fact is due to its forms; but the individual fact is a creature, and creativity is the ultimate behind all forms, inexplicable by forms, and conditioned by its creatures.[29]

The creatures of facts—and data—constitute entities that bring worlds into being. Sense data are productive of new environments, entities, and occasions that make particular modalities of sensibility possible. A social environment then plays a formative part in conditioning and supporting creatures of fact and creatures of data.[30] These are creatures of data because they are involved in creative processes that bring sensing to possibility and that in-form the environments where these modes of sensing have relevance.

It might be useful at this point to step back, briefly, to explain how creativity unfolds within Whitehead's cosmology. "Creativity" is another word for the creative advance, or process, of entities, which explains how entities may be understood both as being and becoming. Creativity is "ultimate" within Whitehead's speculative philosophy, and this approach enables an engagement with entities that moves beyond a fixed subject-object relation to attend to processual and immanent conditions of the formation of entities. In this way, the entities that concresce through a creative advance are "creatures."[31] Creatures are subject-superjects, they are the "conditions" that creativity settles into, since creativity can only be known through its conditions. It should also be noted that creativity is not inherently good. Whitehead expressly develops a "neutral" metaphysics that seeks to explain process, but he does not cast judgment on the ways in which a creative advance settles into creatures.[32]

A process of creaturing data then attends to the ways in which data are not fixed objects gathered through universal criteria but instead are entities through which forms and practices emerge as creatures, and through creaturely processes. As discussed throughout this study, perceiving subject-superjects combine as *feeling* entities through actual occasions. These entities might otherwise be termed creatures, since they are formations of conditioned creativity. Furthermore, the "datum," as Whitehead discusses it, is not simply an external array of objects awaiting conceptual classification by a human subject. Instead, the datum is that which subject-superjects feel. Through this experiencing (and so processing and transforming) the datum, subject superjects are able to generate actual entities, or creatures. As he explains:

> The philosophies of substance presuppose a subject which then encounters a datum, and then reacts to the datum. The philosophy of organism presupposes a datum which is met with feelings, and progressively attains the unity of a subject. But with this doctrine, "superject" would be a better term than "subject."[33]

Whitehead uses Kant's notion of experience as "constructive functioning," but reverses the order of experience. Experience is not always "on its way to knowledge" in the form of "objective content," but rather can be understood in the way that the datum is felt and processed to become superjectal. In this approach, Whitehead inverts the usual Kantian way of understanding experience (i.e., as a subject decoding universal objects) to suggest that objects find "satisfaction" in subject-superjects, which take account of the datum in particular ways.[34]

I suggest we understand Whitehead's Kantian inversion as a process of creaturing data, since it draws attention to the ways in which data are always felt and experienced by and as creatures, which through feeling further give rise to distinct forms of data. A process of transforming the datum into felt experience is a process of creaturing data because what issues through this process are subject-superjects involved in processes of being and becoming creatures. Perhaps in the most concisely stated version of this insight, Whitehead writes, "An actual entity is an act of experience."[35] Feeling the datum is a process of transforming the datum into experience, which concresces as an actual entity or creature. Creaturing is then the description of this process of *feeling the datum,* where creatures are the actual entities formed through creaturing the datum.

If we consider the "data" that digital sensors generate, then these devices might be understood less as technologies for gathering data and more as technologies for processing, transforming, and creaturing data—as a felt form of the datum. While it may be easy enough to query the assertion that more data and more democratically gathered data might lead to action and engagement, an approach to creaturing data suggests that it might be relevant to attend to the ways in which data are taken up, felt, experienced, taken into account, gain relevance, and attain "power" as the process whereby particular perceptions or modes of prehension involve or prevail over others.[36]

Practices of processing environmental data are the routes whereby data achieve "subjective satisfaction" and become relevant to the persistence and further formation of that data. Furthermore, subjects that process and transform environmental data are human *and* more-than-human creatures. This is to say that subjects include a vast array of entities, from pavements to trees to sensors—that form and are formed by creative processes of taking up and transforming data. Environmental data are not simply gathered from environments, as though this process only requires that human subjects discover objective universal data to be communicated. Instead, subjects constitutively enable the becoming and being (which is to say, settlement and endurance) of particular forms of data through the ways in which they experience the datum. The creatures that concresce draw from and express worlds in which those data have relevance. The formation of environments as monitor-able then comprises a key part of how data and facts take hold, since environments are co-created along with the processes of subjects parsing and creaturing data.

CREATURING DATA II: DOGS, PIGEONS, AND EGGS

Sensing technologies are then entangled with and mobilize new environmental monitoring practices and new ways of gathering data. The modes of engagement and spaces through which data are gathered, analyzed, and communicated are central to the emergence of these environmental modes of practice. Citizen-sensing projects are frequently described as data campaigns or as identifying an issue about which more data may be needed in order to effect policy changes. As numerous studies of science and technology have noted, however, data are always embedded within political practices, structures, and institutions that in-form

Figure 6.3. Feral Robotic Dogs website. A project for autonomous dog-robots to sense pollution in environments, developed by Natalie Jeremijenko et al. Screen capture.

everything from how data are delineated and collected to how they are joined up, communicated, and acted upon.[37]

For the remainder of this chapter, I focus on these aspects of environmental data as forming creatures of data. By looking specifically at three projects that engage with computational modes of sensing environments, I consider the relations, practices, and political possibilities that concresce to form these distinct creatures of environmental data. Feral Robotic Dogs, Pigeon Blog, and Air Quality Egg are citizen-sensing projects that largely focus on doing science and environmental monitoring differently through the actors, arrangements, tools and spaces where monitoring is undertaken. At the same time, in pollution-monitoring projects, the gathering of sense data is often closely tied to a/effecting political action and environmental change by addressing how data are generated, collected, and acted upon—as well as creatured through processes of subject-superjects feeling the datum.

Dogs

One of the earliest creative-practice projects to engage with environmental sensing, Feral Robotic Dogs was originally developed by Natalie Jeremijenko in 2002 through the Bureau of Inverse Technology (or BIT) and developed in additional versions and deployments through 2006. The project adapted existing Sony Aibo toy dogs by "upgrading" them with all-terrain bodies and environmental-sensing brains and noses. Ready-made robotic toy dogs with preprogrammed tasks were identified as having more interesting potential uses: these were creatures "awaiting further instructions." The first generation "gamma dog" was proposed to store and transmit environmental data from "any radioactive source" that exceeds EPA thresholds, where the deployment of these dogs in multiples would "provide informational spectacle and conclusive on-data convergence in a given local area."[38] In their development, the semiautonomous gadgets were rerouted to "sniffer" dog mode and fitted with environmental sensors capable of detecting environmental pollutants including volatile organic compounds (VOCs), CO, and methane (CH_4), while providing general indications of air quality.

A number of deployments of the dogs were then developed for sites of likely pollution, including a Former Gas Plant at East 173rd Street Works at the Bronx River in New York, where dogs scouted for volatile organic solvents and polycyclic aromatic hydrocarbons; and at Baldwin Park, Orlando, Florida, where robots were deployed to search for VOCs at "sites of community interest," including a former landfill site that was a proposed site for a middle school. This Florida deployment sought to provide an opportunity for an evidence-driven discussion of the environmental issues facing the community and the opportunity to coordinate diverse opinions and interpretations of the phenomena at hand. As the project description notes:

Because the dog's space-filling logic emulates a familiar behavior, i.e. "sniffing out," anyone can participate and try to make sense of this data in real time without necessarily having the technical or scientific training usually required to interpret data from other sources on the same phenomena. It has the potential to raise the standards of evidence involved [and] promote diverse valid interpretations involved in complex environmental and political processes.[39]

Environmental data were to become *evident* through the movements of dogs that sniff out and map pollutants. This was seen as a way to render environmental data more perceptible and more spectacular, while also changing the possibilities for who can generate and access data and so have the means for contributing to environmental and political debates. Inevitably, processes of monitoring pollutants here and in the projects discussed below are bound up with available sensors that are able to measure specific gases and which creature particular types of evidence. Such versions of data-led environmental citizenship become further informed by the prior investment in sensors developed with specialized technical capacities, often for military, industrial, or scientific uses.

In discussing the Feral Robotic Dogs project, a group of artists and technologists who variously came to work on the project suggested that the meeting of robotics and environmental sensing and mapping could propel activism to new types of encounters, where creative explorations with data gathering and environmental monitoring might create renewed engagements with local environments.[40] Here is a sensing project that speculates about the possibility for participatory and citizen-based data collection in order to create more direct and materialized connections between environmental information and the observers of that information.[41] Through the collection of environmental data, it is imagined that a more immediate and accountable mode of environmental action might also be possible.

Yet this direct connection between data and action could be queried on many levels, since such a translation is not necessarily automatic and in many ways depends on an assumed efficacy that scientific data are assumed to have in the world. The ways in which climate change data, for instance, fail to have an immediate effect on political action may give rise to speculation about whether data necessarily constitute incontrovertible evidence with which to influence and change environmental politics. The failure of data to lead to environmental action might, on one level, stem from the assumed force of a scientifically evidenced and "rational" argument, where decisions made in relation to environmental matters of concern instead unfold through multiple and competing political interests. On another level, however, the ways in which data in and of themselves are meant to be—and may also fail to be—compelling may raise questions about the a/effective registers of data. Is a robotic dog a more a/effective data-creature than a spreadsheet,

Figure 6.4. Pigeon Blog website. A project for trained homing pigeons to carry sensor and GPS backpacks and sense air pollution in Southern California, developed by Beatriz da Costa et al. Screen capture.

bar chart, or policy document? The point here is not to set up a false dichotomy between these data forms, but rather to ask about the ways in which the creaturing of data may be one way to experiment with the modes and practices of environmental citizenship. In these creaturely arrangements of dogs, pigeons, and eggs, new distributions of participation might even materialize. But the exact ways in which these forms of participation in-form environmental politics remain a point of speculation that continues to be explored and taken up in subsequent citizen-sensing projects.

Pigeons

If the Feral Robotic Dogs project deployed air-sensing technologies through a robotic toy to make new technical modalities of environmental monitoring more widely available, the Pigeon Blog project raises the question of how air quality sensing transforms even further when pigeons are the reporters and carriers of sensing equipment. Pigeon Blog, developed by Beatriz da Costa with Cina Hazegh and Kevin Ponto in 2006, was a project that used sensor backpacks fitted to homing pigeons to collect low-altitude air quality readings while the pigeons flew through the frequently polluted skies of Southern California. The sensor backpacks consisted of a combined GPS receiver that provided latitude, longitude, and altitude readings, a dual automotive CO and NO sensor, a temperature sensor, and a purpose-built mobile phone for transmitting text messages. The backpack kit was developed as a miniature unit small enough to be carried by the pigeons so that real-time air quality data could be transmitted and visualized as pollution levels on the Pigeon Blog and within a Google Map visualization.[42]

Situated within Southern California and initially developed in Los Angeles, the project addressed the ongoing problem of air pollution and environmental justice by developing an open-source sensing kit that could be used for "grassroots scientific data gathering."[43] The "Pigeon Blog" project was developed as a response to the limited number of fixed air-monitoring stations that are focused on generating longer-term average data about air quality and which may not necessarily be located in areas of the highest-pollution episodes. By providing the possibility for more local measurements and data about local exposures, the Pigeon Blog project sought to complement if not challenge existing data on air pollution to look at the distribution of pollution on a finer-grained level. This approach was shared with AIR, a 2006 Preemptive Media project (briefly mentioned at the introduction to this chapter), on which da Costa collaborated along with Brooke Singer and Jamie Schulte. Consisting of portable air monitors, AIR enabled urban dwellers to complement coarser and fixed air quality data by collecting local data through individual journeys.[44] Equipped with GPS and coordinated with a database of known pollution sources, the air monitors sensed CO, NOx, and ground-level O_3 at distinct locations and provided real-time visualizations of air pollution levels in relation to an EPA air quality index. By making individual maps of urban air pollution exposure, the hope was that urban dwellers could become more aware and engaged in discussing environmental issues through everyday exposure, individual risk, and neighborhood-level mapping.

In both Pigeon Blog and AIR, citizen sensing is presented as an activist project of sorts. Yet Pigeon Blog in particular does not undertake a typical approach to environmental politics and urban air quality. By enrolling pigeons into the project of sensing air, the Pigeon Blog project questions how to develop a mode of "interspecies co-production in the pursuit of resistant action."[45] The pigeons were sent out as "reporters" to draw attention to the issue of air pollution while providing inventive and more accessible ways of gathering data in order to provoke new possibilities for political action. Pigeons participated in this project in multiple ways, since they are creatures with unique navigational abilities and often fly according to major landscape features, such as highways, and are also a pervasive (if often reviled) bird in urban areas. Pigeons often occupy polluted urban areas and may provide a specific view of low-altitude air pollution in areas of high traffic. Pigeons further act as biosensors, and make available distinct urban experiences through proxy modes of sensing.

Pigeons are also key contributors to creaturing data and environmental participation in ways that move beyond the usual spaces of environmental activism.

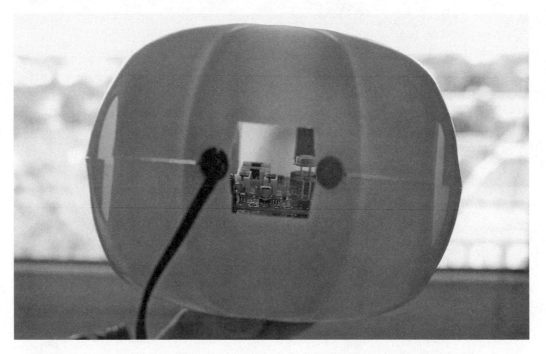

Figure 6.5. Air Quality Egg. View of the Air Quality Egg in its manufactured and saleable form after multiple stages of prototyping. Photograph by Citizen Sense.

Da Costa makes the point that projects such as Pigeon Blog may create new capacities for engaging with environmental information and for mobilizing participation that are not exclusively focused on "how bad things are."[46] While the project set out to provide alternative datasets that might be widely gathered to contribute to more expert approaches to environmental monitoring, in many ways this objective was not achieved. Long-term or even complementary datasets were not generated from the project, and this anticipated outcome even became somewhat incompatible with the project's attempt to experiment with new modes of environmental practice and participation. Instead, Pigeon Blog experimented with the urban, technological, and more-than-human entities that became part of the project of sensing, experiencing, and reporting on air quality. While this project failed to engage with air quality data in the ways initially anticipated it nevertheless arrived at an expanded approach to environmental practice that was less conventionally data driven and more attentive to the ecological modalities of citizenship that might materialize with such a distributed approach to sensing environments.[47]

Eggs

As earlier experiments into environmental sensing, Feral Robotic Dogs and Pigeon Blog tested the ways in which new and distributed modes of participation across human and more-than-human modalities might shift the possibilities for political engagement in air quality. These projects continue to influence citizen-sensing projects currently underway, which have proliferated not least through the increasing availability and affordability of sensor technologies. Air Quality Egg is one project in this area that has sought to connect up maker communities in developing digital devices to enable citizen sensing of air quality.[48] Developed as a "community-led" project, where the community is largely comprised of creative technologists located in New York, London, and Amsterdam, the early prototype version of the egg project consisted of a Nanode sensing platform that detected CO and NO_2, which the project creators identified as key air pollutants. Housed in a rapid-prototyped egg-shaped shell, the air-sensing apparatus was initially developed at workshops in New York City and Amsterdam, further tested at the second Citizen Cyberscience Summit in London in 2012, and subsequently gained considerable backing on Kickstarter, which allowed the device to be manufactured for wider use.

As the eggs developed from prototypes to manufactured sensor kits, they could then be ordered from electronic hobby suppliers. Citizen-owned eggs were in turn mapped in an online platform and shown to be located in the United States, Europe, Australia, and Japan, and variously gathering measurements that were uploaded to a Xively data platform. While the Air Quality Egg project is ostensibly focused on air quality monitoring, it is more centrally located within technical communities that are driven to experiment with the technology of sensing devices. These communities typically engage in what has been referred to as "participatory-sensing" experiments by bringing the technical functionalities of sensors to more immediate points of encounter through setup, testing, modification, and online tutorials for upgrading devices.[49] In the process, many questions have arisen regarding how these technologies actually work and the extent to which the data they generate are accurate. Comments on the community forum related to the Air Quality Egg raised a heated debate as to how viable it is to monitor air quality accurately with such a device, given the scale, lack of calibration, and coarse instrumentation of the metal oxide sensors it uses (which are typically used for automotive functions). To what extent are egg-gathered data useful and accurate? And what if the egg fails to function in the first place (as was the case in several of the prototyping workshops)?

In a project video, however, commentators on the project suggest that the accuracy of data is not a primary issue, since sense data might provoke environmental concerns that could be followed up with more thoroughly scientific study. Here, the focus is on a community of egg users and developers inclined toward testing devices, which might raise further calls to action, even if the trajectories from local, sporadic, and somewhat momentary datasets to the influencing of environmental science, policy, and behavior are not entirely clear.

At one point in the project video, which captures the testing of the egg during the London Cyberscience event in the spring of 2012, a participant remarks, "the chicken is not ready," to refer to the back-and-forth attempts to have the egg setup, calibrated, and ready to gather measurements. In the context of creaturely data, this project seems to be an entity in-formation where sensor-led technical kit is the assumed impetus for galvanizing environmental issues and action. The environments that concresce here are highly in-formed by computational modes of sensing and acting. Data gathered through electronic sensing is seen to be the force that propels perceived possibilities for activism, but the force of data emerges less through the accuracy of data and more through the process of a technical community making a device that can draw attention to data practices as potentially political engagements. Data are creatured in the Air Quality Egg through a device ready to hatch and give rise to new modalities of data-driven activism. But the modes of participating in making devices and generating data may be much different entities and occasions, arguably, than the modes of participating in environmental activism, which may be a legitimating subtext for the egg but not the primary focus in this tech-led and maker-community approach to participation.[50]

EXPERIMENTING WITH ENVIRONMENTAL CITIZENSHIP

The three projects discussed here share a similar approach to environmental sensing as a more democratic engagement with data gathering in order to influence environmental politics. Yet beyond these similarities, much different entities and modalities of citizenship come together in these projects. The Feral Robotic Dogs project tests deployments in landfills and the site of a proposed middle school, making the point that from these sites new communities of interest might emerge to influence environmental debate. Data are rendered as a more haptic and materialized experience, something demonstrated through a fleet of sniffing robotic dogs. The Pigeon Blog seeks to make urban air quality visible through a more-than-human engagement, which at once redistributes environmental participation while creating a more experimental approach to sensing urban environments. And the Air Quality Egg focuses on developing an Internet of Things approach to creating a worldwide sensor network, where new devices and the data they generate lead to new possibilities for participation and the formation of communities.

Yet in each of these projects, the translation from environmental-sensing experiment to citizen-based engagement with environmental issues remains unclear. Such issues are not uncommon within more grassroots modes of citizen sensing and environmental monitoring. As the London Air Quality Network points out, monitoring air quality on a DIY-level may not be as easy as it first appears. This is due to the complexity (and expense) of working with precisely instrumented sensors and the questions of accuracy that pertain to sensing projects that use less refined sensing equipment or that are not set in up a systematic way to study environments over time. Yet the dogs, pigeons, and eggs of these projects are not gathering data at the level of scientific study but are making a case for the development of complementary data sets to inform what is monitored and how it is brought to attention in order to be acted upon.

At this level of action, additional questions arise as to how environmental sense data may influence environmental politics and actions. In an earlier human–computer interaction research project, Common Sense, which tested the deployment of sensors for measuring air quality on street sweepers, the project participants arrived at the observation that environmental community organization is actually the critical factor in order for data generated through sensor deployments to be relevant, meaningful, and actionable.[51] In fact, community environmental organizations have in some cases been rather skeptical of the extent to which more data from computational sensors will necessarily facilitate more effective action. In this way, some researchers question the assumption that more localized and data-led processes of environmental observation and monitoring do actually enable greater environmental participation.

While the uses and effects of citizen-sensing data, particularly as gathered through digital technologies, are arguably still in formation, this situation raises challenging questions about the types of creatures that might concresce through citizen-sensing data. On the one hand, since the relevance of scientific datasets is something that citizen-sensing practices would ideally like to harness, there may be an overreliance on transferring scientific rationales to citizen-sensing datasets. On the other hand, the assumed effectivity of scientific datasets for mobilizing political action also remains unquestioned in this move, since more, or more accurate, scientific data does not always (if ever) incontrovertibly lead to political change, as the ongoing inertia around London's air pollution and the exceedance of EU guidelines clearly indicates.[52]

The EPA's report on new types of monitoring technologies and practices might be read even as more of a provocation for the types of data practices that are still in a generative or experimental phase and the ways in which the creatures of data are still in formation along with their environments of relevance. Citizen-sensing practices for monitoring pollution are experimental both in the technologies of environmental monitoring and data gathering and in the practices and social environments within which these data might have relevance and become creatures of data. In other words, citizen-sensing practices are in-formation as experimental practices that test not just how environmental monitoring data might be differently gathered but also how such data might be mobilized within distinct environments of relevance, and to what (political) a/effect.

Environmental sense data gathered without a clear link to community projects may not have the anticipated effects of facilitating greater participation in environmental matters of concern. Yet, to varying degrees, some of these projects do experiment with the methods, techniques, communities, modes of participation, sites of monitoring, and evidential modes of activism and politics that might materialize as new entities and processes for engaging with environments and environmental issues. These experiments with a/effectivity and practice bring openings—as well as further controversies—to approaches to environmental politics and participation that might be investigated further. Monitoring data—as typically conceived—might not be the critical unit for mobilizing environmental citizenship and action; and a gadget-led process for engaging with politics may not be the most definitive answer to developing new modes of environmental engagement. However, these citizen-sensing projects raise the question of what other experiments might emerge that open up the possibility for new types of environmental politics and new modes of collective participation.

Within this space, the modes and practices of data—the creaturely entities in and through which data manifest and give rise to worlds—are arguably an area yet to be fully explored, since data are so frequently presented as the abstract and dematerialized evidence of environmental facts. But the modalities, materialities, and *creatures* of environmental data may be one way of experimenting with monitoring practices as sites of environmental engagement, where affectivity and the relevance of social environments become critical to considering the effectivity of data.[53] The creatures that concresce here are not only those of environmental sense data but also those of environmental citizenship. Distributed and more-than-human modes of participation contribute to air pollution and its monitoring. From computational sensors to moving air masses, manufacturing and transport, vegetation and animal bodies, temperature gradients and topography, and economic inequality and real estate, as well as policy and modeling, a number of entities converge in the project of experiencing, participating in, and experimenting with sensing air pollution.

PARTICIPATION AS INVOLVEMENT

The engagements that are made possible through the expanded arrangements of environmental data could then be understood as modes of togetherness, or *involvement,* as Whitehead terms it, of environmental problems and events. This is a way of saying that environmental data are participatory, yet not simply as an articulation of humans using and connecting across gadgets or even of objects having agential force. Instead, data are participatory as technological concrescences that make distinct modes of engaging with environments take hold and persist while shaping the actuality and the possibility of environmental politics and imaginaries. The concrescences of environmental monitoring and environmental data encompass more than a concatenation of actors, since the participation of these multiple entities is also a mode of prehension that describes the ways in which "actual entities involve each other."[54]

Involvement is a distinct and processual meeting of entities formed through encounters—as well as absences (hence, positive and negative prehensions). Involvement constitutes more than an assemblage, arguably, since the character and effect of relations are integral to the ways in which entities concresce to produce practices, facts, and subjects-superjects. Involvement is a process that further signals the concrete making of worlds (and environments)—the being and becoming of those worlds, as they endure and change. How then might environmental data be understood not simply as the end *result* of monitoring practices but as the *ingression* or mode of participation and even potential involvement, which mobilizes and concresces monitoring practices in particular ways?[55]

Some atmospheric scientists have suggested that low-cost digital sensors may, on the one hand, encourage more democratic engagement with environmental issues. But, on the other hand, these sensors might not produce data that are as "accurate" as higher-end instrumentation.[56] While data are differently creatured across scientific and citizen-sensing practices, they are also productive in another way, since these realms and hoped-for uses are not mutually exclusive but begin to in-form each other as arrangements of environmental politics and practices of citizenship. While the "accuracy" of citizen-sensed data might not compare to the data gathered through scientific techniques, it might alternatively constitute different processes for creaturing data and for experimenting with environmental citizenship. Such an approach also raises the question of whether an exclusively scientific approach to environments and environmental data is the only way in which environmental politics might have legitimacy and effect. If we keep in mind Whitehead's famous aphorism that science is necessary but not sufficient, then how might environmental citizenship unfold into a wider set of practices and approaches to *creaturing* data?[57]

This chapter has suggested that by attending to the particularities of how data are processed and transformed into experience—by attending to the creatures of data—it might be possible to open up new considerations for how environmental data become relevant, the environments that ensure this relevance, and what the a/effect of these data might be. If we return to the EPA report on the emergence of new monitoring practices and technologies, the point made about how new types of data might yield new types of data practices remains a salient point for this discussion. The report suggests that while legal and compliance-focused data may still largely be produced through scientific processes, new "personal" and "qualitative" uses of data might emerge through digital sensors and smartphones that have as-of-yet not fully determined functions or effects. If current air monitoring is largely focused on legal compliance and atmospheric science uses, then what environmental engagements might come together from different data practices and arrangements? What might citizen-sensing practices cause to "take hold" in relation to air pollution and air quality? And how do these practices exceed the instruments and devices of sensing to encompass a more creaturely distribution of experience?

As discussed through the three projects that mobilize dogs, pigeons, and eggs in a project of gathering environmental data along more citizenly engagements, environmental monitoring always involves more than its instruments, at that same time that data is never reducible to a universal category. The creatures of data that emerge in these projects are entities that concresce as distinct occasions of environmental engagement across experiencing subject-superjects. The "data" gathered here are not simply in service of visualizing environmental phenomena, whether through finer-grained or mobile modalities. Instead, these creatures are the "consequent reasons" for attending to, processing, and transforming data in these specific ways.[58] While scientific datasets might be understood through particular consequent reasons that appeal to the objectivity of air pollution data, mobilizing these same reasons a priori for undertaking citizen-sensing projects may in fact restrict the possibilities that citizen-led monitoring might provide.

If data is de-creatured, as it were, how does this apparent universality of data obscure the arrangements that lend effectivity to data? At the same time, a case could be made here for reconsidering what counts as "raw data." Rather than excising it entirely, raw data might also be considered to be a distinct effect and creature of data. As Whitehead has remarked in relation to Kant, "objective" conditions are simply a way of creaturing those data—as objective.[59] Furthermore, the objectiveness or givenness of data is not an absolute substance or universal condition, but rather is an actual occasion or entity that has settled in this way to become a fact, but which may also change. If we reconsider raw data in this way, then data are less an absolute condition and more a way of creaturing data as a particular resource, where "rawness" sets particular practices and effects in motion. Within scientific practice rawness then is important as a condition of data upon which additional operations are made.[60] But this creaturely inflection of data need not influence or speak for all other modalities for creaturing data nor the many practices that might be set in motion as ongoing inheritances of creaturely data.

In this context, what does it mean to "sense" or experience air pollution with computational sensors? Monitoring air pollution with digital sensors is not just a way of obtaining a "result" or fact about a particular environment but is also about the ways in which data are creatured and mobilized, the social environments that concretize and allow those facts to have relevance, and the additional attendant data practices that might come together to generate a/effects. Creaturing data is an approach that asks how we might consider much more than the "facts" gathered, since the extended social environments, practices, and speculative relations required to bring facts into a space of relevance are crucial to the creatures of data that materialize. Creaturing data is a way of attending to the processing and transforming of environmental data. This is not simply a matter of attending to the extended capacities of generating data but instead involves considering the creatures of data, the entities and situations that form and take hold, whether to solidify, experiment with, or change environmental practices and politics. These creatures, as Whitehead (following James) has reminded us in the epigraph to this chapter, settle into "a democracy of fellow creatures," where the shared experiences of air, pollution, and possibilities for engagement might even bring us into inventive modes of solidarity.

III

Urban Sensing

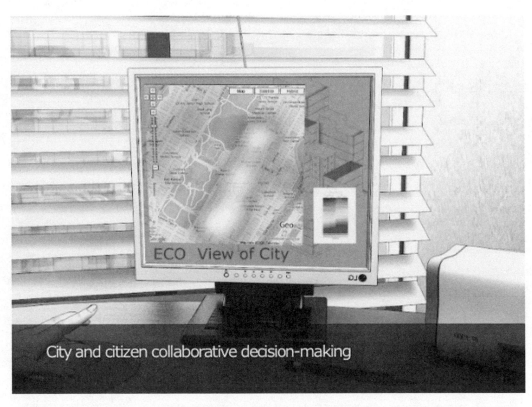

Figure 7.1. Connected Sustainable Cities project video. A collaborative and speculative smart city project developed across the MIT Mobile Experience Lab and Cisco. Screen capture.

7

Citizen Sensing in the
Smart and Sustainable City

From Environments to Environmentality

Cᴵᴛɪᴇꜱ ᴛʜᴀᴛ ᴀʀᴇ ɪɴꜰᴜꜱᴇᴅ ᴡɪᴛʜ and transformed by computational processes seem to be the object of continual reinvention. While informational or cybernetically planned cities have been underway since at least the 1960s,[1] proposals for networked or computable cities began to appear as regular features in urban-development plans from the 1980s onward.[2] From designing for the plasticity of urban architecture to envisioning the city as a zone for technologically spurred economic growth, digital city developments have remade urban spaces as networked, distributed, and flexible sites for capital accumulation and urban experience.

More recent and commercially led proposals for "smart cities" have focused on how networked urbanisms and participatory media might achieve "greener" or more efficient cities that are simultaneously engines for economic growth. Smart city proponents commonly make the case for the necessity of these developments by signaling toward trends in increasing urbanization. While cities are centers of economic growth and innovation, they are also, smart city advocates argue, sites of considerable resource use and greenhouse-gas emissions and are therefore important zones for implementing sustainability initiatives. In these proposals decaying or yet-to-be-built infrastructures are identified as sites of prime smart city development. Smart cities are presented as a neatly packaged way to meet these generalized challenges, thereby ensuring that future cities—whether retrofitted or new—are more sustainable and efficient than ever before.

Although cities infused by digital technologies and imaginaries are not a new development, their implementation to achieve sustainability directives under the guise of smart cities is a more recent tactic for promoting digital technologies. In many smart city proposals, computational technologies are meant to synchronize urban processes and infrastructures to improve resource efficiency, distribution of

185

services, and urban participation. Digital technologies, and specifically ubiquitous computing, have become a recurring theme in articulating how sustainable urbanisms might be achieved; yet the intersection of smart and sustainable urbanisms is an area of study that has yet to be examined in detail, particularly in relation to what modalities of urban environmental citizenship are emphasized or even eliminated in the smart city.

This chapter addresses another aspect of the becoming environmental of computation through the becoming environmental of power that unfolds within smart cities projects.[3] To elaborate upon this particular focus on the becoming environmental of computation *and* power, I take up the emergence of the smart city as a sustainable city by looking at one particular case study, the Connected Sustainable Cities (CSC) project developed by MIT and Cisco within the Connected Urban Development (CUD) initiative. The CSC aspect of the project consists of design proposals developed between 2007 and 2008 by William Mitchell and Federico Casalegno in the MIT Mobile Experience Lab working in conjunction with Cisco CUD. The Cisco CUD initiative was a partnership initiated in 2006 in response to the Clinton Global Initiative for addressing climate change. Pairing with eight cities worldwide, from San Francisco to Madrid, Seoul, and Hamburg, CUD ran until 2010 and has influenced Cisco's ongoing project Smart + Connected Communities, which continues to produce smart city plans, from development underway in Songdo to proposals to develop a "Sustainable 21st Century San Francisco."[4]

Situating this design proposal within a range of smart city projects that include sustainability in their development plans, I examine how this speculative and early smart city project proposes to achieve more sustainable and efficient urbanisms through a number of ubiquitous computing scenarios to be adapted to existing and hypothetical cities. The CSC project proposal bears strong resemblances to many smart city developments still underway and, with its connection to Cisco, one of the primary developers of network architecture for cities, is an influential demonstration of smart city imaginings. Many of the tools developed through the CUD project consist of planning documents, white papers, eco toolkits, multimedia demonstrations, and speculative designs meant to guide smart city development.[5] As an important but perhaps overlooked part of the process of promoting smart cities, these designs, narratives, and documents have played a key role in rearticulating the smart city as a sustainable city. However, this chapter focuses on these proposals not simply as *discursive* renderings of cities but as elements within an urban computational *dispositif,* or apparatus,[6] which performs material–political and environmental relations across speculative designs, technological imaginaries, urban development plans, democratic engagements through participatory media, and networked infrastructures, many of which are folded into present-day urban development plans and practices, even when the smart city is an ever-elusive project to be realized.

Smart city plans and designs, as proposed and uncertainly realized, articulate distinct materialities and spatialities as well as formations of power and governance. By considering Foucault's concept of *environmentality* in this context, I examine the ways in which the CSC project performs distributions of governance within and through proposals for smart environments and technologies. I emphasize this aspect of Foucault's discussion of environmentality in order to open up and develop further his unfinished questioning of how environmental technologies as spatial modes of governance might alter material–political distributions of power and possible modes of subjectification.[7] Revisiting and reworking Foucault's notion of environmentality not as the production of environmental *subjects* but as a spatial–material distribution and relationality of power through environments, technologies, and ways of life, I consider how practices and operations of citizenship concretize that are a critical part of the imaginings of smart and sustainable cities. This reading of environmentality in the smart city recasts who or what counts as a "citizen" and attends to the ways in which citizenship is articulated *environmentally* through the distribution and feedback of monitoring and urban data practices, rather than as an individual subject to be governed.

The primary way in which sustainability is to be achieved within smart cities is through more efficient processes and responsive urban citizens participating in computational sensing and monitoring practices. Urban citizens become sensing nodes—or citizen sensors—within smart city proposals. This is a way of understanding citizen sensing not as a practice synonymous with citizen science but as a modality of citizenship that concretizes through interaction with computational sensing technologies used for environmental monitoring and feedback. In this context, I take up the proposals for smart cities as developed in the CSC project to ask: What are the implications of computationally organized distributions of environmental governance that are programmed for distinct functionalities and are managed by corporate and state actors that engage with cities as datasets to be manipulated? Which articulations of environmentality concretize within sustainable smart city proposals and developments when governance is performed through environments that are computationally programmed? And when sensing citizens become operatives within urban computational systems, how might environmental technologies delimit citizen-like practices to a series of actions focused on monitoring and managing data? Might this mean that citizenship is less about a fixed human subject and more about an operationalization of citizenship that largely relies on digital technics to become animate?

REMAKING SMART CITIES

As might be gathered from the multiple literatures and projects directed toward smart cities, there are numerous interpretations for what even counts as a smart city.[8] It could involve new media districts or automated infrastructures equipped with networked digital sensors, it could refer to the correspondence between online and offline worlds, or it might encompass augmented urban experiences made possible through mobile devices. While earlier research on computational urbanisms may have focused on the relationship between the digital and physical city and the ways in which "virtual" digital technologies might respatialize or represent physical cities,[9] increasingly these approaches have transformed into the ways in which cities are now being remade and marketed through both software and the material infrastructures of digital technologies.[10] Ubiquitous computing remakes cities, rather than displacing or virtually representing them, by generating considerable amounts of data to manage urban processes, as well as by directly embedding devices in urban infrastructures and spaces.

"Smartness," while a generalized reference to computational urbanisms, increasingly refers to urban sustainability strategies that hinge on the implementation of ubiquitous urban computing, or the "fourth utility," as Cisco has termed it.[11] In an industry white paper, "A Theory of Smart Cities," IBM authors involved with the Smarter Planet initiative suggest that the term "smart cities" derives from "smart growth," a concept used in urban planning in the late 1990s to describe strategies for curtailing sprawl and inefficient resource use, which later changed to describe IT-enabled infrastructures and processes oriented toward such objectives.[12] This recurring theme within government and industry white papers on smart cities addresses the ways in which networked sensor technologies are meant to optimize urban processes and resources, including transport, buildings, electricity, and industry, and make them more efficient. Sensor-operationalized and automated environments perform a distinct version of sustainability, where efficiency is the overall goal that influences the merging of economic growth with green objectives. Indeed, smart cities are frequently identified as a hoped-for source of considerable new revenue generation; and in a report funded by the Rockefeller Foundation, the Institute for the Future suggests that smart cities are likely to be a "multi-trillion dollar global market."[13]

The current wave of smart and sustainable cities projects proposed and underway includes numerous proposals located throughout the world that bear similar objectives, plans, and designs related to economic growth through smart and sustainable computational urbanisms. From Abu Dhabi to Helsinki, and from Smart Grids in India to PlanIT Valley in Portugal, many urban development projects are guided by the implementation of networked sensor environments that are marketed through the logics of efficiency and sustainability. Smart city projects are often set up as public–private partnerships between multinational technology companies including Cisco, IBM, and Hewlett Packard, along with city governments, universities, and design and engineering firms. Proposals may involve retrofitting urban infrastructures in New York or London; developing new cities on greenfields in Songdo, Korea, or Lake Nona, Florida; or intensifying network utilities in midsized cities like Dubuque, Iowa, as test sites for networked sensor applications. The focus here is on the ways in which smartness influences articulations of urban sustainability. But rather than fix a definition of the smart city, I work between suggestions that the ways in which informationalized cities are mobilized can be indicative of political and economic interests[14] and that digitally in-formed cities may be figures that continually change in their imagining, implementation, and experiencing.[15] Although smart cities could be rather generic and universalizing in their approach to urbanism, many smart cities also emerge through the materially and politically contingent spaces and practices of urban design, policy, and development, while also forming commitments to specific—if speculative—urban ways of life.

REMAKING CITIZENS IN SMART CITIES

The computational technologies proposed and developed in smart city projects are meant to in-form urban environments and processes, along with the interactions and practices of urban citizens. Citizen-sensing and participatory platforms are often promoted in smart city plans and proposals as enabling urban dwellers to monitor environmental events in real time through mobile and sensing technologies. Yet proposals focused on enabling citizens to monitor their activities convert these citizens into unwitting gatherers and providers of data that may be used not just to balance energy use, for instance, but also to provide energy companies and governments with details about everyday living patterns. Monitoring and managing data in order to feed back information into urban systems are practices that become constitutive of citizenship. Citizenship transforms into citizen sensing, embodied through practices undertaken in response to (and communication with) computational environments and technologies.

Citizen sensing as a form of engagement is a consistent, if differently emphasized, reference point both for development-led and for creative-practice engagements with smart cities. DIY projects propose citizen involvement through the use of participatory media and sensing technologies, and these citizen-sensing projects stress the difference between grassroots and more large-scale smart city developments. Yet an interesting confluence of imaginaries and practices occurs at the point of tooling up citizens, even to the point of "alter[ing] the subjectivity of contemporary citizenship" by enabling urban dwellers to use sensing technologies to interact with urban environments.[16] What subjectivity is this, and might computational environments be one place to turn to consider how (and where) this subjectivity and citizenship is altered? In other words, when urban processes and architectures shift through ubiquitous computing deployed for efficiency and sustainability, how do urban material politics and possibilities for democratic engagement also transform?[17] My interest in these modalities of citizen sensing within smart cities is not to denounce these proposals and projects as tools of control, which might form a typical technological critique, but rather to understand more precisely the ways in which computational materializations distribute power through urban spaces and processes. As Foucault has suggested, rather than attempt to imagine a space free of power it may be more productive to consider how power is distributed as a way to critique modes of governance by imagining how it might be possible not to be governed quite so much—or in that way.[18]

ENVIRONMENTALITY

I take up questions about transformations in urban process, form, and inhabitation in order to analyze in greater detail the ways in which the environmental technologies of ubiquitous computing influence urban governance and citizenship. "Environmentality" is a term I use to describe these urban transformations, which I revisit and rework through a reading of Foucault's unfinished discussion of this concept in one of his last lectures in *The Birth of Biopolitics*. Foucault signals his interest in environmentality and environmental technologies as he moves from a historical to a more contemporary and neoliberal consideration of biopolitics in relation to the milieu and environment as sites of governance. Here, he suggests the subject or population may be less relevant for understanding the exercise of biopolitical techniques, since alterations of environmental conditions may become a new way to implement regulation.[19] Foucault's discussion of environmentality emerges from an analysis of criminality, where in one example he considers how approaches to regulating the supply of drugs may have had a greater impact on conditions of addiction in comparison with strategies that have targeted individual addicted users or populations of addicted users. Working less with an explanation and more with an open-ended suggestion of what he sees as a growing trend toward *environmental* governance rather than subject-based or population-based distributions of governance, he notes, "Action is brought to bear on the rules of the game rather than on the players, and finally in which there is an environmental type of intervention instead of the internal subjugation of individuals."[20] Moving beyond this example, Foucault gestures toward a broader notion of environmentality where influencing the "rules of the game" through the modulation and regulation of environments may be a more current description of governmentality, above and beyond direct attempts to influence or govern individual behavior or the norms of populations. Behavior may be addressed or governed, but the technique is environmental.

Foucault closes his lecture by indicating that in the following week he would examine in greater detail these questions of environmental regulation. However, he does not develop this strand of thought further, and instead, his six pages outlining his approach to environmentality are included as a footnote in *The Birth of Biopolitics* lectures.[21] Consisting more of an unanswered question than a theoretical roadmap, Foucault's discussion of environmentality ranges from a historical analysis of the governing of populations to a consideration of more contemporary modes of governance that may have been unfolding or already underway at the time of his lecture. While his specific concept of environmentality remains a footnote to his discussion of neoliberal modes of governance, it is a provocation for thinking through the effects of the increasing promotion and distribution of computational technologies in order to manage urban environments. In what ways do smart city proposals for urban development articulate and enact distinctly environmental modes of governance, and what are the spatial, material, and citizenly contours of these modes of governance?

The use of the term "environmentality" that I am developing and transforming based on the biopolitics lectures is rather different from the ways in which it has often been taken up based on Foucault's earlier work, from the making of environmentally aware subjects for the purposes of forest conservation in India,[22] to the use of environmentality as a term to capture the "green governmentality" of environmental organizations.[23] Environmentality as a concept does offer up ways of thinking about governance toward environmentalist objectives. But it is important to bear in mind the translations that are made across environmentality and *environmentalism*. Foucault's analysis of environmentality does not directly pertain to environmentalism as such, but rather to an understanding of governance through the milieu.[24] In fact, Foucault's interest in environmental modes of governance touches on strategies of "environmental technology and environmental psychology,"[25] fields that could include designing survival systems or shopping-mall experiences.[26] Environmental modes of governance are also as likely to emerge from the failure to meet environmentalist objectives. Events such as Hurricane Katrina, as Massumi suggests in his analysis of environmentality, generate distinct modes of crisis-oriented governance that emerge in relation to the uncertainty of climate change—a condition of "war and weather" that sets in motion a spatial politics of ongoing disruption and response.[27]

BIOPOLITICS 2.0

Foucault's discussion of environmentality, however abbreviated, addresses the role of environmental technologies in governance and in many ways relates to his abiding attention to the *milieu* (no doubt an influence from Canguilhem) as a site of biopolitical management. Biopolitics, or the governing of life, as he analyzed it in its late eighteenth- and nineteenth-century formations, was concerned with "control over relations between the human race, or human beings insofar as they are a species, insofar as they are living beings, and their environment, the milieu in which they live."[28] If we further take biopolitics to include those distributions of power that influence not just *life,* but also *how to live,*[29] then how are *ways of life* governed through these particular environmental distributions? Indeed, the phrase "ways of life," which Foucault deploys to discuss biopolitical arrangements and distributions of power, is taken up by Revel to suggest that biopolitics is a concept that is not exclusively concerned with "control," as perhaps has been overemphasized through readings of Foucault's earlier work, but that focuses on the spatial–material conditions and distributions of power that are characteristic of and relatively binding within any given time and place.[30] "Ways of life," or "life lived," is a biopolitical concept and approach that also moves beyond understandings of life as a given *biological entity* (this reading of biopolitics may have more to do with Agamben's work on biopolitics and bare life)[31] and instead suggests that *ways of life* are situated, emergent, and practiced through spatial and material power relations. Such a concept does not describe a totalizing schema of power but points to understandings of how power emerges and operates within ways of life, as well as suggesting possibilities for generating alternative ways of life.

A different formation of biopolitics emerges in the context of environmentality, since biopolitics unfolds in relation to a milieu that is less oriented toward control over *populations* and instead performs through environmental modes of governance. In order to capture and examine the ways of life that materialize within the CSC smart city proposal, I use the term *biopolitics 2.0* (with a hint of irony) to refer to the participatory or "2.0" digital technologies at play within smart cities and to examine specific ways of life that unfold within the smart city. Biopolitics 2.0 is a device for analyzing biopolitics as a historically situated concept, a point that Foucault stressed in his development of the term. The 2.0 of biopolitics captures the situatedness of this term, which includes the proliferation of user-generated content through participatory digital media that is a key part of the imagining of how smart cities are to operate; it also includes the *versioning* of digital technologies through the transition of computation from desktops to environments,[32] whether in the shape of mobile digital devices or sensors embedded in urban infrastructure, objects, and networks—something that is captured by the term "City 2.0," which circulates as a parallel term to the smart city.

The biopolitical milieu concretizes material–spatial arrangements in which and through which distinct *dispositifs,* or apparatuses, operate. The apparatus of computational urbanism can be analyzed through networks, techniques, and relations of power that extend from infrastructure to governance and planning, everyday practices, urban imaginaries, architectures, resources, and more. But this "heterogeneous ensemble" can be described through the "nature of the connection" that unfolds across these elements.[33] In his discussions on biopolitics, the apparatus, and the milieu, Foucault repeatedly suggests that the ways in which relations are articulated are key to understanding how modes of governance, ways of life, and political possibilities emerge or are sustained.

Computational monitoring and responsiveness characterize the "nature of the connection" across environments and citizens in smart cities. Biopolitical 2.0

relations are performed through the need to promote economic development while addressing impending environmental calamity, conditions characterized by an *urgency* that Foucault critically identifies as being crucial to the historical situation of the apparatus and, consequently, to the operation of biopolitics.[34] Within smart city proposals and projects, cities are presented as urgent environmental, social, and economic problems that the digital reorganization of urban infrastructures is meant to address by increasing productivity while achieving efficiency. By drawing together Foucault's understanding of how power might operate environmentally and biopolitically, I shift the emphasis toward understanding urban spaces and citizenship within relational or connective registers, with an emphasis on the computational practices and processes that are meant to remake and influence smart city ways of life. In reading and contextualizing these aspects of Foucault as less focused on disciplined or controlled subjects or populations, I also bring environmentality into a space where it is possible to consider how smart cities qualify environmentality by recasting what counts as "the rules of the game."

To say that smart cities might be understood through a biopolitics 2.0 analysis is not so much to suggest that digital technologies are simply tools of control as to examine how the spatial and material programs that are imagined and implemented within smart city proposals generate distinct types of power arrangements and modes of environmentality that entangle urban dwellers within specific performances of citizenship. But within these programs for computational urbanism, the processual and practiced ways of life that unfold or are proposed to unfold inevitably materialize in multiple ways. The "rules of the game" that Foucault

Figure 7.2. Connected Sustainable Cities project scenario for sensors detecting pollution in Curitiba, from the *Connected Sustainable Cities* pamphlet written by Mitchell and Casalegno. Screen capture.

described as central to environmentality might need to be revised as a less static or deterministic rendering of how governance works. Smart city design proposals, on one level, establish propositions and programs for how computational urbanisms are to operate; but on another level, programs never go according to plan and are never singularly enacted. Environmentality might be advanced by considering smart cities not as the running of code in a command-and-control logic of governing space but as the multiple, iterative, and even faltering materializations of imagined and lived computational urbanisms.

CONNECTED SUSTAINABLE CITIES

Working at this juncture of environmental modes of governance, environmental technologies, and sustainability as they are operationalized in smart cities, the CSC project within the CUD puts forward a vision for a near future of ubiquitous urban computing oriented toward increased sustainability. The project proposal materials advocate the smart city as the key to addressing issues of climate change and resource shortages, where sustainable urban environments may be achieved through intelligent digital architectures. The CSC design proposals and policy tools, as well as the core visioning document—*Connected Sustainable Cities*, by Mitchell and Casalegno—develop scenarios for everyday life enhanced, and even altered, by smart information technologies, which "will support new, intelligently sustainable urban living patterns."[35]

Within the CSC design proposals, the technology that most operationalizes smart environments and the programmed interactions between city and citizens is ubiquitous computing in the form of "continuous, fine-grained electronic sensing" through "sensors and tags" that are "mounted on buildings and infrastructures, carried in moving vehicles, integrated with wireless mobile devices such as telephones, and attached to products."[36] Sensor devices are distributed throughout and monitor the urban environment. The continual generation of data provides "detailed, real-time pictures" of urban practices and infrastructures that can be managed, synched, and apportioned to support "the optimal allocation of scarce resources."[37] Digital sensor technologies perform urban processes as a project of efficiency, where environments are embedded with computational technologies that provide urban management and regulation.

Like many smart city proposals, the CSC sites are made smart through several common areas of intervention largely oriented toward increasing productivity while enhancing efficiency. A video lays out the rationale for the project and the core areas it addresses, including platforms developed to aid commuting, home recycling, self-managing one's carbon footprint, facilitating flexibility in urban spaces, and collaborative decision-making as model areas in which improved efficiency by means of digital connectivity and improved visibility of environmental data may save resources and lower greenhouse gas emissions. While many of the applications envisaged in the proposal are already in use within cities, from electronic bicycle rental schemes to smart meters for managing energy use, the project suggests a further coordinated dissemination of sensor technologies and platforms for achieving more efficient urban processes.

In the CUD project video and CSC design document, urban design and planning proposals take place not at the scale of the master plan but at the scale of the scenario. From Curitiba to Hamburg, the episodic urban patterns addressed in these designs and policies include urban services, eco-monitoring toolkits, and speculative platforms intended to achieve smart and "seamless" automated living. Yet in many cases the urban interventions take place in a hypothetical city or in a specified city that is rendered sufficiently general as to be receptive to computational interventions within a universalized language of the everyday. In a design scenario sketched out for "managing homes" in Madrid, numerous capabilities are proposed to make homes more efficient. Mobile phones are GPS-enabled to communicate with sensor-equipped kitchen appliances, so that a family dinner may be cooked by balancing location and timing. The home thermostat will similarly sync with GPS and calendars on mobile phones, so that the home is heated in time for the family's arrival. The organization of activities unfolds through programmed and activated environments so as to realize the most productive and efficient use of time and resources. In the Madrid scenario, monitoring residents' behaviors in detail through sensors and data is essential for achieving efficiency. With this information, environments are meant to become self-adjusting and to perform optimally.

The CSC efficiency initiatives promise to "streamlin[e] the management of cities," lessen environmental footprints, and "enhanc[e] how people experience urban life."[38] By tracking locations and daily activities, smart technologies present the possibility that dinners will self-cook and homes will self-heat. These "enabling technologies" perform new arrangements of environments and ways of life: "smart" thermostats couple with calendars, locations, and even "a human body's 'bio-signals,'" and "skin temperature and heart rate" may be monitored through sensors to ensure optimum indoor temperatures. Similarly, communication with kitchen appliances is proposed to occur through "Toshiba's 'Femininity' line of home network appliances." These technologies ensure the home will be warm, safe, and provided with the latest recipes.[39]

The importance of the everyday as a site of intervention signals the ways in which smart city proposals are generative of distinct ways of life, where a "microphysics of power" is performed through everyday scenarios.[40] Governance and the managing of urban milieus occur not through delineations of territory but through enabling the connections and processes of everyday urban inhabitations within computational modalities. The actions of citizens have less to do with individuals exercising rights and responsibilities and more to do with operationalizing the cybernetic functions of the smart city. Participation involves computational responsiveness and is coextensive with actions of monitoring and managing one's relations to environments, rather than advancing democratic engagement through dialogue and debate. The citizen is a data point, both a generator of data and a responsive node in a system of feedback. The program of efficiency assumes that human participants will respond within the acceptable range of actions, so that smart cities will function optimally. Yet programs for efficiency that are multiply distributed will inevitably be multiply enacted across human and more-than-human registers, so that smart bicycles are left in creeks and sensing devices are hacked to surreptitiously monitor domestic environments or intervene in them. This smart city proposal raises questions as to how these orchestrated ways of life would be actually lived, thereby rerouting programs of efficiency and productivity.

PROGRAMMING CITIES

As specifically rendered through smart technologies, the motivating logic of sustainability becomes oriented toward saving time and resources. This in turn informs proposals for how to embed smart technologies within everyday environments in order to ensure more efficient ways of life. Monitoring is a practice enabled by sensors, and so it becomes a central activity in articulating the sustainability and efficiency of smart cities. The sensing that takes place in the smart city involves continually monitoring processes in order to manage them. The urban sense data generated through smart city processes are meant to facilitate the regulation of urban processes within a human–machine continuum of sensing and acting, such that "the responsiveness of connected sustainable cities can be achieved through well-informed and coordinated human action, automated actuation of machines and systems, or some combination of the two."[41] Humans may participate in the sensor city through mobile devices and platforms, but the coordination across "manual and automated" urban processes unfolds within programmed environments, which organize the inputs and outputs of humans and machines.

"The programmed city" is a speculative and actual project that has been critical to the ongoing development of ubiquitous computing but which has also demonstrated the complicated and uncertain ways in which programmable environments are realized.[42] Programming as described in the CSC document has multiple resonances, signaling the architectural sense of programming space for particular activities as well as the programming of urban development and policy and the computational programming of environments.[43] Within smart city proposals, programming of environments is a way in which the "nature of the connection" within the computational *dispositif* is performed across a spatial arrangement of digital devices, software, cities, development plans, citizens, practices, and more.

The notion of programming, while specific to computation, is further coupled with notions of what the environment is and how it may be made programmable. As discussed in the introduction to *Program Earth,* some of the early imaginings of sensor environments speculate on how everyday life may be transformed with the migration of computation from the desktop to the environment.[44] While many of these visions are user focused, environmental sensors also transform notions of how or where sensing takes place to encompass more distributed and nonhuman modalities of sensing.[45] The programming of environments is then perhaps one of the key ways in which "the milieu" is now best described as "the environment," where the postwar rise of the term "the environment" typically corresponds with more cybernetic approaches to systems and ecology,[46] as well as referring to the conditions in which computation can operate.

A growing body of research in the area of software studies now focuses on the intersection of computation and space, making the point that computing— often in the form of software or code—has a considerable influence on the ways in which spatial processes unfold or even cease to function when software fails.[47] While software is increasingly in-forming spatial and material processes, I situate the performativity of software within (rather than above or prior to) the material-political-technical operations of the computational *dispositif,* since programmability necessarily signals more than the unfurling of scripts that act on the world in a discursive architecture of command-and-control. Software is also not so easily separated from the hardware it would activate.[48] Instead, as I suggest here, programmability points to the ways in which *computational operations unfold across material–political situations,* even at the level of speculative designs or imaginings of political processes (where computational approaches to perceived urban "problems" may in-form how these issues are initially framed *in order to be computable*), while indicating how actual programs may not run according to plan.

The computational articulations of governance and citizenship within the CSC proposals are uncertain indicators for how urban practices might actually unfold, even when processes are meant to be automated for efficiency—but it is exactly the faltering and imperfect aspects of programmed environments that might become sites for political encounters in smart cities. Some smart city initiatives are finding that the less "modern" political structures of city councils, for instance, do not make for easily compatible smart city development contexts. Urban governance may be divided into multiple wards or councils across and through which the seamless flow of data and implementation of digital infrastructures may be complicated or halted. "Realizing programs of action" within software development "is complicated and contested," as Mackenzie notes.[49] Code is also not singularly written or deployed but may be a hodgepodge of just-effective-enough script written by multiple actors and running in momentarily viable ways on specific platforms. Beyond the realm of software development, I argue that the smart city is another realm in which programming does not unfold as an easy execution of code. A change to any element of the code, hardware, or interoperability with other devices may shift the program and its effects. When code is meant to reprogram urban environments, it also becomes entangled in complex urban processes that interrupt the simple enactment of scripts.

The CSC proposals demonstrate the ways in which the programmed environments of the smart city give rise to—and even require—distinct urban materialities in order to be operable. The several modalities of sensing and programming that emerge within the CSC documentation are expressive of programs to sense and monitor in order to manage and regulate the material processes of the smart city environment, from the circulation of people and goods to processes of participation, all of which are seen to interconnect through the "digital nervous system" of the smart city.[50] In the CSC scenarios the metabolic circuit of inputs and outputs that is made optimally efficient simplifies the processes necessary to transform urban materialities—through electronicizing, tagging, and monitoring—in order to make them programmable and efficient. Yet ubiquitous urban computing would require a considerable outlay of materials and resources in order for cities to operate in these modalities. Urban materialities are then doubly elided through the dematerializing logic of digital technology, since automation, improved timing, and coordination seem to minimize—and even eliminate—the resource requirements and wastes of smart cities; electronic technologies also seem to have no resource requirements, whether in their manufacture, operation, or disposal. Resource requirements and material entanglements are apparently minimized through the improved flow offered by smart technologies.

Digital technologies—and the digital apparatus—are generative of processes of materialization that do not so much elide materialities as transform them through computational modalities.[51] The uneven and material ways in which computation unfolds within cities breaks with this kind of frictionless understanding of how computation might seamlessly perform a set of efficiency objectives. Smart cities could be characterized largely by the gaps and accidents of computational technologies, which are also part of the "experience" of how these devices and systems perform and are implemented.[52]

PROGRAMMING PARTICIPATION

The infrastructures at play in the CSC vision partially consist of grids and services remade into smart electrical grids, smart transport, and smart water. But they also consist of participatory and mobile citizen-sensing platforms through which urban dwellers are to monitor environments and engage with smart systems. Participatory media and environmental devices facilitate this more sustainable city by enabling forms of participation that are compatible with it. The smart infrastructures and citizen-sensing platforms in the CSC project enable monitoring practices while structuring responses that regulate or recalibrate everyday practices. Sustainable transit options become more viable through the deployment of "urban citizenship engagement points" that allow for personalized planning of bus routes, carpooling, and bicycle rental.[53] Energy contributions may be made at the intersection of smart transit systems or architectural surfaces and mobile monitoring devices. Urban spaces may be easily reconfigured or adapted to allow working and networking in any location at any time and to facilitate the "intensification of urban land use." The way in which these practices are activated occurs across the programs embedded within urban environments and mobile devices. Digitally enhanced infrastructure and citizens are articulated as corresponding nodes, where technologies and strategies for environmental efficiency become coextensive with citizen participation—and "changed human behavior."[54]

While additional design scenarios address traffic in Seoul and work-anywhere-anytime proposals for Hamburg, as well as coordinate public transit in San Francisco and use mobile platforms to organize daily health monitoring, one scenario based in an unnamed North American urban location focuses on "taking personal responsibility" through the narrative of a love contest between two male friends vying for the attentions of an eco-female.[55] This scenario demonstrates how "the biggest variable in sustainability"—that is, "human behavior"—may be monitored and advanced effectively through ICT applications. The male competitors in this scenario engage in logging their daily travel plans online to produce carbon footprints for comparison, installing a home monitoring system to measure electricity use, and monitoring water use to create a water budget. As the scenario outlines:

> Monitor, monitor, monitor . . . that's a lot of what both men do. They realize that the key to winning Joan's heart is to show her they're making the right decisions, and that means they need a lot of clear information that is meaningful—and actionable.[56]

Monitoring behavior and generating data are seen to be the basis for making sound decisions to advance everyday sustainable practices. Programs of responsiveness are critical to the ways in which sustainable practices are designed to emerge in this smart city proposal. In order for these schemes to function, urban citizens need to play their part, whether by partaking in transport systems or by generating energy through their continual movement within urban environments. Urban environmental citizens are responsible for making "informed, responsible choices."[57] Yet these proposals explicitly outline the repertoire of actions and reflections that the smart city will enable, in which the sensing citizen becomes an expression of productive infrastructures. Mitchell and Casalegno stress the benefits of informed participation in urban processes facilitated by participatory media and ubiquitous computing—technologies that, they argue, make a heightened sense of responsibility possible.[58] Urban citizenship is remade through these environmental technologies, which mobilize urban citizens as operatives within the processing of urban environmental data; citizen activities become extensions and expressions of informationalized and efficient material–political practices. Citizens who sense and track their own consumption patterns and local environmental processes have a set of citizen-like actions at their disposal, enabled by environmental technologies that allow them to be participants within the smart city.

The balancing of smart systems with citizen engagement is typically seen as a necessary area to address when considering the issues of surveillance and control that smart cities may generate. As the previously cited Rockefeller-funded report suggests, global technology companies such as IBM and Cisco may have a rather different set of objectives than "citizen hacktivists," and yet both of these companies have vested interests in contributing to emerging smart city proposals.[59] Digital technologies are seemingly liberating tools, allowing citizens to engage in ever more democratic actions; and yet, the monitoring and capture of sensor data within nearly every aspect of urban life vis-à-vis devices deployed by global technology companies suggest new levels of control. But could it be that this apparent dichotomy between sensing citizen and smart city is less clear-cut? In many ways, participatory media are already tools of variously restricted political engagement,[60] while smart urban infrastructures never quite manifest (if at all) in the totalizing visions presented.

The sensing citizen is an expression of the ideal mode of citizen participation in smart city visions, rather than a resisting agent to them. Sensing citizens are the necessary participants in smart cities—where smart cities are the foregone conclusion. Dumb citizens in smart cities would be a totalitarian overshoot, since they would be entities subject to monitoring without participating in the flow of information (a situation that will be addressed in chapter 8). The smart city raises additional questions about the politics of urban exclusion, about who is able to be a participating citizen in a city that is powered through access to digital devices. Yet the participatory agency that is embedded within smart city developments does not settle on an individual human subject, and citizenship is instead articulated through environmental operations. Within the CSC proposals there exists the possibility that—given a possible failure or limitation of human responsiveness (a lack of interest in participating in the smart city)—the system may operate on its own. In these scenarios, due to a lack of "human attention and cognitive capacity" as well as a desire not to "burden people with having to think constantly about controlling the systems that surround them," it may be relevant to deploy "automated actuation," the project authors suggest. This would mean that urban systems become self-managing such that "buildings and cities will evolve towards the condition of rooted-in-place robots."[61] Citizens would be figures responding within the program of environmentality. However, the smart city program is able to operate independently by sensing environments as well as actuating them and intervening in them to the point where environmental technologies may override citizens if they do not perform according to preset functions—or the rules of the game.

Processes of regulating urban environments within smart city proposals do not require internal subjugation as such, since governance is distributed within environments that default to automatic modes of regulation. Here is a version of biopolitics 2.0, where monitoring behavior is less about governing individuals or populations and more about establishing environmental conditions in which responsive (and correct) modes of behavior can emerge. Environmentality does not require the creation of normative subjects, as Foucault suggests, since the environmental citizen is not governed as a distinct figure; rather, environmentality is an extension of the actions and forces—automaticity and responsiveness—embedded and performed within environments. Such a situation could be characterized as what Deleuze calls the making of "dividuals," a term he uses to describe the fluid entity that emerges within a "computer" age.[62] For Deleuze, automation is coextensive with a deindividualizing set of processes characterized by patterns of responsiveness that rely less on individual engagement and more on the correct cybernetic connection.

Working transversally with this concept, however, I would suggest that smart city proposals signal less toward the elimination of individuals absolutely, since the "citizen" is an important operator within these spaces. Rather, the citizen works through processes that might generate *ambividuals:* ambient and malleable urban operators that are expressions of computer environments. While the ambividual is not an expression of a cognitive subject, it does articulate the distribution of nodes of action within the smart city. Ambividuals are not singularly demarcated or erased but variously contingent and responsive to fluctuating events, which are managed through informational practices. This resonates with Foucault's suggestion that one characteristic of environmental technologies is the development of "a framework around the individual which is loose enough for him to be able to play."[63] But I would suggest that who or what counts as an ambividual is not restricted to a human actor in the smart city, since the articulation of actions and responses occurs across human-to-machine and machine-to-machine fields of action.

CITIZEN SENSING AND SENSING CITIZENS

A final point of consideration that emerges within smart city and citizen-sensing frameworks is the extent to which environmental monitoring leads to actionable data. Smart city infrastructures are projected to operate as a self-regulating environment, but the monitoring technologies that are meant to enable efficiencies within these systems are less obviously able to generate efficiencies or action within citizen practices. In a CSC scenario demonstrating the types of urban environmental citizenship made possible within the green and digital city, proposals are made for residents of Curitiba to experience enhanced and synchronized mass transit options while monitoring and reporting on air pollution at these nodes. Citizen reporting and community engagement are amplified by virtue of ICT connectivity. Through these monitoring and reporting capabilities, positive changes follow as a result of increased information and connectivity: gather the air pollution data, report to the relevant political body, and environmental justice will be realized. These activities and concerns are presented as universally applicable, in that anyone may have cause to monitor and collect pollution data and diligently forward this on to relevant governmental parties. The ambividual actions "coded" into these processes do not presuppose a particular subject, since a fully automated sensor may equally perform such a function. Rather, these programs of responsiveness allow for a fully interchangeable procession of human-to-machine or machine-to-machine data operations.

A similar trajectory is typically envisaged for self-regulating citizen activities: information on energy consumption will be made visible, a correcting action will be taken, and balance to the cybernetic-informational system will be restored. In these scenarios environmental technologies monitor environments and citizens, while citizens monitor environments and themselves. Citizens armed with environmental data are central democratic operators within these environments. But the "governing" contained within cybernetics may not neatly translate into the governing of environments.[64] It may be that the very responsiveness that enables citizens to gather data does not extend to enabling them to meaningfully act upon the data gathered, since this would require changing the urban "system" in which they have become effective operators. Similarly, dominant, if problematic, narratives within sustainability of continued growth through improved efficiency and ongoing monitoring typically do not mobilize an overall resource or waste reduction (what is well known within energy discourse as the "rebound effect"). Strategies of monitoring and efficiency might co-opt urbanites into modes of environmentality and biopolitics that leave modes of neoliberal power unexamined, since the aim of realizing sustainability objectives through citizen engagement is a worthy pursuit.

Foucault's broader interest within the biopolitics lectures is in how neoliberal analyses are brought to bear on governance and subjects, such that economic logics of efficiency in-form what may have previously been understood through social or noneconomic modalities.[65] Environmentality describes the distribution of governance within environments as well as a qualification of governmentality through a market logic that would implement efficiency and productivity as the best guiding principles for urban ways of life. Individuals become governable to the extent that they operate as *homo economicus*,[66] where governance unfolds as an environmental distribution of possible responses made according to the criteria of efficiency and maximum utility.

The transformation of citizens to data-gathering nodes potentially focuses the complexity of civic action toward a relatively reductive if legible set of actions. Participation in this smart and sustainable city is instrumentalized in terms of remedying environment issues through efficiency and devices that will harvest and connect up information to arrive at this outcome. Yet the informational- and efficiency-based approach to monitoring environments raises more questions about what constitutes effective environmental action than it answers. In order for such instrumentalization to occur, urban processes and participation directed toward sustainability, in many ways, must be programmed to be amenable to a version of (computational) politics that is able to operate on these issues. The modes of sensing as monitoring and responsiveness presented within many sensor-focused and smart-focused cities projects raise the question of whether a citizen might be more than an entity that emerges within the parameters of acceptable responsiveness.

FROM NETWORKS TO RELAYS, FROM PROGRAMS TO WAYS OF LIFE

The smart sustainable city vision discussed here is presented as a technical solution to political and environmental issues—an approach that is characteristic of many smart city projects. While the CSC and CUD project proposals are developed as conceptual-level design and planning documents, many of the questions raised here about how smart cities and citizen-monitoring projects organize political participation and the imagining of urban environmental citizenship are relevant for considering the proliferation of projects now taking place in these areas, both at the level of community engagement and through urban policy and development partnerships.[67]

As I have argued, sustainable smart city proposals give rise to new modes of environmentality as well as biopolitical configurations of governance through distinctly digital *dispositifs*. Given Foucault's focus on the historical specificity of these concepts and the events to which they refer, it is timely to revisit and revise these concepts in the context of newly emerging smart city proposals. The environmentality, biopolitics 2.0, and digital political technologies that unfold through many smart city proposals are expressive of distributions of governance and operations of citizenship within programmed environments and technologies. A biopolitics 2.0 emerges within smart cities that involves the programming of environments and citizens for responsiveness and efficiency. Such programming is generative of political techniques for governing everyday ways of life, where urban processes, citizen engagements, and governance unfold through the spatial and temporal networks of sensors, algorithms, databases, and mobile platforms that constitute the environments of smart cities.

The environmentality that emerges through proposals for urban sustainability within the CSC project and many similar smart city projects involves monitoring, economizing, and producing a vision of digitalized economic growth. Such smart cities present ways of life that are orchestrated toward sustainability objectives characterized by productivity and efficiency. The data that develop through these practices are generative of practices of monitoring environments and activities, while activating environmental modes of governance that are found within the jurisdiction of public authorities as well as technology companies that own, manage, and use urban data. From Google Transit to Cisco TelePresence, HP Halo, and Toshiba Femininity, a range of environmental sensor and participatory technologies function in the CSC and other smart city scenarios that are tools of neoliberal governance, and are operated across state and nonstate actors.

I have emphasized how Foucault's interest in environmentality can be advanced in the context of smart cities to consider how distributions of power within and through environments and environmental technologies are performative of the operations of citizenship—rather than of the individual *subjectness* of citizenship. The environmentalist aspects of the smart and sustainable city are not contingent on the production of an environmentalist or reflexively ecological subjectivity, and the performance of smart urban citizenship occurs not by expanding the possibilities of democratically engaged citizens but rather by delimiting the practices constitutive of citizenship. The "rules of the game" of the smart city do not articulate reversals, openings, or critiques of urban environmental ways of life. Rather, practices are made efficient, streamlined, and oriented toward enhancing existing economic processes. And yet, within this approach to environmentality through smart cities, what we might take as the rules or program of the smart city game might be understood less as a deterministic coding of cities and more as something that might unevenly materialize in practices and events. While design proposals put forward a persuasively singular case for the smart city program, inevitably multiple smart cities emerge through the circulation and implementation of this program.

Pushing Foucault's notion of environmentality even further, I suggest that his concept of the "rules of the game" might be recast in the context of smart cities less as rules and more as programs—here of responsiveness—that delimit and enable in particular ways but that also unfold, materialize, or fail in unexpected ways. If urban programs are not singular and are continually in process, then environmentality might also be updated to address the ways in which programs do not go according to plan, and work-arounds might also develop. Such an approach is not so much a simple recuperation of human resistance as a suggestion that programs are not fixed, and that in their unfolding and operating they inevitably give rise to new practices of urban environmental citizenship and ways of life that emerge across human and more-than-human urban entanglements.

This approach to ways of life is important in formulating not a simple denunciation of the smart city but rather a proposal for how to attend to the distinct environmental inhabitations and modalities of citizenship—and possibilities for urban collectives—that concretize in smart city proposals and developments. Subjectification, which Deleuze discusses as an important concept in Foucault's work, is ultimately concerned not with the production of fixed subjects but rather with the possibility of identifying, critiquing, and even creating ways of life.[68] Smart city projects require an attention to—and critique of—the ways of life that are generated and sustained in these proposals and developments. Critique, as articulated in a conversation between Deleuze and Foucault, can be an important way in which to experiment with political engagements and form "relays" between "theoretical action and practical action."[69] From this perspective the ways of life proposed in the CSC scenarios might serve as provocation for thinking through how to experiment with urban imaginaries and practices in order not to be governed *like that*. If we read biopolitics 2.0 as a concept attentive to the ways of life that are generated and sustained within smart cities, and if this computational apparatus operates environmentally, then what new relays for theory and practice might emerge within our increasingly computational urbanisms?

Figure 8.1. TechniCity MOOC, student contribution of video analytics for detecting urban events for crowd and public-space management. Screen capture

8

Engaging the Idiot in
Participatory Digital Urbanism

A LONGSIDE PROPOSALS FOR smart and sustainable cities, a number of DIY and participatory urbanism projects use digital technologies to generate new modes of urban engagement. As discussed in the last chapter, urban infrastructures are increasingly embedded with computational sensor technologies that are intended to automate urban processes and facilitate urban efficiencies. From tracking transport journeys and updating bus arrival times to traffic cameras and cycle-hire schemes—as well as monitoring air and water quality, river and sea levels, energy use, and waste—urban infrastructures, services, and functions are being remade or newly introduced through the sensor-actuator exchanges of digital urbanism. However, these developments are not just about the automation of urban life at the infrastructural level: they also include collecting new forms of input from citizens engaged in participatory sensing. Such smart city developments in-form modes of urban engagement through interaction with sensor technologies, smart phones, digital devices, apps, and platforms that are meant to coactivate urban functions.

This chapter considers in more detail the participatory urbanism and sensing projects that are underway or have been prototyped, as well as the broader context of the literature, training, courses, and gatherings that are essential to learning how to become a contributing citizen in this participatory urbanism framework. From web- and app-based social media initiatives such as SeeClickFix (United States) and FixMyStreet (UK) that enable urban dwellers to crowdsource and report on urban infrastructure in need of repair to civic apps that are meant to facilitate access to government services, a range of citizen-sensing initiatives now encourage urban engagement through digital devices, platforms, and infrastructures.[1] Often referred to as practices of DIY, participatory, or open-source urbanism, these modes of participation contribute to the development of what is meant to facilitate digitally improved functionalities and experiences within contemporary and anticipated cities.

As discussed in the last chapter, participatory urbanism projects are a necessary response to and development within smart cities proposals. Digital imaginaries for increased participation are often continuous with the sorts of exchanges that smart cities would enable, since interaction requires the use of smart phones, data collection, monitoring platforms, and assorted devices and tools that are meant to facilitate participation. Sensor-based and digitally enabled modes of DIY and participatory urbanism have been proposed as grassroots strategies for articulating new types of commons and democratic urban participation, as well as strategies integral to smart city development proposals. By focusing specifically on the use of citizen-sensing applications for environmental monitoring and urban sustainability, I analyze the distinct modes of participation and urbanism that are expressed in these projects. Two questions that I address in this chapter include: How do citizens become sensors in participatory digital-urbanism projects? And how are cities cast as computable problems so that sensing citizens can act upon them? The first question considers the specific capacities of citizens and publics that are operationalized through digital practices dependent upon urban environmental sensors. In other words, what types of urban participation do projects such as updatable maps for street repairs, air quality sensors, or platforms for tree planting activate or enable? The second question continues discussions from the last chapter and is attuned to the ways in which urban problems are broken into computable tasks.[2] So this question further asks: When addressing the "problem of the city," which modalities of urban politics are potentially made more problematic? And what other practices might be created through an approach that specifically seeks to trouble the dynamics of DIY digital urbanisms?

While the smart city is often broadly identified as a combination of networks and sensors, this hardware-software view of the smart city leaves unexamined the types of participation that might unfold within these new or revised urban settings. An implicit assumption within many digital-urbanism proposals and projects is that urban participants will engage with programs of participation as planned—that they will become an extension of computational logics and exchanges and will readily perform as citizen sensors and citizen actuators. I examine the ways in which participation is articulated within digital- and participatory-urbanism projects. But I further focus on the ways in which participation does not always unfold as expected, and may even be a site of disruption—intentional or otherwise. The ways in which programs of participation do not go according to plan was signaled in the previous chapter on smart cities as a possible site where politics emerge and are invented, where the "rules of the game," as Foucault puts it, might generate encounters not just with governance as it is planned, but with governance as it is interrupted or rerouted. It is these possible interruptions and reroutings that I take up here.

To advance a discussion of the ways in which participation proliferates beyond the "rules of the game" and, in so doing, provokes political encounters and inhabitations, I take up Stengers's discussion of cosmopolitics and participation, where she asks how it might be possible to attend to the role of the "idiot," or those who would typically be seen to have nothing to contribute to the "common account" of how to approach political problems.[3] In her proposal, the idiot challenges a notion of participation and politics that easily settles into consensus. This is not the idiot as a simplistic form of insult—as in a dumb or stupid citizen, the simple counterpart to the smartness of the smart city. Instead, the idiot or the idiotic is someone or something that causes us to think about and encounter the complexities of participation and social life as something other than prescribed or settled.

Stengers draws on Deleuze's conceptual persona of the idiot to consider how the idiot "slows the others down" specifically by resisting "the consensual way in which the situation is presented and in which emergencies mobilize thought or action."[4] Resistance here is not a matter of searching after what is true or false, but rather is a way of attempting to reroute what counts as important. But this is not a strategy for trading one agenda for another, since the cosmos- within the cosmopolitical is an "interstice," following Whitehead—a space of unknowing "constituted by these multiple, divergent worlds and to the articulations of which they could eventually be capable."[5] Such spaces and engagements characterized by unknowing and divergence mean that which counts as "political" can never be assumed or finalized. As Stengers outlines, the idiot captures not the "common good" or cosmopolitanism of Kant, but rather signals toward cosmopolitical registers of hesitation and uncertainty, as well as the generative political encounters that can arise from such uncertainty.[6]

A growing body of literature deploys the idiot as a figure and process of engagement.[7] Adopting Stengers's elaboration on the idiot, Mike Michael discusses the patterns of "overspill" and typologies of misbehavior that might typically be disregarded by social scientists attempting to facilitate and study public engagement. Absence, incapacity, refusal, disruption, distraction, and irony are examples of ways in which participation does not unfold according to researchers' plans, but instead irrupts through various idiotic registers that transform the agenda and outcomes of participation.[8] While other discussions of the idiot variously focus on processes of individuation and the making of subjects in relation to new media,[9] my use of the idiot in this discussion of participation in the smart city engages most centrally with Stengers's version of the idiot as a figure that cannot be articulated through a fixed subject position, not even if it is one of inversion. Instead, the idiot as understood here is a troubling and transformative agent within participatory processes who cannot or will not abide by the terms of participation that are meant to facilitate and enhance democratic engagement.

Using Stengers's figure of the "idiot," which has further resonances with forms of non- or sub-citizenship as understood through the Greek definition of the idiot as a noncitizen or private individual, I suggest that disruptive, confused, and thwarted actions fall outside of the usual delineations of what counts as participatory digital urbanism. The becoming environmental of computation in relation to participatory urbanism then involves the ways in which these programs do not go according to plan. In addition, seemingly illegitimate contributions challenge us to consider how cities hold together and unfold as sites of political engagement, how participants often contribute as disruptive agents, and how sensor-based digital technologies, platforms, and networks organize participation (as well as inclusion and exclusion) in the city, whether smart or otherwise.

After first reviewing the rise of civic apps and platforms, I turn to discuss a range of both practical and theoretical approaches to digital participatory urbanism and draw out a discussion of how participation unfolds and to what a/effect. I then discuss in more detail two specific examples of participatory urbanism. I first address the ways in which digital and participatory citizens are often in need of training in order to be able to operate within digital cities and through digital exchanges as "smart" citizens. A number of smart city training opportunities exist in courses and events. Through an account of my experience as a student on a Massive Open Online Course (MOOC) on smart cities, I discuss the ways in which students were instructed to learn about and participate in the smart city, including a task to crowdsource examples of urban sensors for a course forum. I focus specifically on the ways in which departures from this participatory platform occurred and how these departures potentially sparked specific types of idiotic encounters in attempting to understand how sensors influence cities and citizenship.

From this example, I then discuss a second example of FixMyStreet, a UK-based platform that allows citizens to identify and report problems with street spaces. In my discussion of this platform, I am interested to consider both what a street becomes when it is the focus of efforts to fix it, such as reporting potholes and illegally dumped rubbish, and also what modes and task flows of citizenship and participation are enabled or fall outside the boundaries of legitimate participation. FixMyStreet relies on particular types of citizen-led reporting, most often undertaken through that composite sensor-apparatus of the smartphone. But it also captures multiple instances of grievances logged that cannot be easily dealt with through this platform. Such "reports" might be considered idiotic, since they slow down the assumed ways in which citizens are meant to participate in maintaining streets and instead raise open-ended questions and complaints that reveal how many types of street-based concerns and politics are not easily amenable to "fixing."

From these two examples and the modes of participation that they activate, I ask how the figure of the idiot may provoke different approaches to thinking about and creating participation in the digitally equipped and sensor-based city. Many current projects and proposals for smart cities and digital participation assume versions of politics and citizenship that are relatively untroubled and solution-oriented. Citizens need only train up and gain capacities in order to contribute to digitally enabled urban processes. With access to the latest sensor-based platforms and infrastructures, new levels of citizen-to-citizen, citizen-to-government, and citizen-to-city interaction are assumed to unfold. But the idiot troubles these communicative and political arrangements. It instead indicates how computational approaches to cities, citizenship, and politics may give rise to a faltering and hesitating set of practices that do not advance an unproblematic approach to participation, but rather throw it into question.

PARTICIPATORY URBAN ACTIONS

An extensive list of applications and platforms could be made of projects that are variously situated in the area of participatory, open-source, and DIY digital urbanisms. From Code for America to the New Urban Mechanics with their Adopt-a-Hydrant scheme, as well as FixMyStreet, CitySourced, Maker Cities, and Urban Prototyping, a number of projects fuse participation with sensors, apps, infrastructure, devices, software, events, and even manifestoes to articulate and put in motion a 2.0 version of urban citizenship. In the course of reviewing this developing area of practices, I have trawled through websites and used apps that would make me more civic and participatory, and I have visited "meetup" events, as well as signed up for online training sessions in the form of MOOCs and webinars, taken participation surveys, and attended tech demonstrations, hackathons, and fairs. Participatory applications and initiatives have in many ways settled into these formats, where an emphasis is placed on co-creating technologies and services in settings where there is a relatively high enthusiasm for the possibilities of new technologies but often a relatively underexamined approach to what counts as participation and what types of urban politics are activated through these digital engagements.

While I focus here on digital and sensor-based modes of participatory urbanism, it also bears mentioning that there is a long-standing tradition of participatory urbanism projects that span art, architecture, design, and feminist politics. There has been a recent resurgence of these projects that is somewhat parallel if distinct from digital participatory urbanisms. These projects include the Canadian Centre for Architecture's exhibition and catalogue, *Actions: What You Can Do with the City*, the *Spontaneous Interventions* United States pavilion at the 2012 Venice Biennale, and work by creative practitioners such as L'atelier d'architecture autogérée (aaa) and Public Works that focus on creating urban contexts for citizen- and community-led transformation and use of spaces.[10] In another way, many participatory urbanism projects that are not sensor-based often adopt computational metaphors and platforms to describe and organize practices, from Wikicity to Tactical Urbanism,[11] where digital networks and commons are established as tools for achieving greater urban participation. It is then worth noting that there is a much wider stream of participatory urbanism projects underway that runs alongside and at times mutually influences or diverges from sensor-driven approaches to cities.

Similar to this wider context of participatory urbanism projects, many digital participatory urban initiatives have developed through a stated concern with civic responsibility and through an interest in remaking government "from below." Code for America (of which there are regional chapters, as well as a similar but nonaffiliated European version, Code for Europe) is a project established in 2010 that made fellowships available for coders to develop civic-minded software freely available through the software repository GitHub, as well as platforms to encourage greater participation in urban life. The rationale for these projects is that if the urban-computational "system" is put together in the best way, then government may run more efficiently and better address the needs and concerns of citizens.[12] Projects that have developed through the Code for America initiative include an Adopt-a-Hydrant platform that was developed in collaboration with the New Urban Mechanics and prototyped in Boston in 2012. With this platform, members of the public could locate a nearby hydrant, identify the hydrant on a map, and adopt and take responsibility for clearing away snow when the hydrant became buried in winter snow storms. This platform was adapted to several other applications, including Adopt-a-Siren for maintaining tsunami sirens in Hawaii and Adopt-a-Storm Drain for clearing drains. From hydrants to potholes, parking spaces, and animal services, within these projects there is an attention to the mundane and even "bureaucratic" role of governance. Code for America seeks to take up and transform the bureaucratic aspects of governance through code, apps, and platforms, hackathons, GitHub repositories, and open data. Infrastructure is "adopted" in order to maintain it. Data are harvested in order to compare understandings of air pollution and exposure. Coding is undertaken in order to achieve new efficiencies. And citizens participate through computational registers that reroute the practices and responsibilities of local government—where citizens shift from agents with "voices" to agents with "hands."[13] Coding, and the hands that would undertake this practice, are seen as a way to "fix citizenship," since as Pahlka notes, "We're not going to fix government until we fix citizenship."[14]

"Civic Apps" are a similar area of development, which has at various times been held up as the next vital improvement in urban life and "public services provision." As noted in its 2013 forecast, the UK-based innovation think tank the National Endowment for Science, Technology and the Arts (Nesta) predicted civic apps would develop rapidly and so shift the ways in which urban participation unfolds. Code for America Commons was identified as one of these examples, "which operates like a community-driven app store to share technology for public good."[15] The solving of urban problems was not the only area Nesta identified as notable for development, since SeeClickFix, for instance, operates by rewarding users with "civic points," a system that has turned to operate for profit within "thousands of communities." Through ad and software sales, a civic platform had morphed into a profitable venture, as well as a private service that could be sold back to municipalities in order to manage "customer-response services."[16] Many civic apps are in fact consumer-focused or oriented toward the quantified self and allow urban dwellers to locate the best coffee spots and bars, as well as have better access to cabs or public transit, while monitoring their exposure to air pollution or keeping track of miles walked.[17] These are "simple tools" that are meant to aid in "navigating public life and which make it easy to take part in."[18]

Civic apps are then productive of new economies and political economies of participation and are not simply articulations of digital and democratic engagement. As Ulises Mejias notes, however, participation in these networks offers up information, but at the same time the user of these platforms becomes "the product being sold," where participation is "not coercive in a straightforward manner" but is organized to undergird particular economic exchanges and to reinforce particular modes of sociality.[19] Participation through these platforms is then most typically aligned with digital economies where user-citizens provide the data-material that often generates profits for tech companies but less frequently contributes to substantive resources for urban communities or citizens. Users and participants of sensor-based digital platforms provide sensor data that influence, if not benefit, particular types of technological and urban economies. Participation in networks requires the free labor of participants, but the networks are owned, controlled, and operated by companies that collect data in ways that are not typically transparent or contributory to advancing more democratic urban engagement or more equitable economies.[20]

Sensors and sensor data then have effects in the ways in which participation is organized and the uses to which it is put. Participation needs to be organized in order to be activated with and through sensor exchanges that contribute to the amassing of sensor data. A task flow is activated within sensor-based cities that break down participation into executable tasks, thereby making participation a computable problem. Such computational logic may in-form problems of participation with or without sensors, since the task-flow approach to cities may migrate from an application having to do with sensor-regulated traffic flow to a more general method for managing pedestrians.

Participation further unfolds through both active and passive registers. "Participatory sensing" is a term that has variously been used to describe the ways in which people can record environmental phenomena "through sensors built into mobile phones."[21] The sensors in mobile phones, including image, location, sound, direction, and acceleration, have been identified as allowing for participatory and citizen-led modes of sensing that can be used to "record images, motion, and other signals, automatically associating them with location and time."[22] Described as "a new collective capacity" that is developing and that allows for analyzing "invisible" aspects of life, sensors in smartphones are different and yet at times complementary to infrastructural sensors in cities used to organize transport, energy, and more. In fact, sensors in smartphones are often the primary site of interaction and activation for citizen-led participatory sensing. Here, participation in many ways is both facilitated and yet delimited by the capacities of smartphones. Examples of modes of participatory engagement with smartphones then extend to tagged images of community assets, images of safety hazards, apps for travel monitoring, and maps of running routes. As one group of researchers working in this area suggest, "Participatory Sensing's *innovation* lies in how people can use today's technologies to observe, document, and act on issues that matter to them. Participatory Sensing's potential *power* comes from the already widespread adoption of the technologies by people across so many demographics."[23]

While a clear trajectory is laid out from identifying what to sense, gathering this sensor data, and then moving forward to action, the dynamics of participation within this progression remain somewhat unclear. How will participants first decide where to focus their attentions, and how will they further know how best to document their concerns? And within what context will they be able to make claims or advance actions on the basis of sensor data they have assembled? The more overtly participatory aspects of sensing, in many ways, might remain somewhat gestural, since this is less an exercise of democratic urban citizenship and more a case of citizens participating by becoming sensors in order generate urban data. In this respect, participatory sensing and passive data collection may have more in common than at first suspected. Passive data collection generally entails citizens having to do very little, other than turn on their smartphones or other sensor devices, such as wearables, to collect location data or other environmental details. Participation occurs here in a passive register because it does not require input from the human user and it takes place by users simply being equipped with smartphones, where sensing takes place in background registers without further citizen-based translations into urban engagement.

Within the area of passive data collection, civic apps are also identified by Nesta as enabling new modes of participation, where citizens might contribute to larger data stores through enabling their vehicles to send data about speed bumps, for instance, or by contributing location data or personal health data via smartphones or wearables.[24] In these participatory applications, sensor engagements and sensor data are not only located in the urban environment and implanted within infrastructure but also integral to the technologies that people carry such that humans become sensors and actuators, passively facilitating the detection and reporting of urban problems such as potholes. Sensors may be located across infrastructures, smartphones, and vehicles, but what is most striking about participatory and passive modes of sensing is the way in which citizens perform as sensors, such that humans and sensors may even undertake interchangeable functions. Participation is in-formed through these exchanges—not as a technologically determined activity, however, as much as a concretization of urban dwellers, cities, infrastructure, everyday life, sensor devices, databases, big data economies, and emerging practices of civic-ness.

The new entities that concretize through sensor-participation raise the question of whether urban citizens and participation are as much in the prototype stage as sensors. Both the practices that human-users are meant to develop and adapt to, as well as the possibilities for more autonomous sensing devices, can be said to be in development in this urban-computational loop of participation, sensing, data, analytics, and action. What are these new capacities of citizenship and urban politics that are meant to be activated vis-à-vis participatory digital urbanism? As Mark Andrejevic has pointed out, the rise of ubiquitous computing has actually contributed to "a heightened form of passive interaction: the gathering of detailed information in an increasingly unobtrusive manner."[25] Within ubiquitous-computing scenarios, participation unfolds most readily where humans interfere the least with inputs and outputs and where citizens become sensing nodes or perform in ways that are readily computable. Such an approach could be described as making humans "easier to use," as Trebor Scholz has noted in relation to digital labor.[26] But this ease of use does not necessarily clearly indicate new capacities of citizenship.

In these examples, computation and participation coincide in order to enable new modes of citizen engagement, and yet in the process there is a remaking of the processes of participation, cities, and citizenship toward computability. Urban life is articulated through a series of computational problems that can be solved or enhanced through participatory platforms and programs. Citizens achieve participation through using these platforms to perform urban functions (effectively becoming actuators of coded sensor actions), and at a presumably higher level by writing programs in the first place, which code urban life for particular forms of participation. Yet within this range of actions and technologies, participation remains relatively unquestioned as a practice. The ways in which coding, for instance, may facilitate particular types of urban participation are assumed to be positive and unproblematic contributions to urban life. Civic apps are tools for achieving "public life" or the "common good." As Stengers suggests in relation to the idiot, "The idea is precisely to slow down the construction of this common world, to create a space for hesitation regarding what it means to say 'good.'"[27] Although cities are regularly cast as sites in need of urgent attention, Stengers's cosmopolitical proposal mobilizes the idiot as an entity that does not deny the "emergency" but at the same time cannot accept the usual ways in which the emergency—here the problem of rapidly growing cities in the face of resource shortage and climate change—has been framed and operationalized.

With the idiot in mind, it then becomes possible to ask in what ways do problem-solving code, reconfigured service economies, and participatory sensing for data collection reshape how participation unfolds? Public life and participation become articulated as data-collection sites and practices, as crowdsourced actions for solving definable problems, and as new business opportunities that emerge from these same ventures. Civic apps attempt to remake and reinvigorate civicness by framing urban problems so that they are actionable, and actions are easily performed through smartphones or digital platforms. The program of participation here is not just a means to achieve efficiency but is a way to individually and collectively frame and solve urban problems, whether through simply defined actions or passive contributions of data that would facilitate the management of urban functions. This is the well-known Leibnizian dream of making all problems solvable by rendering them in terms of computation, such that should any dilemmas arise, a clear program may be in place so that one only needs to say: "Let us calculate!"[28] But in making urban participation actionable, and by articulating the conditions through which collectives might form and have effect, digital modalities of participation also delineate distinct forms of engagement that break down into tasks, modes of training, and voluntarization of local government services (when in many cases, these services may be under stress due to budget cuts).

The approach to making urban problems computable might be further situated within the growing area of "urban science."[29] Transportation problems, contributions to civic process, and land use are examples of issues that might be best dealt with through gathering, analyzing, and managing data sets otherwise characterized as "big data." Beyond infrastructures equipped with sensors that contribute to the store of big data, participatory sensing and data collection are also meant to contribute to large datasets that are intended to make cities more efficient and easier to manage. Participation occurs through individual contributions that scale up to aggregate functions, where the action and possibility of participatory sensing is most pronounced when gathered into databases that are mined and managed by distant actors, whether tech companies or city governments. Participation in this sense requires a certain deferral, where one's contribution of sense data becomes a resource for other urban actors. While participatory sensing is meant to put tools into the "hands of citizens," in reality most citizens will not have the time or resources to mine data in order to influence urban processes or to effect urban political change.

The contribution that participatory sensing and data collection make to big data might in one way be described through a topological logic of digital infrastructures that Matthew Fuller and Andrew Goffey discuss as a "faith in small numbers." As they write, "The effects of small numbers are pressed upon us as exemplars of the instability of global systems and of the power of the individual to effect real social change."[30] Small numbers—and by extension here, the small numbers of participatory sensing that add up to big data—are the operators whereby urban citizens are meant to be able to make change, whether through individual contributions or the collective amassing of data. Possibility occurs through computability, and so sensors at infrastructural and participatory levels are the devices that facilitate the computability and possibility of urban processes. Sensor data that feeds into big data promises to achieve distinct types of functionality and connection through enumeration and processing. Participation through sensor-based devices is then contributing to modes of urban life and urban engagement that are characterized by computability as the register of urban politics and social life.

Read/Write/Execute

While the participatory and civic aspects of these apps, platforms, and modes of digital DIY urban engagement differ, they in various ways could be described as projects that take up an approach to developing "read/write" urban practices. Read/write, as another computational term that has been mapped onto participatory digital urbanism, identifies cities as hitherto primarily read-only spaces, which may become write-able, or participated in, through computational practices. Participating with digital or sensor-based devices allows urban inhabitants to modify cities, as they might modify computer code. Participation may even become a "script" to rewrite, where urban inhabitants may contribute to the formation of new modes of experience.[31]

Following the read/write trajectory, many approaches to facilitating urban participation that proceed by making the city open to modification focus on citizens' abilities to add stories to places, to create alternative urban experiences, or to open up other ways of perceiving and engaging with the city that might at the same time be more democratically oriented. And yet, a recurring question arises with many of these projects as to how participation might be narrowed when it is necessarily routed through a program of engagement that requires input and output, sensing and actuating. If not to solve urban problems, then participatory practices are still bound to computational modalities that would appear to open up digital participation to a wider set of inputs but which continue to restrict the potential of inputs to set registers of computational recognition. Whatever falls outside recognition will simply not compute.

The read/write mandate is not just about citizens writing—it can include a consideration of how cities influence this process by "writing" back. Saskia Sassen has written in her discussion of open-source urbanism that more consideration might be given to moments when cities talk back to smart technologies. In other words, she suggests we attend to the ways in which technologies might be "urbanized" through cities reworking these devices and systems. In this way, urban life might be less prone to the simplifications that occur when attempting to fit cities into the parameters of smart technologies.[32] How does a city variously interrupt, resist, or reroute these attempts to solve urban problems or facilitate digital participation? Sassen suggests cities do this through their "incompleteness," whereas many of the proposals for "intelligent cities" make urban spaces into "closed systems."[33] Practices that "urbanize" technology would then make space for this incompleteness and would resist closure in favor of multidirectional openings in urban life.

One question that arises from this proposal to let the city "talk back," however, is in what ways this talking is registered. Sassen identifies sites where smart and sensor-based technology may be undergoing forms of urbanization, and these are generally places of considerable privilege, from "the Network Architecture Lab at Columbia, the SENSEable City Lab at MIT, and much of the work gathered at the *Design and the Elastic Mind* exhibit at MoMA."[34] A second question that emerges from Sassen's proposal is the extent to which the urbanization of technology still requires a communicative and computational logic. Although the city might "talk" back, if it is to urbanize technology it must do so within particular modalities. Despite multiple critiques of primarily communication-based approaches to publics, cities, and citizenship,[35] many of the ways in which participation unfolds continue to involve the exchange of "messages" that have, as an implicit trajectory, some sort of debate-and-consensus-based dialogue as the condition for urban life.

And yet, despite the persistence of a certain communicative logic, Sassen implicitly raises the question of who is reading and writing, and she extends this dynamic into "the city." If the city is talking back, and is able to interrupt the closed logic of intelligent technologies, might this exchange also infuse the city with idiotic capacities? In this project of participating within cities of sensors, where the city might talk back, distributions of participation are also by no means located exclusively within human agency. To what extent do the multiple nonhumans inhabiting cities contribute to these idiotic capacities, whether traffic lights or bus stops or road signs? The failure of sensor systems might be one way in which the idiotic capacities of digital participation concretizes. If participation through sensor-equipped technologies is a path to citizenship or to exercising citizenship, then do sensors and sensor-enabled infrastructures also become enrolled in the project of citizenship—of enabling, disabling, or otherwise rendering this program incoherent? The becoming environmental of computation then becomes as much a process of rupture and disconnect as an all-encompassing program of computability.

Writing is a practice that quickly morphs into hacking, moreover, since these are modes of intervention that seek to remake the "civic process." Read/write becomes read/write/execute, in an ideal approach to participatory digital urbanism. Hacking, and an approach to urban sites as executable, is further articulated as a "civic duty," since it is not enough simply to gather open government data and produce crime maps, for instance.[36] Instead, computationally enabled modes of participating in cities must remake or hack the very ways in which we participate in urban life.

Yet what aptitudes are required in order to be a citizen who knows how to "write" or even to "execute" in digital modalities or, in other words, to participate through computational means? In some cases, new modes of training have emerged in order for citizens to learn techniques that might be useful for participating in the digital and sensor-based city. These courses include offerings on MOOCS, hackathons, urban prototyping sessions, meetups, and more. Urban Prototyping is one such initiative, initially based in San Francisco and which subsequently traveled to other locations including Singapore and London. Funded through the Gray Area Foundation for the Arts (GAFFTA), the initial 2012 Urban Prototyping festival in San Francisco focused on "transforming public space through citizen experiments."[37] Drawing on a sort of tactical urbanism approach, projects spanned from the digital to the analogue, including glowing crosswalks, air pollution monitors, digital sidewalk cinemas, and public urinals. The objective of the first festival was to "make cities better, faster," where prototyping could move to replication and adoption by multiple cities in any number of locations.[38]

The Urban Prototyping event I attended in April 2013 in London at Imperial College consisted of part seminar series and part hackathon, with an emphasis throughout on identifying and addressing impending urban problems that could be made solvable through digital modalities. The possibility to "change the world through digital technologies" was a feature of the seminar and hackathon events, where usually inaccessible data would be available for hacking, mash-ups, game development, and more, which might lead to developing technologies and platforms for "citizens to overcome the serious challenges that our society faces."[39] Within this Urban Prototyping event, questions proliferated about how digital technology might be reforming the relations between citizens and government. One discussion led to the question: Who will control the city: Team Architecture or Team Computing? Another discussion came to imagine a time when government would cease to play a central role in public life: How would citizens become more resilient in the absence of government? Deskilling and reskilling were terms often deployed: we might have to unlearn some ways of thinking about cities and citizenship in order to develop new practices. Other questions that arose included: In what ways does citizen engagement tend toward the optimization of systems, often at the expense of privacy? Is digital citizenship worth it, in this case? Is it possible to develop a "peer progressive" model of urban engagement? How might it be possible to make crowdsourcing more representative, so that it is not simply a collection of contributions made by those who have a vested interest? And perhaps most perplexing of all: How might it be possible to motivate people to participate in these new modes of digital citizenship, since this proved to be a difficult challenge, even when participation projects were gameified to be apparently more readily engaging. Prototyping projects developed during the hackathon were attempts to experiment with these issues of urban citizenship and engagement via digital modalities.[40] But these multiple questions about participatory digital urbanism loom large, sparking considerations about how prototyping can, if ever, move toward new modes of urban citizenship and engagement.

Open-source urban prototyping, as Alberto Corsín Jiménez has argued, might be a technique for digital experimentation that keeps urban possibilities open, particularly in relation to urban infrastructure.[41] Yet in order to experiment within digital modalities, skills need to be gained; and experimentation requires considerable levels of education ("deskilling and reskilling"). Prototyping draws attention to the situations in and through which experimentation unfolds. What would be another way of considering how technologies concresce with urban situations, and are not just free-floating tools that would solve free-floating problems? If technologies are put to the test in these contexts, then participation becomes articulated through actual registers of engagement rather than as hypothetical platforms and gestures toward the common good. Idiots and idiotic encounters might even proliferate in these encounters and activate new approaches to the project of participation in the digital and sensor-based city.

LEARNING TO PARTICIPATE IN THE SENSOR-BASED CITY

As discussed in the previous chapter on proposals for smart cities and in the introduction to this chapter, smart cities effectively require smart citizens rather than dumb ones in order to make this version of digital intelligence operable. Smart citizens need to play their part in the smart city. But in what ways do smart cities effectively program what counts as smart citizenship? And what divergences might occur within this delineation of a smart and participatory citizen? A smart city is usually defined as some version of a highly communicative urbanism comprised of networks and sensors. Here, smartness is a feature of enhanced communicability, where new modes of communicative participation constitute new capacities for citizenship and citizen participation. Sensors and networks are the fundamental building blocks in this version of the smart city. But are there critical traits of citizens, or modalities of citizen participation that are necessary to smart cities? If we follow one historical thread of what counts as a citizen, this requires that one have a voice or be able to communicate within public and urban forums. By extension, to be a citizen in the smart city would require that one has the ability to have a voice and participate within the communicative registers and exchanges enabled through digital technologies.

To explore this area further, I turn to two examples of participatory urbanism that intersect with sensor-based cities. The first example involves the discussion of my experience participating in a TechniCity MOOC on smart cities. The second example attends to a participatory urbanism platform, "FixMyStreet," that collects reports for fixing urban streets. As I discuss below, these examples raise issues about the characteristics and practices of participatory urbanism as it intersects with participatory sensing and of how computationally and sensor-enabled and cities are meant to facilitate, encourage, and advance urban participation. While both of these examples are largely focused on online participation, as the wider array of participatory urbanism examples indicates, online and offline worlds are entangled so as to redirect attention toward modalities of participation rather than attempt to delineate the apparently virtual or physical locations in which participation might unfold.

TechniCity

In February 2013 (and again in February 2014), I enrolled as a student in a MOOC, "TechniCity," which had the stated aim of exploring the ways in which technologies—specifically computational networks, mobile devices, and sensors—are changing cities. The point of the course was to examine "from a critical viewpoint" the "sweeping transformation" taking place in how cities are designed, modeled, and engaged with. In this eight-week course, the weekly lectures consisted of presentations by "thought leaders," including discussions on sensors and smart grids as well as data and participation from academics, creative practitioners, tech gurus, and Silicon Valley technologists. Suggested readings included academic papers on apps for urban planning, industry white papers on smart cities, and discussions of citizen sensors.[42] A number of platforms and course forums intersected with the lectures and readings to encourage student participation and contributions to the course. From a "virtual salon" to MindMixer, SiftIt, Spotify, LinkedIn, Twitter, and Facebook, as well as Google hangouts, the course was multiply networked and made use of social media as part of its educational remit.

I include this MOOC example since there are numerous ways in which smart cities and sensing applications are delivered through and as educational and training encounters. It seems citizens have a lot to learn in order to become smart and participatory, and so need to be educated and skilled up in order to participate in the smart city. What are these forms of instruction, what do they teach, and how do they purport to make us smarter (or at least smart enough) and more participatory citizens? In what ways might one—by not participating—find oneself to be idiotic or on the fray in relation to this mode of *becoming smart*? How does this practice of training up for the smart city resonate with Deleuze's suggestion that, in a computer-oriented society, continuous training takes the place of schooling?[43]

Are we almost always in need of further training in the smart city, where computation sets the terms of instruction?

As an admittedly haphazard student in this course, I watched lectures on sensors for managing traffic, and I reviewed lesson plan materials on how to scrape Google Street View. I did not contribute to forums as actively as the course suggested I should, but I did review others' input and wonder at the range and spread of contributions. I specifically learned from the lecture delivered during week two of the course about the "ladder of participation," a figure taken from a classic 1969 urban participation text written by Sherry Arnstein.[44] This ladder-based figure of citizen participation moves from the lowly depths of manipulation and therapy to the more enlightened stages of "citizen power," which includes partnership, delegated power, and citizen control. While Arnstein's paper puts forward this model as a way to suggest different stages and modes of citizen engagement, she is as interested in making the ladder a relatively expanded and nondefined space, where she suggests the rungs on the ladder might be just a few of many more, and that there may be 150 or more stages or spaces of citizen involvement, for instance. In the TechniCity course, the ladder is presented in a rather linear progression, however, from control to enablement, where the more engaged realms are inevitably facilitated by digital technology. In fact, it could be argued that, in the course and in this specific lesson, digital technologies were presented as facilitating these higher reaches of citizen engagement.

And yet, I wondered while pacing through this lesson, what happens if you are less adept at using digital technologies, or what if these devices do not realize their promissory aims and instead tend to lead to the usual inertia that often accompanies urban political problems? What aptitudes and resources might be required in order to be a citizen who might be located at the upper rungs of this ladder of participation? What happens if you fall off the ladder, or if you are not able to climb the ladder in the first place?

Momentarily stepping off the ladder of citizen participation, we might also consider what space there is for the idiot in this typology of engagement. If citizen engagement always requires that people are present in a forum and expressing themselves, then to what extent do idiotic presences challenge the progression from control to empowerment that digital technologies would facilitate? This space might be characterized as cosmopolitical, following Stengers, where "political voices" are less able to "master the situation they discuss" because "the political arena is peopled with shadows." As Stengers further writes, "This is a feeling that political good will can so easily obliterate when no answer is given to the demand: 'Express yourself, express your objections, your proposals, your contribution to the common world that we're building.'"[45]

A core tenet of social media is the project of expressing yourself. Sensor-based urbanisms would unfold not just through the expressive practices of individuals continually pinging and posting messages but also through the voluntary or default sharing of patterns of urban inhabitation. But these various ways of expressing oneself, ostensibly in service of an optimized urban experience, might as easily become sites of nonengagement, idiotic disruptions, or points of interference. The idiot in this way does not offer up a "program" of participation, expressive or otherwise, but instead has "far more to do with a passing fright that scares self-assurance, however justified."[46] Not only is Stengers's provocation to consider the figure of the idiot not a ladder of empowerment, such a proposal also suggests that we might attend to the ways in which participation is always a diverging rather than an easily or necessarily unifying set of engagements.

Rather than assume that the participatory program will necessarily be followed exactly as instructed, I suggest it is productive to consider examples of how these programs splinter into idiotic contributions. Alongside other students working with the lesson plans and materials, I examine one forum that asked students to document the types and locations of sensors that influence everyday life, which were crowdsourced for discussion by course participants. The forum

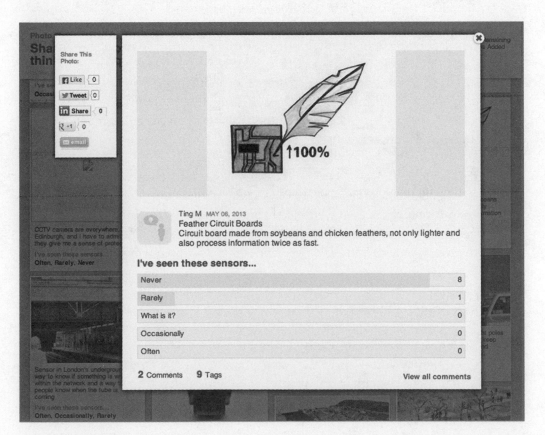

Figure 8.2. TechniCity MOOC, student contribution of "sensor." A "feather circuit board" offered as an urban sensor example, which most contributing students had never previously seen. Screen capture.

consisted of examples of sensors in the city, uploaded by course participants and available for commenting and voting by other course participants, who could indicate, "I've seen these sensors," with categories to tick including "never," "rarely," "occasionally," "often," and "what is it?" Sensor examples ranged from smart bus stops to lampposts and parking spots to environmental sensors to detect urban heat islands and seismic activity, but many examples were curious for the questions they raised about what is a sensor in the city, and what might its capacities be, particularly in relation to enhancing participation. I discuss three of these contributions to the sensor forum as examples of idiotic contributions to the TechniCity MOOC.

Society of Machines

The first sensor-forum example is drawn from an ID card uploaded by a MOOC participant to demonstrate sensor-activated access, and consists of a magnetic card that would pair with an electronic card reader. A staff card for a school psychologist named Jane Doe, this card was tagged as having been seen "often," "occasionally," and "rarely" by other course participants. The card and contribution signal toward sensors as actuators, as granting or restricting access to locations depending upon the status of the card bearer. While not necessarily an idiotic contribution in itself, this example uploaded by a MOOC participant bears comparison to an earlier vision of a computer-modulated city in an example discussed by Deleuze. In his Postscript text, he suggests that particular machines coincide with particular societies.[47] He then relays an anecdote from Guattari in order to demonstrate his point. He writes:

> Félix Guattari has imagined a town where anyone can leave their flat, their street, their neighborhood, using their (dividual) electronic card that opens this or that barrier; but the card may also be rejected on a particular day, or between certain times of day; it doesn't depend on the barrier but on the computer that is making sure everyone is in a permissible place, and effecting a universal modulation.[48]

The electronic card, as it intersects with computational urbanisms, is not a simple example of sensors embedded in urban contexts but rather raises questions about how programmed machines might enable or disable access to urban pathways. Idiocy might manifest in the form of programs that inadvertently restrict when they should allow access or in the form of those who would find themselves unable to traverse the city at any point, in an idiotic deferral to sensorized environments. As Christian Ulrik Anderson and Søren Pold write in relation to the issue of access and cities, "The digital urban and scripted space is at once a functional, aesthetic and political space. It manifests itself not as a grand spectacle but most often as a space where one can log-in (or be left out)."[49] Access modulates participation in the digital city differently, and the possibilities for engagements across political, functional, and aesthetic registers in many ways depend upon the distributed conditions of access.

From these computerized encounters, questions arise as to how participation is scripted, prevented, or rerouted in the sensor-based smart city. The program of participation is here more than a set of rules that are abided by, since programs rarely go according to plan, and may even be characterized by their accidents. Guattari, together with his electronic card, is participating in the sensor-based city, but if he does not have access he can become idiotic through the same technologies that would ordinarily make him a smart and participating citizen. Deleuze thus suggests that Guattari is an ancillary part of the card-to-urban-access interaction, where he becomes a dividual.

Political engagements here occur across human and nonhuman registers. A human might have notional access, but at the same time defers the possibility of access to sensor environments—in this case, cards and infrastructure—that embody the material conditions and politics of access. Of course, it would be possible to invoke the well-known examples of the actor-network theory of Bruno Latour, with his discussion of seat belts and speed bumps, or Madeline Akrich and her discussion of objects and scripts, which could be another way to talk about programs set in motion by objects that are not simply the result of human intentionality through use.[50] In this approach, contra Deleuze, the agency of the card expresses a distribution of action that cannot be located primarily in a human subject. The card may even force actions and responses in ways that give rise to different practices on the part of the card bearer. What Deleuze describes through Guattari is an example of an interrupted or broken program of participation, where the object-script that would facilitate participation can become a locus of control, differently articulated politics, or a machine society that unfolds in distinctly computational ways.

An idiotic encounter then emerges at the point of attempting to gain access and having that access restricted, of querying this restriction, and of being made to inhabit urban sites in ways that are orchestrated by smart technologies contra the intentions of users—technologies that might even be, in some ways, "smarter" than their users. Idiot-ness might here be described as a distributed condition that encompasses would-be urban citizens, access protocols, cards, and sensors, as well as urban spaces and infrastructures. Following Lucy Suchman, we might say that "users" (and their possibility to become idiotic) are distributed across machines, people, conditions of access, programs of engagement, and more.[51] At the same time, as Andrejevic suggests, within environments that increasingly depend on ubiquitous computing for conditions of access, "unwired humans will come across as singularly unintelligent, non-conversant and incomprehensible."[52] A world of ambient intelligence makes a human without sensing capabilities potentially idiotic. Far from being a naked citizen contributing to urban life through an inclination toward the common good, the digital citizen requires an extensive arrangement of resources in order to participate and be sensible in the sensor-based city.

The Private Life of the Public Idiot

The second sensor occasion charts how an idiot has at various times been defined as someone unable to participate in public life—where someone cannot be a citizen if he or she is not able to contribute to public forums. But in current circumstances that might be characterized by the excessive production of (previously private) data, we could also distinguish the idiocy that emerges when private lives are made baldly public, with intimate details made available everywhere all the time, so that this distinction collapses such that the public-ness of citizenship is no longer a defining trait. Another example of a student contribution added to the sensors-in-the-city forum of this MOOC then includes a billboard that reads:

> Michael-
>
> GPS Tracker—$250, Nikon Camera with zoom lens—$1600, Catching my LYING HUSBAND and buying this billboard with our investment account—Priceless.
>
> Tell Jessica you're moving in!
>
> -*Jennifer*[53]

From Google Glasses to GPS tracking and exposé billboards, the idiot appears at this newly blurred set of intersections between public participation and private lives where an "open" city and ensuing excess of data are made available through sensors tracking and reporting the banalities and infractions of everyday life. Is spilling the details of one's private life into public forums a mode of participatory urbanism? At what point do data even qualify as private, when privacy is an increasingly obsolete concept? Do practices involving the constant monitoring and sharing of personal data for wider distribution constitute idiotic or good citizenship? Or do they challenge the priorities and spaces of citizen participation beyond public and private to sites of contestation, the enrollment of sympathetic supporters, or the airing of grievances, as well as making contributions to the "greater good" through the sharing of intimate data?

With passive data collection, the terms of participation and citizenship in sensor-based cities are such that the default setting is to overstep or disregard the usual distinctions of private and public life. Private life, the usual space of the idiot, comes flooding into public life, the space of the assumed "common good." Here is an idiotic encounter through political renegotiations that occur where egregious information about citizens is deliberately or inadvertently made available. These renegotiations may also be absent, since in order to participate citizens may have no choice but to provide information if they are to participate as able communicative digital citizens, because to unplug or not provide data would be to become an idiot.

While many advocates for sensor-based urban participation suggest that these practices open up new forms of civic agency, sensor-based engagements restructure participation and the processes by which an urban citizen is understood to be making a contribution. Participation need not even be "active," in the sense of exercising volition, but can occur through passive data collection, tracking (self-initiated or conducted by others), wearing devices, and having technologies report on one's behalf. Participatory contributions are not expressions of intent, or even interruption, but rather involve being an available sensor-based and data-producing urban entity. GPS data might be picked up to articulate points about infidelity or illegality, but within the processes of data analytics most of these moments of pattern detection will instead be parsed by algorithms rather than disgruntled partners. But the previously idiotic space of private activities floods into public forums, while reconstituting the ways in which public-ness and public life is designated. The contribution—even if passive—of private data becomes a new way of engaging in, articulating, and sustaining the digital city.

Failure to Compute: 3 + 2 Does Not Equal 5

A third sensor occasion considers the idiot as a person or entity that does not compute. Idiot-ness may emerge through failing to follow instructions or interpreting instructions in a way so tangential to the project aims that moments of befuddlement, dismay, or confusion proliferate. This final list of sensor examples added by students includes images of bacon cooking in a frying pan with a splatter shield, a pistol-shaped hairdryer, and a feather circuit board. One might ask: Where is the city or sensors in this crowdsourced example of urban sensors? Perhaps lodged in the strange interstices of a bacon-splatter tray there is some sensor-actuator response ready to contain popping grease from defiling domicile walls, but the bustling city seems a distant land from this fry-up scene. Possibly these MOOC student-citizens have misunderstood their assignment and are providing deliberately misleading or incorrect data as a way to interrupt the steady flow of contributions to this forum. Or spam bots may have intervened and garbled the contributions of otherwise diligent citizens-in-training.

Whatever the occasion, in the mixing of citizens and sensors of all sorts one could politely say that potentially other-than-smart contributions are made. Maybe lodged in the pistol hairdryer example provided by one MOOC participant a sensor awaits activation for a morning beauty routine. This may gesture toward an urban encounter in waiting, but its relevance to the forum remains a mystery. And a microchip ostensibly assembled from soybeans and chicken feathers seems to offer an organic approach to sensors. Whether speculative, fictional, or the material of one course member's strange electronics experiments, this wholesome sensor option begs the question of what materials will variously construct smart and sensor-filled cities. Is it idiotic to propose that sensors should be biodegradable and organic?

My point in this analysis of participatory engagements with thinking through digital urban participation vis-à-vis the idiot is to attend to these concrete examples that I encountered during my attempt to train up to the challenges of the smart city while learning from other participant contributions. In the process, I questioned what sorts of intelligence were brought together, promoted, or alternatively derailed in attempts to make citizens smart and cities apparently more participatory. The participatory citizen unfolds as a potentially idiotic figure in these sensor examples, which ask us to rethink the settlements at which smart and sensor-based participation would seek to arrive. The idiot is not simply a ruse, but rather a figure that troubles the instructions and assumptions of smartness, of smart cities, citizens, or actions that would course through with efficient connectivity, a cascade of clear decision-making, and problem-solving actions.

As Deleuze has suggested, one version of the idiot may be someone who insists that 3 + 2 does not equal 5. This is an "old" version of the idiot, one who "would doubt every truth of nature." In contrast, "the new idiot has no wish for indubitable truths; he will never be 'resigned' to the fact that 3 + 2= 5 and wills the absurd."[54] Whether old or new idiot, basic computations do not yield expected results. The figure of the idiot (as a "hazy presence") here suggests it may be possible to consider how participation unfolds neither as a simple formula for achieving the common good nor as the dutiful actions of rational actors, but rather through distributed and often disruptive modalities of engagement that reorient accounts of smart, idiotic, or instrumental forms of participation toward other engagements with collective experience, plural realisms, and unexpected potential for creative advance in urban situations.

In the updated 2014 version of the TechniCity course, the final lecture, "The Future of the TechniCity," ends with the cautionary explanation and statement:

> This is our last week together! We end the course thinking about the future of technology in our cities. We'll explore wearable technology, mobile money, and more. But most importantly we have a cautionary note that technology isn't the be all end all solution for all of the world's problems.[55]

As Mackenzie established in his Simondonian-influenced study, *Transductions,* "technicity" as an abstraction is in fact a way of articulating relations so that technological analyses do not focus on technologies as apparently given and do not impute deterministic agency to devices.[56] Instead, technicity draws attention

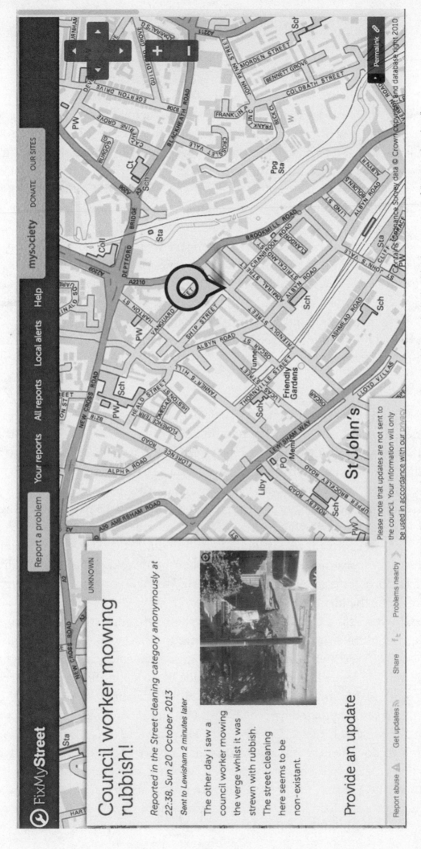

Figure 8.3. Council Worker Mowing Rubbish! submission to FixMyStreet platform to document urban problem to be solved in the category of "street cleaning." Screen capture.

to the shifting and transductive relations that emerge through extended technological engagements and arrangements. Here is a different version of technicity, one which demonstrates that even while the instructors of this TechniCity course caution against seeing smart city technologies as simple solutions, solutionism may in fact be the primary type of relation that sensor-based technologies tend to articulate. But this is also exactly where idiots (whether human or nonhuman) proliferate within these solution-oriented approaches to digitally equipped citizenship. This is also where politics might be best be identified, understood, and practiced—as a reworking of digital participatory urbanism. By attending to these idiotic encounters and presences, as Stengers writes, such a cosmopolitical approach might "protect us from an 'entrepreneurial' version of politics, giving voice only to the clearly-defined interests that have the means to mutually counterbalance one another."[57] Instead of a politics that attempts to sidestep hesitation and divergence in pursuit of an unquestioned common vision of the city, a cosmopolitical proposal would see these idiotic encounters as an ongoing condition of engagement.

Citizens Fixing Streets

If anything, digital programs for participation raise as many questions as they offer answers about how engagement in a sensor-based city might unfold. While held up as a solution to the problem of the city, participatory and sensor-based technologies inevitably create their own obstacles, diversions, and problems that do not necessarily allow for an easy passage to a more participatory urbanism. The second participatory urbanism project that I discuss is particularly illuminating for these issues, which it makes evident in relation to how urban problems are identified, categorized, reported, and acted upon. FixMyStreet is an online platform that citizens can use to report problems with urban spaces and infrastructure. The problems are in some cases reported to local councils, which might then decide to fix the identified problems.

FixMyStreet is a UK-based service developed through mySociety, a nonprofit group established in 2004 that uses digital tools to make governments more accountable.[58] FixMyStreet was launched in 2007 as what might be classified as a "civic app." In line with the distinction made by Nesta in the above discussion on civic apps, FixMyStreet occupies a niche in the participatory digital city ecosystem that is also populated by SeeClickFix (an American-based version of the platform), as well as Citizens Connect and FillThatHole (a UK-based platform for cyclists to document potholes), platforms that are specifically oriented toward enabling citizens to report problems with urban streets and infrastructure.[59]

In my investigation into the FixMyStreet app and platform, I focused on the borough of Lewisham within southeastern London, an area with a diverse population of varying socioeconomic circumstances. Far from the wealthy western side of town, Lewisham residents may engage with street fixing in ways that may be particularly attentive to the at-times precarious state of infrastructure here. But because this platform and app work through universal categories of concern for street conditions, from abandoned vehicles to dumped rubbish to graffiti and road defects, the differences in urban reporting may show up most clearly at the point of written and visual problems logged.

The FixMyStreet platform and app are relatively easy to use, and only require that one first choose a borough, identify the location of the particular report on a Google map, assign the problem a category, add any additional textual or visual detail, and include one's individual details with email. The report will then be made live, and is then typically communicated to the local council, which may or may not decide to "fix" the problem reported. Problems reported over the past two years in Lewisham range from rubbish dumped on streets to dog fouling, missing utility covers, rubbish- and rat-filled derelict phone booths, dead or nuisance animals, potholes and more potholes, and missing or incorrect signage, as well as a general concern with the condition of infrastructure and an impending sense that if particular problems are not fixed then future consequences could be disastrous.

When problems are fixed, a green icon will appear with the text, "fixed." When problems are left to linger, a grey box will appear with the text, "unknown." Problems most likely to be fixed include clearing dumped rubbish and fly tipping, covering missing utility holes, and dealing with any infrastructural issues that might lead to litigious action. Problems that are often left to linger include potholes, phone booths, dead animals, and anything that constitutes criminal activity as opposed to an urban-space issue. In August 2009, one person logged that a Christmas tree had been abandoned. In April 2013, the problem was reported as fixed. A commentator then wrote, "4 years later? Are you sure it didn't just rot down?"[60] The timing and accountability of fixing can be thrown into question when problems linger in an unknown state and when fixes surface that seem to be attributed more to neglect than to the active remedying of problems. By questioning the process by which problems are fixed, this Christmas-tree commentator introduces an idiotic mode of participation, slowing down and asking how these problems go away, if ever.

Problems that remain unknown, unfixed, and unaddressed are many, a situation that raises questions as to how effective this platform is for empowering citizens to communicate with councils or to have their grievances heard and addressed. One report notes with alarm, "Council worker mowing rubbish!"[61] Here, rather than pick up litter on a grassy right-of-way, a council worker took the more expedient route of mowing through plastic bags and bottle tops in order to shred them to a finer and potentially less noticeable bit of debris. The person logging this report takes this as an indication of general neglect for street cleaning, and no doubt also as a sign of a lack of care for the neighborhood. But in many ways such an observation verges on the idiotic, as it does not define the task in such a way that it may be easily fixed. Rants, complaints, and general observations of neglect: these are not computable problems but rather could be characterized as participatory noise that runs through this platform.

As the FixMyStreet site indicates, problems are typically reported to the local council or "relevant local body," such as a transport agency. The overview notes that the site's purpose is for the reporting of "physical problems," and other platforms may be more appropriate for problems not within this category. However, even problems that might be clearly identified as "physical" do not always feed through to solutions. For instance, one site user complains about a recurring problem with rubbish overflowing at one particular intersection, which eventually leads to "rats in the rubbish." While there seemed to be a slight improvement after reporting the rubbish situation, the problem continued, so that the user writes, "Please don't suggest 'Do it yourself': I've offered, still waiting for a reply from Lewisham Council."[62] There is a sort of ambivalence here around DIY—the user is taking the time to report a street problem while also indicating a willingness to

Figure 8.4. Council Worker Mowing Rubbish! submission to FixMyStreet platform, with example of photograph uploaded by anonymous reporter of "street cleaning" problem. Screen capture.

engage with the problem if communication with the council can be established, and yet the user also expresses a frustration with DIY and a sensed lack of accountability on behalf of the council. DIY here is not a simple pathway of citizen empowerment, but a space of hesitation. It is unclear who should or will take responsibility for the rat and rubbish problem, and so the FixMyStreet platform registers this as an idiotic exchange of sorts, where the problem is not solved, the "common good" is not readily advanced, accountability and empowerment are not clearly articulated, and urban engagement remains a faltering undertaking.

Within the FixMyStreet platform, the street becomes an object of citizens sensing and identifying problems, reporting these problems for repair, and being variously impressed or disappointed at the resolutions achieved or inaction that results. While citizens perform sensing functions, the sense data they collect and report does not necessarily lead to actuating solutions. The street further becomes the site of a particular type of problem making. Physical problems that readily fit within preidentified categories and local council chains of action stand the best chance of being addressed, while problems that fall outside this scope may be rendered as idiotic—as not being addressable within the logic or terms of the platform. Even when physical problems with streets and infrastructure are coded into tasks, solution-based actions are not readily realized. Who is responsible for the city? This is a question that remains unanswered in many of the FixMyStreet platform exchanges.

RETHINKING THE SENSORING OF PARTICIPATION

FixMyStreet, as with many other platforms and participatory- and passive-sensing projects, relies on a certain task flow in order to make problems identifiable and operable. Participation in the digital city then becomes a matter both of "instrumenting the citizen" and of breaking down urban problems into computable tasks. These tasks, as identified in one IBM paper on citizen sensors, typically involve selecting an urban event to report, collecting sense data, and reporting and analyzing data collected, most often through a smartphone interface.[63] Sense data collected usually includes that which smartphones can sense, including images, location, altitude, acceleration, temperature, and direction. There are multiple sensors in a smartphone, and this microecology of participating user with sensing devices in relation to a city full of problems waiting to be solved becomes the basis for the task flow of participation in the digital city.

Citizens then perform sensing tasks that are continuous with a sensor-filled city and its sensor devices. In this way, a proliferation of sensors does not necessarily need to be the determining driver of sensor interactions as such, since citizens are also sensing within computational logics through an established chain of tasks. Citizens become interchangeable with sensors, they even *become* sensors. Citizens monitor themselves and environments, environments self-monitor, and in this loop of gathering data for participation and democratic urban action, urban sensing networks do more than simply record and model existing patterns of urban life—they also enable and form ongoing possibilities for urban engagement. The analysis of sensor data, whether gathered through participatory or passive means, leads to the performance of new actions and modulations, thereby informing the "behavior" of citizens and further possibilities for participation.

While much attention is deservedly directed toward analyzing sensors for their capacities of surveillance, this particular analysis has not focused on surveillance-based concerns in relation to participatory sensing and urbanism. However, as is especially clear in the context of participatory urbanism, surveillance emerges not only as a project of real-time observation but also as a store of multiple banal details and data about everyday life that can be mined for patterns and even turned into anticipatory and predictive engagements. Surveillance in this respect is about capacities yet to come as much as current documentation in the production of new datasets, whether crowdsourced data on potholes, passively sensed data from individual journeys, or participatory sensing in the form of eager urbanites annotating, coding, and hacking urban space. Sensor- and computation-based approaches to urban life are then as much descriptive gatherings of current events as they are productive expressions of new practices, ways of life, and modes of politics.

Within these task flows undertaken by instrumented citizens, participation is far from an easy and straightforward project of generating and analyzing urban data for optimal urban functioning. Instead, participation becomes characterized as much by multiple idiotic encounters—across human and nonhuman sites of engagement—as it does by purported new modes of democratic engagement. A considerable amount of literature on digital participation focuses on the ways in which people connect, social movements form, and (even revolutionary) actions unfold.[64] There is also a significant body of literature that closely examines the networks within and through which participation takes place and which draws attention to the inevitable power relations that in-form what sorts of participation might even be possible in the first place.[65] Social and digital media are generally advanced as tools that address the problem of participation, even fomenting revolutions. Alongside the extensive set of participatory digital urbanism projects, there are a number of literatures that also draw out—critically, theoretically, and practically—the ways in which digital participation is unfolding and might continue to develop. Clearly, there is a multiplicity of terms and practices circulating that carry the descriptor of participation. My point here has not been to undertake a comprehensive review of these practices since this work been done elsewhere in various ways.[66]

Moving laterally from these studies interested in the capacities of social media to spark actions, I instead attend here to the ways in which participation does not go according to plan. In this sense, I align this analysis with research that might question or critique the democratic engagements that digital media might facilitate.[67] Other studies, such as the collection on *DIY Citizenship* edited by Matt Ratto and Megan Boler, collect together more empirical encounters with DIY citizenship, digital and otherwise, to consider how more active and making-based forms of engagement might transform citizenship.[68] Kurt Iveson suggests that, rather than ask whether digital urban platforms enable particular types of engagement and citizenship, we should instead ask: "'What is the vision of the good citizen and the good city that they seek to enact?'"[69] Similarly focusing on politics as that which occurs at sites and moments of disruption, Iveson notes that participatory platforms do not *de facto* lead to politics or political engagements, and may in fact have depoliticizing effects. His point of inquiry is then to ask whose version of the good city we are enacting through participatory technologies of engagement. But a more Stengerian focus on politics and disruption might, vis-à-vis the idiot, ask not about what vision of the good city or good citizen is being enacted, but instead query any project that would assume its own goodness. The idiot always asks, "But what about . . . ?" thereby disrupting an unproblematic marching toward a common good. As the idiot reminds us, there is always more at stake.

Civic Media and Tactical Media: Re-versioning Participation

In critiques of the smart city a number of writers and tech gurus make a point of calling for greater attention to the role that "grassroots" or "bottom-up" citizen engagements can play in giving rise to a more human, just, and equitable set of digital city developments.[70] And there are many projects underway that would attempt to respond to this challenge, which may variously be seen as projects of disruption, democratization, and enablement. Numerous projects have sprung up that attempt to experiment with alternative approaches to digitally oriented and sensor-based urban participation and at times also to disrupt the smart city rhetoric that is often primarily issuing from technology companies. Projects in this area span from Spontaneous Intervention projects shown at the U.S. Pavilion during the 2012 Venice Biennale, including the San Francisco Garden Registry, a platform for identifying used and underused garden space in the San Francisco region; PetaJarkata, a crowdsourced and Twitter-based project for enabling residents of Jakarta to report and respond to recurring flood events in Jakarta; and the City Bug Report, a project by Henrik Korsgaard and Martin Brynskov that allows citizens to report on bugs in the smart city and that makes space for the "messy" social and political encounters that inevitably unfold in cities.[71] Many of these projects continue to work within a citizen-sensor-urban-problem dynamic but at the same time open up other spaces for participation that encounter the disruptions and possible openings that might make for other urban political engagements.

While the point here is not to identify "good" and "bad" examples of participatory digital urbanism, these strategies could be described as re-versioning participation—of working within a prevailing set of approaches to participation and experimenting within alternative possible outcomes.[72] Many of these projects unfold through a sort of "tactical urbanism meets tactical media."[73] They are experiments with urban technology, participation, and citizenship. While some offer up different ways of thinking about urban inhabitations and practices, others are driven by a technology-for-technology's sake ethos—of making the city computable in order to test the capacities of sensor devices. All of these projects raise the question of what role these tactical projects might play in seriously recasting the increasing digitalization of urban spaces and processes.

Writing on tactical media, a form of digital disruption that is in some ways the uneasy forerunner to more integrative forms of participatory media, Rita Raley suggests that tactical media should not be evaluated exclusively for its assumed effectivity. She suggests, "The right question to ask is not whether tactical media *works* or not, whether it succeeds or fails in spectacular fashion to effect structural transformation; rather, we should be asking to what extent it strengthens social relations and to what extent its activities are virtuosic."[74] In this passage, Raley responds to an argument made by Geert Lovink and Ned Rossiter, who critique tactical media for its fleetingness and ability to be co-opted back within the very systems and arrangements it would interrupt and reroute. Recognizing the validity of their point, Raley nonetheless suggests that focusing on effectiveness may be the wrong criterion with which to evaluate tactical media. Virtuosity, as an alternative criterion, enables events to have their own momentum and purpose, thereby sidestepping a teleological agenda that says tactical media are only as good as their impacts.[75]

If we were to extend this approach to an analysis of participatory urbanism and sensing, we could then suggest that participation may not necessarily be characterized by preplanned outcomes, but rather it may rework processes of engagement toward making, changing, and mobilizing situations differently. Yet there remains an abiding question as to whether even virtuosic participatory media fold back into digital media that circumscribe politics as computable problems. With all of these projects, it may then be useful to keep in mind the idiot, the one who "does not compute," and who, by not abiding by the terms of participation, forces us to reencounter the problem and the politics of urban engagement.

A COSMOPOLITICAL ECOLOGY OF PARTICIPATION

If we were to return at the end of this chapter to a consideration of all the multiple sensors that are embedded in urban spaces and that facilitate access to urban infrastructures, or that are transported by urban citizens in the form of smartphones, we might make a list of technologies that spans from automated traffic lights to smart bus stops to bicycle-sharing schemes, increasingly smart energy grids and screen-filled urban environments, not to mention innumerable CCTV cameras trained on urban spaces. Urban spaces are increasingly sensorized, and urban engagements are also in-formed by these sensor-based modalities. As I have suggested in this chapter, sensors are not just about the proliferating hardware and associated software that would automate urban infrastructural functions but also about the changing character of urban engagements. These engagements shift within the technomaterial context of sensorized cities and also through the computational logics that are put into play where citizens become sensors and perform sensor-like functions in relation to computable urban problems.

Participatory actions within sensor-based cities most often consist of monitoring environments and reporting data for networked analysis. Action—a response to the urban problems identified—is often assumed to flow from this monitoring data. The intelligence that would actuate and solve urban problems is made to be continuous across citizens and digital urban infrastructures, an intelligence that is narrated as being bound up with democratic engagement and sustainable action. If citizens—and urban algorithmic networks—have more data about urban life, then urban experiences are meant to be optimized and made more participatory and intelligent.

But these modes of intelligence and participation, as I have suggested here, are as likely to be generative of idiotic encounters as participatory ones, and the actors in these digital exchanges—whether human or nonhuman— are as likely to produce idiotic non sequiturs as they are rational advances to enhanced urbanization. The distributions of multiple sensors used to manage urban processes as well as encourage participation are sites productive of idiotic engagements. These idiotic sensor occasions include encounters that take place in interactions with the sensor city, in sites of data production and circulation, in the contributions made and interpretations given to participatory platforms. Opportunities for idiotic participation also proliferate across distributed human and nonhuman arrangements and engagements.

With the idiot brought into a more considered part of the dynamics of participatory digital urbanism, perhaps cities could begin to be understood less as technical problems in need of fixes. Drawing on Stengers not just for her discussion of the idiot but also for her Whitehead-inspired discussion of how to make the field of problem making more inventive,[76] I suggest the computable problem of contemporary cities here gives reason to reconsider how this problem has been cast. By reconsidering the field of the problem of the computable and sensor-based city, it might finally be possible to reinvent participatory practices that unfold within digital urbanism encounters.

Proposals for sensor-based and digital urban participation are often narrated through a vision of the good city and the smart citizen. But as the idiotic encounters

I have discussed demonstrate, these projects also raise questions about who participates, how urban problems are identified, and what participation is meant to accomplish. Citizens become sensors and urban problems become computable in many of the projects I have discussed here. Yet it may be necessary to consider how the idiot disrupts consensual visions of goodness and instead troubles participation in ways that reroute the rules of the game. As Stengers writes, "One has to be wary of individual good will. Adding a cosmopolitical dimension to the problems that we consider from a political angle does not lead to answers everyone should finally accept. It raises the question of the way in which the cry of fright or the murmur of the idiot can be heard 'collectively,' in the assemblage created around a political issue."[77] Stengers suggests that the idiot especially forces us to attend to the concrete conditions of problems. If urban environments are under stress in one way or another, these problems would then need to be attended to in their specificity and not as conditions conducive to solutions propagated by universal information architectures. Such a specific (cosmo-)political ecology of problems is then an important part of attending to urban conditions. These specific conditions ensure that we cannot proceed through "blind confidence" or "good intentions" but rather must "[build] an active memory of the way solutions that we might have considered promising turn out to be failures, deformations or perversions."[78] This chapter has described such a catalog of failures and deformations through a few encounters with the idiot at sites of participatory digital urbanism. This list could be extended. But it is also an opening and invitation to consider how participation and the problem of the city might be reinvented by attending to the diverse inhabitations that break with the program of digital urbanism.

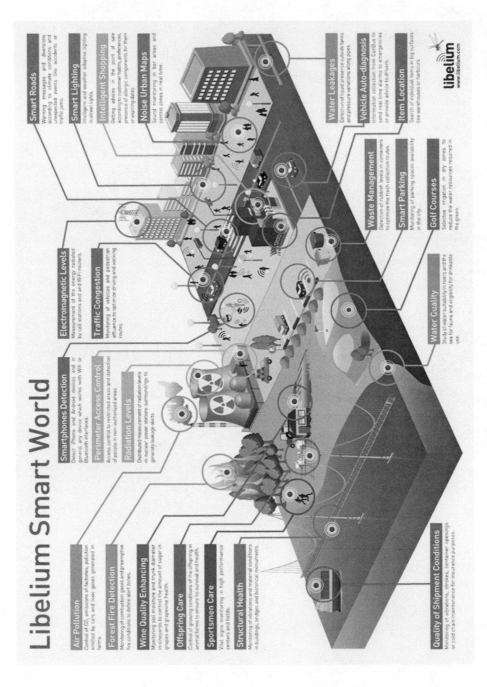

Figure 9.1. Libelium Smart World. Libelium's infographic comprising Smart Cities, Internet of Things, and other sensing applications. http://www.libelium.com/top_50_iot_sensor_applications_ranking. Copyright Libelium Comunicaciones Distribuidas S.L. 2013.

9

Digital Infrastructures of Withness

Constructing a Speculative City

Pₐʀᴛɪᴄɪᴘᴀᴛɪᴏɴ ɪɴ ᴛʜᴇ ᴅɪɢɪᴛᴀʟ ᴄɪᴛʏ, as discussed in the last chapter, is far from a clear and simple project of using digital technologies or platforms for achieving common good. The contributions and pathways of participation unfold along multiple often-circuitous routes, while modes of participation are also entangled with more-than-human entities that may disrupt or prevent engagements. In the emerging smart city—as it is now being implemented—digital sensors are not simply scattered nodes of computational hardware but are also bundled into urban infrastructures in ways that remake these systems and the practices that they animate. The digital infrastructures that variously constitute the smart city, from smart grids to connected devices and smartphones, are also sites where participation is organized and potentially altered and rerouted.

As discussed in chapter 7, many proposals for smart city projects still exist at the speculative or near-future stage. Yet some initiatives have translated into alterations to the built environment, including new services as well as retrofitted urban infrastructures and processes. This chapter looks at sensor technologies and smart city initiatives as they have been implemented and discusses how these distributed modes of environmental sensing influence urban experiences. I specifically discuss sensor projects and smart cities technologies developed in London, a city that is frequently referred to as one of the smartest worldwide, and which has a number of current and ongoing initiatives to implement smart technologies.

I am interested to understand digital infrastructure in these emerging smart city initiatives, not as something that can be defined at the outset, but as something that comes into being through concretizing technological arrangements as well as distinct ways of inhabiting these infrastructures. While it would be possible to make a straightforward list of the typical ways in which this "fourth utility" is unfolding, I suspend these assumptions about what exactly constitutes smart and digital infrastructure in order to examine more closely how digital infra-structure actually manifests through technologies and practices. Simondon makes the point that an abstract technology is not necessarily a cognitive model that is implemented, but rather is a set of dynamic changes that occur in any given tech-nocultural system to make possible the concretization of particular technologies. In considering the smart city as one such abstract technology, I suggest it is vital to attend to the specific environments in and through which specific smart cities are able to take hold.[1]

Furthermore, Simondon suggests that as technologies become concrete they do not become more fixed, but rather become *more indeterminate*.[2] Abstraction, as it turns out, may constitute a more static rendering of technology than its con-crete instantiations, where multiple ways of materializing, practicing, inhabiting, and intersecting with technology can occur. Processes of constructing a specula-tive smart city constitute ways of working out and transindividuating the entities that populate smart cities and the ways in which they relate and connect up. This chapter turns to consider the transformations that occur as smart cities migrate from an abstract and even speculative set of technologies to more concrete mate-rializations. In relation to implemented sensor technologies, I ask: What are the processes that these infrastructures instigate and sustain? How do they at once individuate and join up cities and citizens? What are the capacities of these infra-structures and what modes of inhabitation do they facilitate?

Part of my objective in asking these questions is to consider the processes whereby digital infrastructures become environmental in the city. This focus takes up the emphasis placed throughout this study on the *environmental* aspects of computation, but here particularly attends to existing urban infrastructures. By "becoming environmental," I also address much more than a process of spatial-izing digital technologies, since as I have discussed throughout this study, envi-ronments are not merely established backdrops against which the activities of human and more-than-human entities unfold, but rather are involved in distinct processes of becoming and concrescing along with entities. Environments are then interconnected with entities inhabiting those environments. Environments are also the conditions in which particular entities may take hold—they ensure the success or the lapsing of particular entities and establish further conditions for invention. It is important to continue to extend the environmental aspects of computation in these ways, since it enables a more dynamic and processual under-standing of how environments and digital media concresce to form actual entities and actual occasions.

Environments and environmental computation also constitute situations in which "withness" might be articulated and in-form processes of participation. Withness, following Whitehead, is a concept that signals modes of being and becoming together, of concrescing, such that the possibilities for both urban ontological engagements as well as urban speculative futures are undertaken.[3] Withness raises the question of how we "possess" the world and become together, not exclusively as a matter of intelligence or rational cogitating actors, but as embodied if differently directed creatures in shared worlds. In this chapter, I attend to the ways in which digital urban infrastructures are productive of occasions of withness. Borrowing this term from Whitehead, I suggest here that withness is a way of moving an analysis of participation to the things, entities, and occasions that are brought together within digital infrastructures and articulating how these modalities of withness influence the types of politics that might in turn unfold. Digital infrastructures in smart city developments induce particular types of participation, which are further productive of modes of withness that generate individuations, embodiments, and inhabitations within the very particular environmental–computational worlds of smart cities.

As discussed in the last chapter, much attention has been directed toward the ways in which digital technologies might activate distinct types of action and engagement, where by virtue of having information and being connected to a seemingly expansive if hazy "community" of fellow users, urban citizens can be better attuned to political projects. But I would suggest that there is much detail that is elided in this characterization of participation, since it assumes a rush to effective action that may never transpire. By attending to the withness of participation, I instead want to identify whom or what particular participatory practices gather together, what these relations or nonrelations might be productive of, and what political contestations are bundled into these groupings. If much participatory literature has attended to the empowerment of individual or collective digital users, this analysis instead considers how a user is only one figure within a wider infrastructural network of participatory and transindividuating politics and action.

While most digital participation focuses on the empowerment of individual users who together make up a collective, or who exercise a self-ness that reinforces notions of doing-it-yourself, here I want to move away *both* from the idea of empowered individuals and from action itself, since the assumption of making citizens and cities productive through optimized actions is so much a part of smart city and digital participation projects that to suspend action in particular is a way not only to ask the Stengerian question in relation to the idiot, "What are we busy doing,"[4] but also to attend to ways in which urban life might be characterized other than as the ceaseless unfolding of productive (and often economically aligned) individual activity.

Moving from abstract to concrete instantiations of the smart city, I further consider how to move from programs of participation (as discussed in the previous two chapters) to articulations of withness. Participatory, DIY, and smart urbanism projects are articulations that circumscribe withness in particular ways:

they are commitments, with consequences, to ways of being and becoming together. Control systems communicating with transport networks, CCTV performing video analytics for pattern detection, smartphone apps providing access to urban maps and services: in each of these examples, what in the smart city rhetoric would be described through narratives of optimization, facilitation, and efficiency are occasions of withness that express, constitute, and contribute to the ongoing formation of distinct urban individuations and inhabitations.

How are we *with* the (smart) city, its infrastructure, its other inhabitants, and the many computational devices that would steer us, when emphasis is placed on coordinating flows of movement so that stoppage, disruption, breakage, and jamming are minimized? What is the withness of a ceaselessly flowing city, of a city that never stops, that in its automated efficiency continues to process goods, information, and waste in the small hours of the night? Clearly, to discuss digital infrastructures of witness then also requires attending to infrastructure as process. While in some cases emphasis may be drawn toward the physical aspects of infrastructure, with roads, bridges, sewers, electricity, and telecommunications interconnected as basic organizational structures, at the same time a growing body of research now attends to infrastructure as process, which may become more pronounced in moments of failure and breakdown.[5] Digital infrastructures draw our attention to the ways in which the occasions of cities would not be possible without infrastructural processes. Distinct urban processes are facilitated, enabled, and connected up through these digital infrastructures.

I employ the notion of witness to characterize particular human and more-than-human engagements and concrescences that occur with and through digital infrastructures. I further attend to these infrastructures both as they are emerging as new urban technologies and as they orient urban life toward speculative constructions and additional urban potentialities. In my use of the term "speculation," as discussed throughout this study, I am particularly concerned to draw out the ways in which the future is already present in current instantiations of smart cities. The future is being made with each sensorized traffic intersection that is implemented, with every smartphone-interacting signage that is developed, and with every prototype and policy document that legislates toward making smart urban environments possible.

Speculation in relation to the smart city is an ongoing practice that, complementing the discussion undertaken in chapter 7 of this study, is both an imaginary and a material–political practice. Speculation is not a project of making fictions but rather is a practice of constructing particular trajectories of urban practice and inhabitation. Construction occurs here in at least two senses: of being built, and of forming the conditions in which new speculative urbanisms (and modes of withness) may unfold. The process of constructing a speculative smart city then points toward experimentation and invention. In the conclusion to this chapter,

I consider how digital infrastructures, smart cities, and participatory urbanisms might be advanced through considering modes of withness that are experimental and that, in a Simondonian sense, might become more *inventive* as indeterminate technical arrangements.

AN INVENTORY OF SENSORS IN THE SMART CITY

Even a notional inventory of sensors in London might reveal that this is a city popping with proverbial intelligence, where the number of sensors in place and in operation suggests that the apparently speculative smart city is already in construction. On any given day traversing the city, one might encounter sensors not just in the smartphones that many people carry (devices that are packed with humidity sensors, temperature sensors, a digital compass, an accelerometer, a gyroscope, GPS, a touch sensor, a microphone, an ambient light sensor, an optical proximity sensor, and an image sensor) but also in every bit of urban infrastructure. Here are multiple amalgamations of sensors and networks, which collectively are meant to add up to more intelligent urbanisms.

An inventory of smart technologies might also incorporate where sensors are located, what functions they enable, how they transform urban processes, what modes of engagement they require or facilitate, who is able to access them and the data they generate, and what happens if they fail or go awry. And indeed, even notionally following such an approach in London, one finds that sensors can be located within preexisting utility infrastructures, including electricity grids and increasing numbers of smart meters; in water mains for leakage and crack detection, as well as flow rates and meters; in waste transfer stations and collection bins; in sewers for tunnel monitoring (and, prospectively, for chemical- and biosensing for drugs, bombs, and diseases); and in an increasing array of smart home technologies that moderate the boundaries between home consumption and utility provision.[6]

Multiple sensors and smart networks are already in use in transit systems in London, from pedestrian crossings for managing the density and flow of foot traffic to bus sensors for alerts on current arrival times (which include GPS location sensors; GPRS transceivers; physical bus measurements via odometer, gyrometer, and turn rate sensors; and mobile IP connections for transmitting data from eight thousand buses to the CentreComm command center).[7] There are sensors on the cycle-hire docking stations and on buses to avoid collisions with cyclists. Sensors are in regular use at traffic lights and as cameras for live-traffic feeds.[8] Sensors are further dotted throughout the London Underground on escalators, in tunnels and on lifts, on HVAC control systems, and joined up with loudspeakers and CCTV. In this extended Underground system and sensor network, data on temperature, vibration, humidity, faults, system alerts, and equipment degradation are fed live to a secure cloud for further integration in order to monitor and automate tasks.[9]

Sensors are located on building sites, to monitor vibration and strain from pil-ing, and at tunnel construction sites.[10] There are sensors on parking bays to signal availability in order to discourage drivers from circling to find a parking spot.[11] Innumerable CCTV cameras operate as sensors (as discussed in chapter 2), and video analytics are used for crowd management to detect suspicious activity and identify hazards or accidents.[12] Sensors are also in use for multiple forms of mon-itoring related to health and the environment, including the LAQN air-monitoring network discussed in chapter 6, as well as sensors at Environment Agency sites for monitoring river levels, water quality,[13] and air pollution in relation to industrial activity.[14] Sensors monitor for vibrations and displacement at the Thames Barrier[15] and the wind speed and direction on the London Eye.[16] Multiple sensors are at work to facilitate weather monitoring, including stations from the Met Office as well as smaller weather stations, amateur stations feeding into Weather Underground, and mobile phones set up as weather stations feeding into crowdsourced maps.[17]

Sensors can be found overlapping with existing infrastructures, in some cases forming new networks; sensors are in place to monitor specific temporary uses and events such as construction; sensors are monitoring air and vibration and water levels; and sensors are carried around in smartphones, as wearables, and other portable devices, whether as DIY citizen-sensing tools or monitors for detect-ing specific phenomena. Sensors are also not the only source for data generated to manage urban systems—alternative data sources include both static and dynamic data collected from social media streams, participatory-sensing systems, and pre-dictive and strategic modeling capabilities.[18]

Additional London-based smart city initiatives include Living PlanIT, a proj-ect developed out of Formula One racing technology and put to new use at the London City Airport. This project consists of using data, smartphones, apps, and retail schemes to maximize shopping opportunities for High Net Worth individu-als, as well as a schematic plan for an "urban operating system" for London, together with energy, water, waste, mobility, security, building controls, analytics, and apps.[19] In a somewhat different arena, the Intel Collaborative Research Insti-tute (ICRI) for the Sustainable Connected Cities project is a collaborative venture across Intel, University College London (UCL), and Imperial that includes proj-ects such as "living labs" for testing air quality in London.[20] A video documenting the Intel initiative argues that cities such as London are "reaching capacity," start-ing to "break down," and so need to be "more efficient." There are even more projects than this underway that demonstrate the extent to which London is an active site for experimenting with, prototyping, implementing, and trialing sensor-based smart cities projects.

These sensor architectures and developments are, of course, situated within a larger context of multiple smart cities pilot projects and plans underway in London, from the Smart London Plan developed by the Greater London Authority to the

Sensing London initiatives advanced by the Future Cities Catapult, funded through the Technology Strategy Board.[21] The Smart London Plan, for instance, outlines the rapidly growing population of London, which is expected to expand by one million by 2021, as well as numerous other urban constraints that smart infrastructure is meant to help manage, primarily in the form of "mobile internet applications, the internet-of-things, cloud computing and insights from big data."[22]

With this extensive yet by no means complete inventory of sensors as well as other data streams increasingly informing the urban processes of London, it is still not entirely clear at what point this proliferation of technologies might cross a threshold to constitute digital infrastructure or fully formed smart city developments. As mentioned in the introduction to this chapter, a list of sensor technologies does not necessarily point automatically to infrastructural development and may instead raise the question of to what extent sensors must concretize in order to constitute smart infrastructure. Some of these developments are bundled onto existing infrastructure to remake it either as more efficient or more readily maintained. Other developments enable emergent forms of connectivity that form largely invisible networks within everyday operations. But at what point would this array of sensors tip over into smart infrastructure and smart city-ness? Do infrastructures within cities further need to become their own self-regulating organisms in order to qualify as smart?

London is of course not alone in the proliferation of smart cities projects underway, and (as discussed in chapter 7) farther afield there are notable developments in Santander, which has been dubbed the "smartest city" in Europe;[23] in Songdo, Korea, which developed an entire smart urban development from scratch on a green field site;[24] in Rio de Janeiro, where the famous "control room" has by now circulated as a pervasive image of smart cities in operation;[25] in Dublin, where Intel has created multiple sensor "gateways" and a CityWatch platform;[26] and in multiple smaller cities such as Dubuque, Iowa, and Milwaukee, Wisconsin, where IBM has undertaken an approach that consists of meeting with urban planners and community members to identify and document urban issues that might become sites for smart city intervention and technological development.[27]

While Intel, along with many other advocates of smart cities technologies, have variously suggested that cities are at breaking points and these technologies will facilitate efficiency and added capacity, the cities in which sensor technologies and networks are being implemented represent a wide range of urban circumstances, from the admittedly often-overloaded infrastructures of London to the seaside midsize urbanism of Santander to the relatively low-density urbanisms of midwestern locations such as Dubuque and Milwaukee. These cities are not all alike, but their need for sensor and smart technology has been largely narrated through the pervasive urban problematic of cities at a breaking point due to demographic change, climate change, and rapid urbanization. Perhaps it might

Figure 9.2. Siemens Crystal, a smart and sustainable building in the Royal Docks area of east London. Photograph by author.

be necessary to look more closely at what a concentration of sensors entails within a specific urban site in London.

Meeting Desigo at the Crystal

Seeking to gain a more detailed understanding of how these sensor and smart technologies unfold in built environments, I identified one site in London to learn how the specific urban dynamics in this location have coincided with the development and operation of new digital infrastructures. On the north side of the River Thames, in a shiny-spaceship hulk of a building situated in London's Green Enterprise Zone in the Royal Docks, the Siemens Crystal rises as an icon of both sustainable and futuristic urbanism.[28] The development consists of the Crystal building, which opened in autumn 2012 as a model of sustainable and smart architecture, as well as an exhibition on smart urbanism that documents smart cities as they are unfolding and might continue to develop in the future.

Like most of London, the Royal Docks is an area of intensive development and ongoing real estate speculation. The Emirates cable cars that pass overhead at this site were developed for the 2012 London Olympics, and new housing and office developments circle the riverside grounds of the Crystal. Further development plans are proposed for a new Silvertown Tunnel to connect the north and south banks of the Thames, along with an expanded London City Airport nearby and a £1.7 billion Chinese development in this enterprise zone to establish a base for Asian businesses.[29] Smartness inevitably becomes bundled into these ongoing plans and proposals for urban development and expansion.

I am visiting the Crystal on this particular day to take a "technical tour" of the smart and sustainable building that is meant to also be a showpiece and the centerpiece of the exhibition on smart and sustainable cities. One could say that before I even arrived at the Crystal my journey crossed through several levels of smart technology, from planning my journey online with the London TfL search tool providing real-time travel conditions, to receiving live arrival times for the Docklands Light Railway (DLR) at the station, to the Oyster card RFID reader used to access transport, the CCTV cameras under whose watchful gaze I fell, and the automated driverless running of the light rail trains that transported me to the Crystal. As I later learn from the technical tour, before I had even entered the Crystal building, I had also already entered the first of four levels of smart security by approaching the "SiteIQ" bubble that surrounded the site, a field of cameras and motion sensors working through algorithms and Video Motion Detection (VMD) to parse whether I might be a suspicious entity, present at the wrong time of day, or incorrectly placed or shaped. If I had set off any of the pattern criteria, cameras would be informed to track me and alert security guards to attend to my movements.[30]

Having passed through this layer of security, I enter through the automated doors to sign in for my technical tour. In addition to my visitor pass, I am given a smart card for the exhibition so that I may touch in to gain additional information about each display, while leaving a data footprint at every display I visit. The Crystal, I learn from the tour guide, is a building that is mean to "respond to nature," in its crystalline shape. The structure is supplied with a number of sustainable technologies, from a rooftop that harvests rainwater to be funneled to an underground thirty-thousand-liter rainwater tank to supply water for the building; to an extensive array of solar panels on the rooftop that provides a variable amount of energy, depending upon time of year; to the seventeen kilometers of pipes underground used for the ground-source heat pump and the automated ventilation system that responds to internal and external conditions to cool and heat the building. Sustainability and smartness are once again paired in this digital-urban infrastructure.

All of these technologies are watched over and coordinated by a building management system developed by Siemens named Desigo, which monitors eleven thousand points around the building. A sort of uber-cybernetic control system, Desigo controls the ventilation, lighting, heat, and much more by responding to sensor data provided by the Building Automation Solutions (BAS) sensors located throughout the building. As part of the technical tour, I visit a conference room to learn about Desigo. The Desigo system is projected from a laptop, since it is accessible from any device connected into the network (and much of the operational oversight of Desigo actually takes place not in London, but in Frankfurt, where Siemens is headquartered). Scrolling through the Desigo system, the tour guide selects a conference room on the second floor to zoom into. We are able to see the trench heating and chilled beams that are the basis for the heating and cooling of the building, as well as the forty-one sensors that monitor this particular bit of infrastructure to ensure stable temperatures.

From the Desigo panel, it is possible to set points to change the temperature across the microclimates of the building, to turn lights on and off, to see if windows are closed or open and to operate them from the building management system, to investigate motion sensors that may have been triggered, to observe activity levels in office areas, and to keep track of a log of failures. Desigo also logs in real time the amount of electricity generated from the photovoltaic rooftop installation, which on this particular day is producing between 3 to 6 percent of the building's electricity. A weather station site indicates the local external weather conditions. The system senses external and internal factors such as CO_2, light

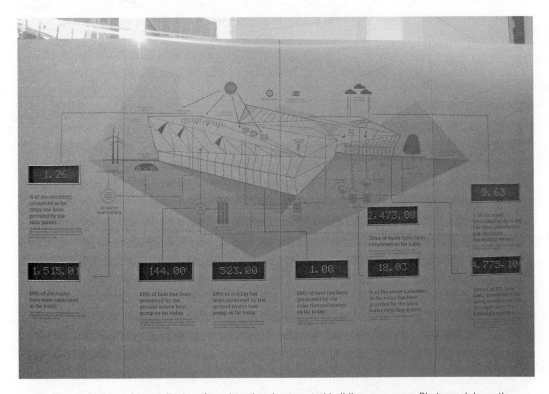

Figure 9.3. Siemens Crystal display of monitored and automated building processes. Photograph by author.

levels, rain intensity and direction, and wind speed and direction, as well as temperature and humidity. Algorithms track sense data to respond to conditions and adjust ventilation in order to maintain comfortable indoor temperatures.

This central automated system is a sort of system of systems, a tool for monitoring and managing the building that largely takes place free from human intervention. While this control system does not need watching over, it can be used remotely—from an app, laptop, or PC—to make adjustments. The building management system largely "takes care of itself," as it senses, adjusts, manages, and maintains a reasonably homeostatic environment. Emerging with this building are orders of smartness, the intelligence of an individual sensor multiplied through a distributed sensor network; and the intelligence of a sensor network multiplied through an automated building system that is self-regulating. Smartness becomes a matter of intelligent automation, of a cybernetic organism that is able to monitor, manage, and adapt to circumstances in real time. This self-regulatory capacity is not just a management tool and ethos for the building, but also for the city, which as the Siemens exhibition would inform me, could benefit from the self-regulating abilities of sensor networks, where the city itself might become a cybernetic organism.

Figure 9.4. *Play the City Game* exhibit, Siemens Crystal. A game for managing the city according to four key indicators for sustainability, which can be altered via a dashboard. Photograph by author.

Play the City Game

After I have completed the technical tour of the building, I turn to spend some time in the Crystal exhibition, which is advertised in the promotional literature as the "world's largest exhibition devoted to urban sustainability."[31] This is a highly technologized and managed version of sustainability, where technological solutions are presented as the key way to help the city balance considerations for the environment, economy, and quality of life.[32] The exhibition begins in the Forces of Change theater, where the "megatrends" of demographic change, climate change, and urbanization are explored as key factors "that determine our future" and are putting pressure on cities, forcing them to adapt in new ways. Here is the recurrent narrative of cities at the breaking point, expressed not just by Siemens, but also by Intel, Cisco, IBM, and a host of other technology companies. Cities are systems, similar to natural ecosystems, the exhibition informs its viewers, and so automated feedback loops are a necessary way to ensure the regulation and management of urban systems.[33]

Following on from this first part of the exhibition to the ground floor, the subsequent part of the exhibition focuses on "creating cities," where importance is placed on understanding the "overall system" in order to manage and address buildings, transport systems, distribution of energy, and water—the vital components in making a city work.[34] A central city icon is located in the center of the exhibition space to signal the importance of this overall view for achieving a "well-designed city." This abstracted urban scene appears in the form of a black tower coursing with an array of colored lights, where the city resembles computer circuitry neatly connected up and pulsing with activity. Here, the city has been remade in electronic form in order to merge with current technological developments.

Planners, the exhibition text and book indicate, are also key to "steering" a city in the right direction.[35] In this part of the exhibition it is then possible to play the game of the future city, and to put oneself "in the shoes of a city manager."[36] This hypothetical city scenario has 3.5 million citizens and $30 billion to administer. In the managing game, the player-planner has forty years and four critical areas to manage: power and water, security, transport, and finance. By adjusting these levers on a dashboard, a status menu indicates how well one is doing with this city management exercise and in addressing urban problems overall. Notably, every urban resource, infrastructure, and problem that is to be addressed must be quantified in order to be made computable. As the booklet accompanying the exhibition notes (citing ex-mayor Bloomberg), "If you can't measure it you can't manage it."[37] Urban life must be enumerated in order to be managed within this cybernetic system. But once measured, the city is meant to emerge as an easily pliable and modifiable system.

Beyond this game of running your own city, there are displays on smart buildings—since buildings consume 40 percent of the world's energy—and the Crystal is an example of how buildings might become more sustainable. There are additional displays on security and crowd control, electricity and the necessity of reconfiguring energy infrastructure to a smart grid, on water and reducing water consumption, on aging and the development of scanning lab tests and genetic profiling to attempt to reduce strains on health care, on measures to improve air quality and reduce climate change–inducing emissions, and on urban transport.

At the end of the exhibition is a Future Life gallery, where New York, London, and Copenhagen are presented circa 2050 as cities where many advanced technologies have been implemented in order to achieve greater sustainability. Future urban scenarios include vertical cities-within-cities that have sprung up to accommodate burgeoning populations, smart grids that respond to fluctuations in renewable energy supply by alerting residents to use energy during a surplus or conserve at peak times, and sensors that make for an aware city—"a seeing, hearing, thinking, feeling, living organism" that is able to respond intelligently to real-time information as it "flows into the city cockpit." All aspects of urban life have been made sustainable, from parks to energy, transport, waterways, and even citizen participation, where "the city responds to the needs of the people." The city also never sleeps, as it efficiently and automatically activates, restocks, recharges, and recycles during the night.

The smart city as built and imagined seems to toggle in this in-between zone, instantiated in some ways but always leaning toward a more complete automation, a more fully self-regulating (and so sustainable) organism that monitors, responds, and adapts in real time in order to achieve the most efficient and optimized balance of resources, time, and money. Sensors becoming networks becoming smart cities all appear to be on a trajectory toward an urban organism that acquires an uncanny intelligence and ability to manage the city as planner, architect, and engineer all rolled into one. But when might this phase-change occur, when sensors and networks take on a life of their own and begin to organize their own automated processes within the city? Smart cities projects seem to suggest that we are all waiting for the moment when the city becomes the ultimate automated organism, where, as Norbert Wiener would suggest, all the sense organs that might constitute individual and assorted automatons participate in and feedback into a larger cybernetic system.[38] Digital infrastructures are part of this accumulation of automaticity, where new structures, organizations, and processes of digital connectivity come together to articulate particular modes of *withness* for subjects (human and nonhuman) and to advance particular ways of "possessing"—and making—worlds.

Leaving the Crystal and its current and future vision of the automated and smart city, I take the Emirates cable car across the river to North Greenwich.

Passing over the waste transfer stations and scrap yards, the bits of real estate in between an industrial past and a still-to-be-fully-instantiated future, I sense a deconcentration of smartness as I surface from the Underground en route to New Cross in South London. Here, far from the sustainable smartness of the Crystal, the connectedness, flow, and seamlessness of the city unravels a bit, where streets are clogged with traffic, housing stock is crumbling, and air pollution is a recurring problem. Among the dilapidated and disused phone booths, convenience shops, and ancient laundromats, I sense that the pacing of this area is less geared toward moving bodies according to some well-tuned and optimized circuitry, since the economies here are of a different and less-privileged sort. How would an automated urban organism deal with this indelible aspect of urban life—that of difference and inequality?

DIGITAL INFRASTRUCTURES OF WITHNESS

As mentioned in the introduction to this chapter, infrastructure is commonly described through resources and what it moves or structures, from energy to water, waste, transport, and communications. When digital infrastructure is described as the "fourth utility," however, it is clear the implementation of urban digital technologies is more than just another piping and cabling job. In this next section, I would like to return to the idea of how an abstract technology or technologies become concrete, and how this process increases indeterminacy. Part of this indeterminacy may actually be understood through the transformations that take place as digital infrastructures take hold and in the numerous entities and relations that proliferate as transindividuations of smart city technologies. These ways of articulating and connecting up entities are constitutive of modalities of withness: of relating, being, and becoming together in the smart city. I now turn to discuss the concretization of digital infrastructures through three modalities of withness—measurement, automatism, and contingency—in order to consider the distinct traversals that digital infrastructures make as they becomes more present in the city. This is a way of saying that we could consider the infrastructural aspects of smart cities not simply as sensors and networks but perhaps also as processes of measurement, automatism, and contingency, which are operations that generate distinct modes of withness. Further to this point, digital infrastructure does not become a matter of hardware and software, physical structure and code, monument and process, but rather of transductive articulations of urban environments, technologies, and inhabitants.

Infrastructure as Measurement

From the TfL network to the Desigo building management system, the sensors proliferating throughout London are assembling into a series of infrastructures engaged in measurement. The rate of vibration, light levels, temperatures inside and outside, air quality levels, and more: the city becomes a set of enumerated data and variables to monitor and manage in real time, as well as to anticipate through predictive modeling. What will traffic density levels likely consist of in two weeks? How will increased temperatures affect air quality? How will energy levels shift with changing temperatures?

While the focus with machine-to-machine communication has largely emphasized the machines or things that are talking to each other, in many ways there are a series of emerging networks and infrastructures that are forming in and through the drive to measure the city in order to manage it. Infrastructure has arguably never been without its modes of calculation, but with digital infrastructures measurement becomes even more infrastructural—it is a condition and resource that enables social processes and organization. Infrastructure in this sense is transversal, characterized not just by a particular resource to be moved and apportioned, whether water or energy, but also by the overlapping, intersecting, and (ideally) interoperable streams of data that allow correlations to be made across previously separate infrastructural milieus.

Or at least, this is the promise. To date, with smart cities as well as the Internet of Things, some commentators note that it is not always clear what is being measured, what should even be taken into account, and how data might be made interoperable.[39] While digital capacities of enumeration folded into infrastructure are meant to be a way of managing urban overload as well as the decay of urban systems,[40] the sensor networks and accompanying data are not yet at the point of delivering on this expectation.[41] Rob van Kranenburg has suggested that "Ambient Intelligence" and the Internet of Things present the problem of deciding which "connectivities we really want as human beings on this planet."[42] Measurement is a primary way in which machine-to-machine and ambient connectivity are unfolding in the smart city and via digital infrastructures. But what does this mode of connectivity involve, exactly? And how does measurement constitute experience (as discussed in chapter 4)?

The measurements that sensors and networks undertake could be described as a process of "taking into account," following Whitehead (and Stengers), where environments are processed through a subject-superject relation. On one level it might be possible to suggest that each individual device is "taking account" by measuring environmental variables, which eventually add up to "big data." If we take seriously Whitehead's invitation to extend subjectness to all entities and not just to humans, then in the smart city we would consider how a location sensor takes account of a moving vehicle, how a light sensor takes account of the sun coming into the windows, how a vibration sensor takes account of the movement of subterranean tunnels, how a sensor network takes account of the multiple streams of data, how an algorithm takes account of distinct patterns in the data, how a program takes account of the conditions of responsiveness in order to implement, actuate, and thereby continue to change and influence the urban environment that is monitored and managed.

But measurement might also signal the ways in which a collective urban-environmental potential has been parsed to be made into data—itemized, quantified, networked, and operationalized—so that individuals and relations are formed through a collective made measurable. This point draws on Simondon's discussion of how relations are not formed through the adding up of individuals to form collectives. Rather, collectives are transindividuated into distinct entities, and it is this mode of parsing collective potential that in-forms individuals and relations. So rather than decide which connectivities are preferred between already enumerated individuals, we might instead attend to the ways in which collectives are turned into measurable entities and individuals, which are further put into relation through infrastructures of measurement.

Infrastructure as Automatism

Infrastructures of measurement are not simply expressions of counting or taking into account, moreover, but are also about engaging in processes whereby that which is enumerated is also capable of becoming *automated*. One of the primary commitments of smart city developments is to automate urban processes so that responses to conditions of overload, situations requiring rerouting, or moments of alarm can be reacted to automatically, through sensors that detect and actuators that respond. Automatism in this way is a key aspect of how Wiener developed his ideas of cybernetics, where "servo-mechanisms" might perform automatic functions in relation to identified triggers. Within these feedback loops, Wiener also integrated humans as sensor-actuators as part of a cybernetic system. The reactions of pilots in warfare, for instance, became one of his areas of focus, where automatism could be extended to a body–machine loop of triggers and actions.[43]

The logic of automatism then points to the ways in which regulation might be achieved, so that urban-infrastructural-human arrangements might unfold more seamlessly, and disruptions might be minimized. In this way, urban systems are managed as cybernetic entities. The smart city implements numerous instances of sensors and humans actuating both other machinic functions, as well as further human actions. Wiener's cybernetic pilot functions through an integrated relationship with a dashboard, where responses emerge in relation to immediate triggers. Cybernetic dashboards are also proliferating in smart cities, where not just the famous control room of Rio but also the more platform-based dashboards that consolidate multiple urban sensor streams become sites for inputting and outputting, for gathering sensor streams in order to respond or actuate in relation to sensor variables. Smart buildings such as the Siemens Crystal has its Desigo control system and dashboard, and London even has several dashboards, including the City Dashboard project, where data from weather stations, tube status, London cycle hire, DEFRA air pollution stations, river levels, Yahoo stock data, traffic cameras, Twitter trends, BBC News, Open Street Map, the electricity grid demand, and even a mood index indicating happiness levels, are amalgamated into both a dashboard and map view.[44] The data made available here are collected from sensor and web feeds, which provide an apparent overview of the city while at the same time indicating a whole range of automated urban processes underway or that might be further managed.

The dashboard-as-platform also demonstrates how the city is becoming a platform. While Internet platforms for social interaction are one aspect of smart city initiatives, platforms also unfold and are distributed across urban space. The city as platform has been a topic of discussion in recent seminars and events sponsored by Microsoft, IBM, and others. Platforms, in these scenarios, are not just Internet-spaces but are also embedded, situational, context-focused applications that map new digital functionalities onto urban infrastructures, processes, and exchanges. Platforms are both localized and distributed in the city and across the Internet. As another layer of infrastructure that enhances the efficiency and timing of cities, digital connectivity and platforms present the possibility of a well-regulated city that becomes sustainable through the enhanced synchronicity and expediency of urban systems. New platforms and connectivities arise to facilitate citizen involvement and monitoring of these processes. Platforms further function as monitoring devices, to be used together with everyday urban practices.

Automatism cuts across these different types and distributions of infrastructure, and a city dashboard or platform signals the many automated urban functions underway. Capturing the potential and processual aspects of infrastructure, the architect Keller Easterling has suggested, "Designing infrastructure is designing action."[45] This statement could easily be read in a deterministic way, where the structures of infrastructure are seen to offer automatic scripts or codes for action. But the statement could also be read less causally and more simultaneously, where infrastructures and actions coincide as entangled and co-emergent processes. A study of infrastructure could very well attend to the actions that are productive of infrastructures, as well as infrastructures that are productive of actions. This raises the question of how actions unfold within and through automated urban infrastructures. Various smart city developments put forward scenarios where the movements, timings, and circulations of bodies and systems will sync up more easily, where crowd control or shopping may be streamlined through apps and code and CCTV, where the city will operate as a real-time organism, where the witness of bodies in cities will be advanced through the seamlessness that automatism enables.

Simondon suggests, however, that automatism is actually quite a low-level and limited way of engaging with and thinking about technology. As someone relatively critical of cybernetics and the master-slave relationship it tends to prop-agate in relation to technology, Simondon suggests that technology is far more inventive when it is open and indeterminate. Limiting technology to "utensil" status is a way to ensure that the technical is also separated from cultural con-cerns, which might otherwise ensure a wider set of engagements with technol-ogy.[46] As Simondon writes, "In order to make a machine automatic, it is necessary to sacrifice many of its functional possibilities and many of its possible uses." Instead, he writes:

> the real perfecting of machines, which we can say raises the level of technicality, has nothing to do with an increase in automatism but, on the contrary, relates to the fact that the functioning of the machine conceals a certain margin of indeter-mination. It is such a margin that allows for the machine's sensitivity to outside information. It is this sensitivity to information on the part of machines, much more than any increase in automatism that makes possible a technical ensemble. A purely automatic machine completely closed in on itself in a predetermined operation could only give summary results. The machine with superior techni-cality is an open machine, and the ensemble of open machines assumes man as permanent organizer and as a living interpreter of the inter-relationships of machines.[47]

Simondon imagines a more engaging set of operations for machines that requires openness, and in particular an openness to human involvement. One could argue that with machine-learning, the participation of humans with machines need not be described in this way—that machines could be open to the unfolding processes of other machines. Or one could argue that Simondon brings us back to a roman-tic version of humanism, where humans in the loop present the ways in which machinic closure may be overcome. But as Simondon's writing indicates through his discussion of the preindividual reserve, what counts as "human" is also not fixed or settled, since machinic engagements also give rise to distinct transindi-viduations of the entities involved.

In a rather different vein, Wiener develops a trajectory of different versions of automata, from clocks to opening doors, and from photocells to computers, which as servo-mechanisms have operated as sense organs that, when coupled to the outside world, translate information into a series of actions.[48] For Wiener, automation occurs across the sensor-actuator exchange. Yet as Simondon's ap-proach to technology suggests, this relationship of automatism would necessarily have to be exploded—and perhaps even von Neumann's computer architecture diagram would need to be redrawn—in order to gather the "responsible and inven-tive" input of (transindividuating) humans into these circuits.[49] Technicity—and by extension, computation—is not always a closed loop of action and reaction, and may just as often open into new orders of indeterminacy, especially as it becomes environmental.

A smart city is meant to join up structures, processes, mechanisms, and gadgets, so that they function more efficiently as some well-timed machine. Infrastructures emerge as a set of relations, but these are not necessarily primarily human relations. Humans are typically flattened into this relationality, not as exceptional actors or beings, but as one more parcel to move along, one more node of information, one more sensor-actuator operation taking place in the city. In this galaxy of machines talking to machines, contra the city game and steering planner of the Crystal exhibition, it is not a human that would cybernetically govern the smart city, but rather an automated system of control where a human operator is one source of input, since many of the protocols and terms of relation are written through the ways in which machines talk (and must talk to) to other machines in order to be interoperable.

Measurement and automatism as modes of connectivity signal the ways in which infrastructures are then necessarily discussed as relations. Writing on infrastructure and the relations it sets in play, urban sociologist AbdouMaliq Simone suggests that the characterization of infrastructure as "in-between" is provocative for considering how cities come together and hold together.[50] We might also read Simone through another Simondonian register, to consider how this in-betweeness might be understood less as the glue between individuals and more as a parsing of collective potential. In-betweeness is about the relationality that occurs through infrastructure. As Simondon has suggested, not only does relationality not precede the act of relating but the terms or individuals related are similarly not fixed in advance but come about through transindividuating processes that give rise to singular and collective human and nonhuman individuals, as well as the conditions of relationality. With this inverted sense of how relation might be seen to materialize, what other sorts of relations might concresce through the digital infrastructure of sensorized and smart cities?

Infrastructure as Contingency

Simondon suggests that as technology moves from abstraction to concretization it becomes more indeterminate and that this process is not linear. As technology concretizes, relations are differently manifested and materialized in ways that might be described as contingent: contingent upon the environments, entities, and technologies that transindividuate together. For instance, contra those studies that would see code as a discursive program to be implemented in urban sites, a Simondonian approach to concretization would instead ask how code transindividuates along with the entities, environments, and relations in and through which code is meant to take hold. Concretization is not the rational implementation of a plan of technology but rather an actual occasion of withness, a particular possession of the world that is less scripted and more generative—potentially even of new types of indeterminacy.

Contingency then is a key way in which the concretization of digital infrastructure can be understood. Infrastructural arrangements might be differently constituted depending upon how sensors concresce into organizing structures. Along with infrastructure, a corpuscular society of sensing entities also concresces. Infrastructure is a condition that enables inhabitations, modes of withness, and worlds to be sustained as modes of being and becoming. But encountering infrastructure as concretization and concrescence is not, arguably, synonymous with infrastructure as assemblage, as previously discussed throughout *Program Earth*. Whereas assemblages might emphasize the adding up of entities, a list of everything that comes together to make infrastructure materialize and operate, a concretization and concrescence of infrastructure places the emphasis differently on the processes and individuations that parse and connect up entities and environments in particular ways, such that the capacities of entities might not even be known in advance of their connecting and relating. Such an approach inevitably emphasizes the contingency and indeterminacy that characterizes such technologies as digital or sensor-based infrastructure.

Perhaps illustrating this point with the most sustained clarity and richness of example is the work of Simone, who focuses on (other-than-smart) cities in the Global South, including Jakarta and Mogadishu. Simone suggests another way of encountering infrastructure as it emerges and is articulated in more contingent and itinerant ways—not as fixed structures, but as provisional, always in process and participatory relations. Simone extends infrastructure to include "people's activities in the city" since, as he writes, "African cities are characterized by incessantly flexible, mobile, and provisional intersections of residents that operate without clearly delineated notions of how the city is to be inhabited and used."[51] By extending infrastructure and its provisionality to people, and people living in particular African cities, Simone draws attention to the distinct if changeable and contingent infrastructures that come together in particular places.

The changeability of infrastructure is something that Mackenzie also captures in his discussion of wireless technology by attending to how "wirelessness" changes with each instantiation of wireless technology.[52] At the same time, the "devices that comprise infrastructure keep changing."[53] Rather than infrastructure necessarily constituting a seamless set of connections, Mackenzie suggests that a whole set of lived and felt transitions take place across infrastructure that are different from the assumed fixed experiences of infrastructure. Instead, wireless infrastructure can be piecemeal, in process, and in need of constant upkeep and repair. While there may a constant hope for infrastructures that are self-administering, the reality is that maintenance is a continual condition through which infrastructure is encountered and experienced.[54] Or, as Graham writes, infrastructure is less stable and more made up of "leaky, partial and heterogeneous entities."[55]

However, in the context of smart cities, it is not uncommon for technical arrangements to be presented as managing diverse urban circumstances toward seamless sets of relations that would be relatively complete and universal across urban contexts. Smart cities developments assume the same modes of withness, where citizens, cities, and technologies, despite their diversity, regularly intersect through a continuous program of sensor-activated cities. Yet urban life is no doubt not the ceaseless undertaking of universal sensor-actuator tasks to be completed, with Tetris-like packages to be dropped in Tetris-like spaces. In the computational architecture of command-and-control automatism writ large over urban environments, there are as many accidents, disruptions, and breakages as there are seamless connections made. People continue to wander and not simply move from home to office to store and back again. Not all urban citizens are hyper-productive economic subjects. Not all urban spaces operate as one more bit of computer circuitry.

If we were to return to Simondon's notion of how the concretization of technology may give rise to indetermination, we could then say that the smart city in some ways commits the error of making the abstraction of infrastructure real without accident or indetermination. The universal visions of smart cities typically assume infrastructures are always the same in their striving for optimization. Even imaginaries for participatory digital urbanism, as discussed in chapter 8, do not typically allow for indeterminacy. These schemes might admit some contingencies in the form of local circumstance, but they move toward the same end point of managing and regulating cities in order to achieve efficiencies and solutions. Yet as Simone reminds us in his writings on the practices of infrastructure, these are characterized by situated contingencies, where people may even tinker with and alter the city and its infrastructure.[56] Contingencies in the smart city may emerge across human and more-than-human registers, moreover, since as the city "plays itself" it no doubt is not simply adapting for optimization but is also generating particular materializations of sensor-spaces, transforming environments through programs of more-than-automatism, and giving rise to proliferating bugs and blockages that are sites for ongoing repair.

CONSTRUCTING A SPECULATIVE CITY

The processes whereby digital technologies become environmental are, on one level, a matter of how technologies are distributed in and among surroundings. But, on another level, and following Simondon, this process does not involve a static arrangement of machines but rather it in-forms the very environments in and through which technologies are distributed such that new environments are made. Weiser's early discussion of ubiquitous computing proposes not just that an array of "tabs" and "badges" and laptops would proliferate but also that these technologies would "disappear" and "weave themselves into the fabric of every-day life until they are indistinguishable from it."[57] The ways in which ubiquitous computing would become environmental here consist of integrating with sur-roundings such that technologies do not require active attention. Rather than seeking to make computing a willful cognitive engagement, Weiser found the success of these computing devices to be exactly their imperceptible integration into experience, where they became infrastructural.[58]

We could say that these infrastructural sensor technologies effectively become the environment with which we would interact, along with the subjects, objects, and the milieus they constitute. As discussed in the introduction to this study, Weiser sought to push computing beyond its presence as a self-contained box—where it merely constituted its own self-enclosed world—and in the process to find ways for it to enhance the world already in existence.[59] But as Weiser's own text intimates, ubiquitous computing would not simply consist of a set of enhance-ments, but rather would transindividuate worlds and environments in new ways, thereby creating other versions of the real. As one example of ubiquitous com-puting elaborated by Weiser indicates:

> In our experimental embodied virtuality, doors open only to the right badge wearer, rooms greet people by name, telephone calls can be automatically for-warded to wherever the recipient may be, receptionists actually know where people are, computer terminals retrieve the preferences of whoever is sitting at them, and appointment diaries write themselves.[60]

Here is a concatenation of actions that would arrive at more seamless actions in the workplace. It reads as a list of new modalities of ubiquitous computing and bears an uncanny resemblance to the Internet of Things scenarios developed in smart cities proposals and applications. Yet as Weiser later indicates, "Neither an explication of the principles of ubiquitous computing nor a list of the technolo-gies involved really gives a sense of what it would be like to live in a world full of invisible widgets."[61] New worlds come into being—digital technologies become environmental, and in turn, new environments emerge as milieus distinctly con-ditioned and transformed through these devices.

Processes whereby technology become environmental and give rise to dis-tinct milieus are then less mediatory and more in-forming and ontogenetic. As mentioned in the beginning of this chapter, making a list of sensor-based digital infrastructure does not necessarily address the distinct ways in which these tech-nologies concretize in and with new environments and environmental conditions. As Weiser's above example of the well-running office indicates, the constitution of environments also involves the constitution of the field of subject-action-event. Digital technologies play a role in making these relations coalesce and concresce in particular ways. Here, digital infrastructures produce medial relationships that are world-making and world-sustaining, rather than mediating across preestablished entities. In these transindividuating infrastructural relations, the conditions and potential for collectives to form and interact arise.[62]

I have been drawing together Simondon's discussion of concretization alongside Whitehead's notion of concrescence in order to return to a consideration of the particular "techno-geographical"[63] milieus that come together along with smart cities technologies. These concretizations and concrescences are also a way of discussing withness, of thinking about how entities become together. In his study of wireless technology, Mackenzie mobilizes a discussion of conjunctive relations to consider "with" as one type of the many ways in which relations unfold. Drawing on James and Nancy, Mackenzie notes that extending the scope of conjunctive relations is a way to think beyond or against merely conceiving of technology through utility or means.[64] It is interesting to note that, for James, "withness" is a lower-level relation, where "the lowest grade of universe would be a world of mere *withness*, of which the parts were only strung together by the conjunction 'and.'"[65] Yet for Whitehead, withness becomes a way to discuss the ways in which worlds are possessed.[66] If we were to take up Simondon on this point, we would also be obliged to note that relations are not able to precede the moments and entities of relating. Withness even constitutes the subjects that would be resident-actors in these worlds.

In this sense, I mobilize withness as a particular way of bracketing off assumptions about what it means for participation to unfold in the smart city. I see withness more as an articulation of processes of participation that involve becoming together, across an extended array of entities, and setting in motion the connections and inheritances that take hold to become something like urban infrastructure, whether solidified as communication conduits or distributed as seemingly immaterial exchanges of practices. In order for infrastructure to exist and persist, it must draw entities into active participation with it. The city is a complex corpuscular society of inheritances that in-form bodies of sense. Withness captures the lived embodiments of participatory experience, while also going further to indicate the ways in which worlds are made and sustained through particular ways of "possessing" the world. In other words, withness describes ways of making worlds.[67] Digital infrastructure is a multiple and more-than-human event, which may be further described as a process whereby multiple actual entities become tuned in to shared registers of withness, along with shared practices for making worlds. In fact, this is how Whitehead describes a community, where multiple entities are effectively resonating within and experiencing a shared register of world-making.[68]

But this is by no means a way in which to render neutral the worlds that are formed. Whitehead for his part recognized the power and force, in the form of persuasion that enabled certain worlds to take hold and not others.[69] Stengers takes up this discussion of power and persuasion within Whitehead to draw parallels to Foucault's discussion of power. Here, it is useful to revisit the analyses made in chapter 7 to bring together both a Foucauldian and Whiteheadian analysis of speculation and power vis-à-vis analyses of infection and persuasion drawn out by Stengers in *Thinking with Whitehead*.[70] As Foucault has noted, the point in discussing power is less to make an effort to outpace or escape its reach but rather to understand its distributions in order to attend to particular formations it enables—or disallows.[71] Stengers takes up Foucault's discussion of power in relation to Whitehead to consider how not just persuasion but also infection and the gaining of a foothold indicate how power is distributed in and through environments and how environments and entities (including facts) are able to persist.

Participation as a concept and practice frequently involves discussions if not disagreements about forms of *agency:* about people becoming more empowered by 2.0 social media or being duped into contributing free labor to exploitative digital economies, about the more-than-human actors and devices that are also participating along with people in digital media exchanges, and about the ways in which agencies of these devices shape the contours of participation. Rather than focus on agency as a key modality of participation, however, I have redirected this consideration of participatory urbanism toward different registers of withness, in the form of persuasion and infection. I have drawn on and adapted these terms from Whitehead and Stengers, who attend to the ways in which entities concresce, a process that arguably has less to do with agency *per se* and more to do with ways in which entities become together and are drawn into processes of mutual—if differing and differently manifested—influence. Influence—persuasion and infection—are decidedly different from agency because they do not involve a subject / object *acting on* another—willfully or otherwise—but rather have more to do with transferences (and Simondonian transductions) that occur across not always clearly delineated cause-and-effect or subject-and-object sites or encounters. This means that rather than search between determinist or constructivist approaches to technology, we might attend to how new entities, relations, and modalities of withness contingent on infection and persuasion generate new technological practices, inhabitations, and ways of life.

Smart cities developments such as those underway and implemented in London raise the question of how modalities of withness are realized, and how different registers of withness concretize along with distinct approaches to "smart" technology. "Our" urban future is differently distributed depending upon how close to the machine "we" are.[72] Those that can speak to it, in its language, stand a better chance of counting and being taken into account as a relevant node in its networks. Those who do not may find they cannot get a foothold in the world the smart city has made and possessed. Possessing a world is not dissimilar to individuating environments and entities. Rather than adding up relations or making lists of technologies, possessions and transindividuations signal processes of taking into account, distributing power, exercising persuasion, in-forming experience, participating, taking hold, and so securing, even if momentarily, an inhabitation, an environment, a mode of becoming.

Withness has been the speculative inhabitation that I have worked with in this chapter, while examining how modes of becoming together are articulated within smart city developments. The lived practices and politics of smart cities, as they are and as they might be, and when addressed through withness, recast questions about what constitutes the environment of the smart city and rework the practices and entities of participation that unfold there.

Withness also points to the ways in which speculative cities are being constructed continually. As Stengers suggests, speculation as a philosophical proposition requires a "'leap of imagination'"[73] that is not as much a matter of projecting abstract ideas of future possibilities as it is a question of how to precipitate and make possible distinct ways of life. In this sense, speculation becomes entangled with persuasion and power, generating a politics of speculation that manifests in particular ways in smart cities projects. It may be that speculation is in fact made possible through distinct modalities of withness—this is not something undertaken alone—and so we might be particularly attentive to the modes of withness that are propagated and sustained in smart cities and sensor-based urbanisms. This might involve querying the programs of withness and participation that digital infrastructures help to sustain.

This chapter ends by suggesting that withness may be generative of experiments. Withness asks how are we thinking with, being with, and becoming with the smart city. It strikes me that many smart city and participatory urbanism projects are missing exactly this more experimental approach to speculation. Rather than open technology to a multiple array of inhabitations, encounters, and modes of withness, these projects most often reduce technology to a "utensil," as Simondon has termed it, or as a project of utility and optimization, rather than of equality, actual quality of life, or even wonder. The conclusion takes up some of these provocations for how to experiment with and speculate toward more inventive environmental inhabitations—computational and otherwise.

Figure C.1. *Forces of Change* exhibit, Siemens Crystal. A video documenting key planetary stressors to be monitored and managed. Photograph by author.

Planetary Computerization, Revisited

In traversing the areas of wild sensing, pollution sensing, and urban sensing, I have sought to build up an account of the environments, entities, and relations that concresce within and through computational sensors. The environmental-sensing projects covered here are situated within a wider array of sensing activities that span from monitoring tides to airflow, as well as observing vegetation and soil conditions. Moving across experimental forests and webcams, migrating animals and climate change, garbage patches and urban air pollution, smart cities, idiotic participation, and automated digital infrastructures, I have discussed how these versions of programmable earths are not singular entities but rather involve the in-forming and multiple activations of distinct techno-geographical environments.

The *becoming environmental of computation* has been the central concept that I have developed to draw out the multiple ways in which earths and environments are programmed and the new entities and experiences that materialize through these processes. But this concept has not been put forward as a way to spatially locate sensor-based media nor to argue for the revolutionary environmental engagements that computational sensors seem to offer. Instead, I have attended to the ways in which sensors become environmental through exchanges and individuations of energy, materialities, subjects of experience, relations, and milieus. Environments are an active part of how actual entities come to concresce, how organisms gain a foothold and endure, and how values are articulated through the unfolding of technical objects. Returning to the ideas put forward in the introduction to this study, we find that programmable earths generate and materialize very particular computable entities, relations, and environments. This is not a simple project of turning the earth into an object or artifact of study, however, but rather a multiply realized set of practices for identifying environments and processes that can be parsed as sensor data, gathered together into analyzable data sets, and operationalized into new formations of environmental engagement and experience.

This could, of course, have been a cautionary tale of how the earth is becoming a highly instrumented control space, where every activity and environment is under observation. While the capacities of environmental sensors to enable an even further degree of surveillance are no doubt formidable, I have sought to develop a slightly different line of inquiry by attending to the ways in which these computational sensor technologies become environmental to create new subjects of experience. The subjects that would be monitored and would undertake monitoring are not static entities but instead are connected up with generative processes for experiencing environments and other entities in distinct ways. This phenomenon further indicates that how environments are felt and acted upon, whether for political or interventionary purposes, is closely tied to the technologies and subjects that prehend or feel environments and have concern for environments.

These environmental–computational operations play out differently across the situations that I have discussed throughout *Program Earth*. Wild sensing draws attention to the ways in which we might remotely monitor environmental processes seemingly in absence of human intervention and toward a greater understanding of the real-time ecological relations that unfold. Yet this section also emphasizes that sensing practices entangle milieus, problems, modes of sensing, and matters of concern with particular formations of environmental study, citizen sensing, and engagement. Pollution sensing draws out the often proxy modes of sensing that organisms and societies of objects express and the entities and relations that proliferate within changing milieus and environments. It considers the new creatures and practices of sensing that concresce through the monitoring and collection of pollution data. And urban sensing attends to speculative and implemented smart city scenarios and technologies to consider how sensing operations attempt to make urban spaces and urban citizens into more manageable, efficient, and responsive entities, as well as the ways in which these programs do not always go according to plan.

Sensing, as each of these sections and chapters indicates, is a practice and technological relation that does not simply detect external stimuli to be processed and turned into manageable content. The program earths and programmed participation that concresce here are less linear and substantialist than this. Instead, these programmed environments draw attention to the tunings and attunings that occur through computational and environmental operations. Ways of feeling and accounting for environments and environmental relations are activated or otherwise delimited through computational sensors, and it is these ways of feeling and the practices and subjects that they sustain that I have sought to make more evident. Environments are not then simply a map of a territory but are a field of resonance and relation that can be drawn into and materialized in the experiences of subjects, whether those subjects are citizen-sensors, soil sensors, moss cams, migrating storks, marine debris in sensor-mapped ocean currents, air pollution–sensing devices, smart buildings, digital infrastructure, or any of the many other actual computational concretizations of environmental-sensing subject-superjects.

Planetary computerization is unfolding apace. As Guattari has suggested (and as was also discussed in the introduction), such computerization might even provide an opportunity for polyphonic engagements across machines, subjects, temporalities, and materialities. Yet while such planetary and computational concrescences are underway, it remains an open question as to whether such computerization might activate the expansive new subjectivities, "creative enchantments," and imaginaries that Guattari had hoped for.[1] As Guattari writes in characteristically profuse form,

> The question that returns here in a haunting fashion is to know why the immense processual potentialities carried by all these computational, telematic, robotic, bureaucratic, biotechnological revolutions so far still only result in a reinforcement of previous systems of alienation, an oppressive mass-mediatization, infantilizing consensual politics. What will enable them finally to lead to a postmedia era, setting them free from segregationalist capitalist values and giving a full lease of life to the beginnings of a revolution in intelligence, sensibility and creation?[2]

Guattari invested a certain speculative energy into the capacities of new technologies. Yet if the current developments related to environmental sensors and the wider landscape of the Internet of Things are anything to go by, then the "reinforcement of previous systems" seems to be the more likely scenario than a transformation in intelligence, sensibility, or creation. Energy meters and security cameras are pervasive, if less than enchanting, technologies in the ever-connected, always-on, searchable and brand-able big-data predictive worlds in the making. In some ways, this is a logical continuation of the seemingly helpful ubiquitous computing world that Weiser imagined:

> When almost every object either contains a computer or can have a tab attached to it, attaining information will be trivial: "Who made that dress? Are there any more in the store? What was the name of the designer of that suit I liked last week?" The computing environment knows the suit you looked at for a long time last week because it knows both of your locations, and it can retroactively find the designer's name even though that information did not interest you at the time.[3]

In a contemporary context, where these once-imaginary computing scenarios have now become standard procedures, such engagements sound potentially less enchanting and more in service of tracking a customer-citizen whose attention is perpetually tuned to consumption.

In this respect, it is also useful to return to Simondon's critique of automation (and by extension, first-order cybernetics) as discussed in the last chapter, where he suggests that automation is quite a limited and uninteresting way to mobilize technology. Instead, Simondon suggests that if we attend to the values articulated and transduced through technology, we might be able to identify opportunities for active experimentation within these cultural and machinic registers as expressions of relationality—and even, potentially, equality and invention. Open technology was a focus of Simondon's in this regard, not in the usual sense of openness that circulates today in relation to "free" hardware and software, but rather an openness that consisted of experimentation with the entities, relations, and collectives that might be individuated through technological engagements.

Such an approach also points toward considerations of how not to fix technology into conditions of inequality but instead to open it into encounters that are more-than-technical encounters with technocultural creativity and response-ability. This would require going beyond engagements with technology as a utensil or instrument, as mentioned in the previous chapter.[4] It would also require going beyond engagements with environments as computational problems to be solved. Instead of remaining firmly embedded within the utensil-problem space, where environmental sensors facilitate increasingly regimented and automated ways of life, another approach might be to consider how environmental computational practices open into experimentation, expanded experiences, and speculative adventures.

PROGRAMMING EARTHS, EXPERIMENTING WORLDS

The problem of automation is closely tied to the processes and imaginaries of programming. Programming, as this work has suggested, is not simply an automatic code or script foisted on the world through a set of discursive enactments. The usual sense of the programmatic often refers exactly to those scripted and less-than-inventive engagements. And so there proliferate antiprograms and counterprograms, seemingly as a way to break out of programs of control. In relation to a discussion of "programmatic statements," Massumi has suggested that breaking with such statements is a way to generate encounters with politics.[5] Programmatic statements might at first appear to be expressions of politics—of a "correct" position to be taken in relation to a certain problematic. But these can actually reinforce deadening certainties that prevent a more engaged and engaging encounter with matters of concern. The point of politics is to go beyond programmatic statements in order to open our inhabitations to more speculative encounters. It is through a break with programmability—both as an automatic code and as a correct position—vis-à-vis experimentation that political inventiveness might unfold.

Programmability, as I have suggested throughout this study, is also a somewhat fraught process, verging on the enactment of scripts for certain outcomes but giving rise to indeterminate environments, entities, and relations. Programmability in the context of program earths is continually unfolding the complex if at times troubling attempts to make environments into spaces of observation, distributed experience, and even automated management. Programming might then be addressed as a technical process that is involved in individuating entities, relations, and milieus. These processes are not a mere list of entities, but rather describe the *collective* articulation of potential, as well as all that is more-than-human in human engagements with technologies.

Participation then becomes an important topic of consideration when attending to the operations of environmental sensors. The citizen-sensing practices that I have discussed throughout this study have demonstrated not just how citizenship, participation, and practices of engaging with environmental problems are bound up with and individuated through environmental sensors, but also how the "citizen" in citizen sensing is an entity that shifts and expands to include other practices constitutive of citizenship, and which might include more ecological ways of accounting for how citizenly engagements unfold. In this way, I have not sought to reinscribe the somewhat pervasive notion that citizen-sensing technologies are liberatory tools for achieving more democratic engagement, or that a heightened ability to work with computational sensors might convey an increased amount of political capability to those few skilled citizens.

Instead, what I am more inclined to consider is that, in the current circumstances, increasingly one is either in the network or out.[6] While Internet dark rooms, off-grid spaces, and antique (read: nonsmart) appliances might be one strategy within an "only dis/connect" approach to the modes of planetary computerization that are now unfolding, such strategies do not realize another space *within* networks so much as opt out of these computational relations altogether. In relation to milieus, both Simondon and Foucault articulate, in a transformed way, Canguilhem's notion that to be simply *in opposition* is to be already defeated. Living entities must extend themselves and make something of their milieus.[7] In this way, Foucault attended to power not by making simple exhortations to resist its influence but rather by noting that power is a force that runs through ways of life, and that it was necessary to reroute power where possible to invent more creative and engaging ways of life. Simondon's point was that a more thoroughgoing engagement with technical objects might materialize through attempts to experiment with other relations with technology, rather than to opt out of these engagements altogether. Technology is not simply to be dismissed as other-than-human,

Simondon has also argued, but rather is part of the collective potential that makes us human in particular ways as it individuates entities and our possibilities of relating. This in turn also has consequences for the extensions and exchanges made with environments and more-than-human organisms.

In drawing out this discussion of how sensor technologies transindividuate environments and entities, individuals, and collectives, I have attempted to build a context in which the *power* of speculation might also become more apparent. This power unfolds through the ways in which propositions are made: What are the computational environments we would inhabit, how do they lure us into becoming together, and what are the processes and practices that give rise to these concrescences? This power also takes hold through the ability to undertake persuasive practices, to infect and in-form the conditions and entities that might activate speculative engagements. The power of speculation might also be articulated across the moments and sites of concretization, where abstract technology takes root, becomes *more indeterminate,* and so potentially is generative of multiple speculations.

In this respect, speculation is as much about propositioning, instigating, and triggering—beyond the usual automated sensor-actuator triggers of cybernetics—toward indeterminacy and openness. As Whitehead has suggested, a speculative project is most interesting when it is involved in "bringing adventures into existence." Rather than demystifying, mapping out, and nailing down everything in a grand gesture of rationalism,[8] adventure seeks to create conditions for both hope and change.[9] A speculative adventure is then an approach invested in experimenting worlds. Less a matter of polemics, adventure and speculation are about making particular things matter—of generating environments and entities that are able to take hold in particular ways because they have exercised power through persuasion (and experience).[10]

Speculative approaches to research and practice are emerging across multiple fields as a way to develop not simply descriptive engagements with topics but also propositions that invent new possibilities for research and practice. One mode of experimentality developed throughout *Program Earth* has involved encountering environmental sensing as a series of propositions for considering how the becoming environmental of computation offers up opportunities for creative and practical as much as analytical engagements.

Citizen-sensing projects often attempt to find ways to broaden the scope of observational practices beyond the sciences exclusively and to make ecological observation more accessible and engaging to a diverse range of participants. Such citizen-sensing projects intend to democratize the collection and use of environmental sensor data in order to increase citizen engagement in environmental issues. The process of gathering and making these observations more participatory is often a way to overcome the relative crisis of environmental engagement in political and cultural spheres: by making environmental change more evident and distributed across sensing subjects, environmental action may also be facilitated. But these practices also tune in subjects, environments, technologies, and multiple other entities into shared registers of sense making.

Program Earth has considered how or whether environmental-sensing and citizen-sensing practices enable expanded engagement not just with the "message" of environmental change but also with understanding environmental concerns at a more intimate level. Through discussing the rise and proliferation of environmental sensing practices within scientific, creative, citizen, and urban applications, I have analyzed the distinct practices of environmental politics that concresce in relation to these technologies, where citizens make use of technoscientific devices at times to reroute the usual spaces of environmental engagement and expertise.

An obvious observation to make would be that citizenship is performed through these sensing technologies. But how does this mode of citizenly practice square with the stabilizations of citizenship to which these practices might refer and/or resonate (or dis-sonate) within democratic contexts? A citizen is not newly emergent with every use of sensors, but these technologies are involved in re-articulating and recasting the materiality, spaces, practices, collectivities, infrastructures, imaginings, abstractions, processes—in other words—concrescences of citizenship. From a perspective informed by Whitehead and Simondon, we could say that citizenship is historically immanent and processual, as well as a site of dynamically articulated collective potential. We could also say that a citizen is not an exclusively human-based subject of experience but inevitably is also part of an extended ecology of attachments.

Citizen sensing, as it is typically conceived, is often positioned as somewhat continuous with citizen science, an activity that might be augmenting science but through more digitally enabled devices. And yet, citizen sensing might also be an expanded way of practicing environmental sensing in relation to pedagogical and political aims.[11] The development of these alternative environmental monitoring practices might, on one level, focus on technologies for engaging with environments, where alternative monitoring practices might be a way to question official versions of events, while also developing tools for continually engaging with environmental issues. Yet, on another level, as practices situated within and generative of milieus, citizen-sensing practices are also inevitably techno-geographical as well as speculative. If more thoroughly engaged with the environmental concerns they would monitor, they might even generate new subjects and politics of experience. Citizen-sensing practices might then not necessarily be data-collection exercises but rather ways of making particular environments and environmental concerns matter and gain a foothold. In this way, citizen sensing might move beyond the unquestioned if rather problematic equation of data-equals-action to engage with the extended milieus, moments of resonance, and points of collective potential that might be forceful sites of investigation and practice.

In discussing environmental sensing and citizen-sensing practices, I have attended to the experimental aspects of these engagements in thinking about the distinct human and more-than-human distributions of experience that are brought together and which take on consistency. However, rather than talk *about* these practices through a discursive analysis that treats them as an object of study, I have considered how an approach attuned to experimentation also seeks to mobilize and invent abstractions along with concrescing practices, materialities, and environments.[12] In other words, I have attempted to write toward the edge of the page, in-between the spaces of thinking and practicing, such that these are not seen as oppositional modalities but rather as continuous in their commitment to experimenting with ways of encountering environments and environmental matters of concern with creative, critical, and political a/effect.

This research finally suggests that new approaches to computation might be developed where digital devices expand beyond automated and user-controlled applications toward more speculative and "open" engagements. One question to bring to any environmental sensing project might then be: How does it give rise to speculative adventures, or otherwise prevent experimentations with worlds? The becoming environmental of computation might be as much a proposition as a fact. Sensorized environments are propositions for particular types of worlds to take hold and for distinct subjects of experience to be sustained. Sensorized environments and citizen-sensing practices put in motion specific ways of feeling worlds and of making particular problems matter. In this sense, citizen-sensing practices materialize not as easy fixes to making environmental engagement more democratic but rather as particular *expressions* of environmental problems, politics, and citizenship. Given that these projects are so actively making worlds, they are also sites that might then be encountered through further speculation.

Simondon might have opted to use the term "invention" over and above "speculation," and at the same time he investigated how processes of transindividuation always left a preindividual reserve that was a site for further potentialities.[13] And yet, invention is also a topic and practice that runs through Stengers and Whitehead. Stengers discusses how speculation involves attending to the "invention of the field in which the problem finds its solution."[14] There is an adventure that is undertaken in formulating questions, and concepts can transform "the way in which a situation raises a problem."[15] A speculative approach can transform questions and the character of experience—it is a conceptual approach that is charged with its ability to shake up and recast the usual approaches to problems.[16]

I end this discussion by considering the theoretical influence of Whitehead and Stengers, who, in a related way, discuss how it might be possible to be *for a world,* and not simply of the world.[17] Our inhabitations are ways of experiencing and, in turn, making worlds. The etho-ecological relations that we articulate along with multiple others require certain environments to take hold and endure in order for those relations and experiences to be sustained. This is a process of persuasion, infection, and power. It is also a process of experimentation and ethics. The milieus that are put into play through environmental sensors are ecologies of amplification—as they connect, they intensify.[18] As computational sensors increasingly take hold and concresce entities, relations, and environments in distinct ways, the question of what sorts of worlds—or program earths—we are involved in sustaining comes into play. While this is not a simple proposal to adopt one particular relation to these technologies, this study suggests that the world-making operations and experiences of sensors might become an area of more intensive—if openly indeterminate—engagement. *Program Earth* is one incomplete and speculative attempt at making an opening into this space.

NOTES

INTRODUCTION

1. McLuhan, "At the Moment of Sputnik."

2. There is an extensive literature across multiple disciplines on "whole earth" imagery. For example, see Cosgrove, "Contested Global Visions"; Jasanoff, "Image and Imagination"; Poole, *Earthrise;* Diederichsen and Frank, *The Whole Earth.*

3. Committee on Scientific Accomplishments of Earth Observations from Space, National Research Council, *Earth Observations from Space,* 15.

4. *New York Times,* "Sputnik's Legacy."

5. Tikhonravov, "The Creation of the First Artificial Earth Satellite."

6. Mack, *Viewing the Earth.* For a discussion of satellites in relation to the televisual, see Parks, *Cultures in Orbit.*

7. The complete text for the Kodak advertisement reads, "The whole earth from a business viewpoint," and details the boundless capacity of satellite sensing technology: "Aerial photographic surveillance started as an art of war. Now it has found work in helping mankind make a better peace with his environment. Kodak Products, for example, monitor dangerous ice on the sea, as well as the health of lakes, and the readiness of hillsides to slide down. Snow fields as fresh water sources are inventoried, as are fishing grounds off continental coasts. Aerial photography also measures social phenomena. Our color-infrared film has been found capable of providing accurate estimates of number of families in areas of high population density. Statistics in the public library may lag behind population shifts. Business decisions require fresh, solid facts. We have customers who can pick economic facts out of the air—from an appropriate altitude. Decision-makers who wish to get in touch with such people should write Eastman Kodak Company." Cited in Horowitz, "Domestic Communications Satellites," 38.

8. Ibid. As Horowitz notes in this article, rather than serving ecological and public interests, primarily, "the principal contractors for NASA's Earth Resources Technology Satellite, launched last July (1972) to identify sources of environmental pollution and monitor mineral resources, are General Electric and Eastman Kodak"; and "one prospective satellite

owner, RCA Globcom, has already noted in its satellite plan to the FCC that it intends to make satellite facilities available to the mining and petroleum industries."

9. For a discussion of how "informational globalism" coincides with "infrastructural globalism," see Edwards, *A Vast Machine,* 23–25.

10. McLuhan, "At the Moment of Sputnik," 49. *Sputnik* was launched October 4, 1957, rather than October 17, 1957, as cited in this instance.

11. Simondon, *Du mode d'existence,* 56.

12. Ibid., 152; Combes, *Gilbert Simondon,* 60; Lamarre, "Afterword," 92.

13. Berkeley, "Edmund C. Berkeley Papers."

14. Welsh, "Sensor Networks, Circa 1967."

15. Weiser, "The Computer for the 21st Century," 98.

16. Ibid., 94.

17. Helmers, "Creating Sensors"; Hsu, Kahn, and Pister, "Wireless Communications for Smart Dust"; Postscapes, "A Brief History."

18. Gross, "The Earth."

19. Ibid.

20. John Parkinson, chief technologist of Ernst & Young, quoted in Gross, "The Earth."

21. Toshitada Doi, chairman of Sony Corporation's Digital Creatures Lab, quoted in Gross, "The Earth."

22. Arthur, "Smartphone Explosion in 2014"; Thomas, "Smartphones Set to Become Even Smarter."

23. IBM, "The Internet of Things."

24. Ibid.

25. Evans, "The Internet of Things." For a discussion of the things in the Internet of Things, see Gabrys, "Re-thingifying the Internet of Things."

26. Bradley et al., "Internet of Everything (IoE) Value Index"; Cisco, "Internet of Everything."

27. Chan, "IPv6."

28. Intel, "What Does the Internet of Things Mean?"

29. For instance, see Scottish Sensor Systems Centre.

30. Wilson, *Sensor Technology Handbook;* O'Sullivan and Igoe, *Sensing and Controlling.* See also Gertz and Di Justo, *Environmental Monitoring with Arduino.*

31. Weiser, "The Computer for the 21st Century," 78.

32. Combes, *Gilbert Simondon;* Simondon, *L'individuation à la lumière.*

33. Whitehead, *Process and Reality,* 31 and 67–88 passim.

34. Stengers, *Thinking with Whitehead,* 163–64.

35. Whitehead, *Process and Reality,* 5–24.

36. Simondon, *Du mode d'existence,* 149, 222; Combes, *Gilbert Simondon,* 62.

37. Neumann, "First Draft of a Report on the EDVAC."

38. As Combes writes, "Simondon's approach entails a substitution of ontogenesis for traditional ontology, grasping the genesis of individuals within the operation of individuation as it is unfolding." See Combes, *Gilbert Simondon,* 3.

39. Hayles, *My Mother Was a Computer,* 17–31.

40. For instance, see Cox, *Speaking Code.*

41. Mackenzie, *Cutting Code,* 169; Gabrys, *Digital Rubbish.*

42. Canguilhem, *Knowledge of Life.*

43. For example, see Foucault, *Security, Territory, Population*.

44. Combes, *Gilbert Simondon*, 78; Lamarre, "Afterword"; Simondon, *Du mode d'existence*, 55–56.

45. Whitehead, *Modes of Thought*, 94, 112; Whitehead, *Process and Reality*, 15, 41.

46. Simondon, "The Genesis of the Individual," 306.

47. Combes, *Gilbert Simondon*, 4.

48. Ibid., 19.

49. Turner, *From Counterculture to Cyberculture*; Höhler, "'Spaceship Earth.'"

50. Heise, *Sense of Place and Sense of Planet*.

51. Edwards, *Vast Machine*, 25.

52. Ibid., xix.

53. Guattari, *Schizoanalytic Cartographies*, 11–12.

54. Ibid.

55. Ibid., 6.

56. McLuhan, "At the Moment of Sputnik," 49.

57. For example, see E. Odum, *Ecology*; H. Odum, *Environment, Power and Society*; Bateson, *Steps to an Ecology of Mind*.

58. Haraway, *Simians, Cyborgs and Women*, 164.

59. There is a rich literature that explores the ways in which cybernetics and information theory have influenced understandings of and approaches to ecologies (and systems, more generally). See, for instance, Haraway, "The High Cost of Information"; Taylor, "Technocratic Optimism"; Elichirigoity, *Planet Management*; Martin, "Environment c. 1973"; Bowker, "How to Be Universal."

60. For an overview of the area of these media-ecological distinctions, see Fuller, *Media Ecologies*.

61. See, for instance, Hörl, *The Ecological Paradigm*.

62. Guattari, *The Three Ecologies*.

63. See Fuller, *Media Ecologies*, 4.

64. Massumi discusses the "becoming-environmental of power" in relation to Foucault's exposition of environmentality. I take up this specific use of becoming-environmental of power in chapter 7 of *Program Earth* in relation to smart cities. Here, however, I draw on Whitehead and Simondon to develop an expanded notion of the *becoming environmental of computation* through the individuations and concrescences generated through environmental sensors. See Massumi, "National Enterprise Emergency"; Foucault, *The Birth of Biopolitics*.

65. Gabrys, "Atmospheres of Communication."

66. Gabrys, *Digital Rubbish*.

67. "Civic science" is a term from Fortun and Fortun, "Scientific Imaginaries and Ethical Plateaus." "Street science" is a term used by Corburn in *Street Science*. See also Burke et al., "Participatory Sensing."

68. For example, see Goodchild, "Citizens as Sensors."

69. For example, see Paulos, Honicky, and Hooker, "Citizen Science"; Aoki et al., "A Vehicle for Research."

70. The most commonly referenced platform for environmental data is one that has been in continual transformation. First taking the form of Pachube, as developed by Usman Haque, this platform was subsequently developed as an open yet commercially

based structure in the form of Cosm and has now become a more commercial and subscription-led platform rebranded as Xively. The development and migration of this platform from an open community to a commercial enterprise is a topic that could be researched and analyzed at length. However, there is not room here to deal in depth with how environmental sense data are aggregated, presented, and made accessible—or commercialized—in online platforms.

71. Key citizen-science research within science and technology studies includes (but is not limited to): Irwin, *Citizen Science*; Irwin and Michael, *Science, Social Theory and Public Knowledge*; Jasanoff, "Technologies of Humility"; Ellis and Waterton, "Environmental Citizenship in the Making"; Wynne, "May the Sheep Graze Safely?"

72. Bratton and Jeremijenko, "Suspicious Images, Latent Interfaces."

73. The work of Kim Fortun is relevant here in thinking about how the "informating of environmentalism" takes place. See Fortun, "Biopolitics and the Informating of Environmentalism." At the same time, I read Fortun's work alongside Simondon, who in a less epistemological register considers how in-forming and in-formation are processes of exchanging material and energy, of taking form, and of giving shape to individuals and environments. See Simondon, *L'individuation à la lumière*, and Combes, *Gilbert Simondon*, 5–6. I discuss this concept in more detail throughout this study, particularly in chapter 4.

74. Whitehead, *Adventures of Ideas*, 176. Latour and Stengers take up Whitehead's notion of concern to discuss "matters of fact" and "matters of concern." For instance, see Latour, "Why Has Critique Run Out of Steam?"; Stengers, "A Constructivist Reading of *Process and Reality*."

75. Mackenzie, *Wirelessness*; Greenfield and Shepard, "Urban Computing and Its Discontents."

76. Hayles, "RFID"; Kitchin and Dodge, *Code/Space*; de Souza e Silva and Frith, *Mobile Interfaces in Public Spaces*.

77. Dourish and Bell, *Divining a Digital Future*.

78. Ekman, *Throughout*.

79. Foth et al., *From Social Butterfly to Engaged Citizen*.

80. Boler, ed., *Digital Media and Democracy*; Jenkins and Thorburn, eds., *Democracy and New Media*; Delwiche and Henderson, eds., *The Participatory Cultures Handbook*; Ratto and Boler, *DIY Citizenship*.

81. Castells, *Networks of Outrage and Hope*; Earl and Kimport, *Digitally Enabled Social Change*; Feenberg and Barney, *Community in the Digital Age*; Rainie and Wellman, *Networked*.

82. Dean, Anderson, and Lovink, *Reformatting Politics*; Lovink, *Networks without a Cause*; Raley, *Tactical Media*.

83. Approaches to the more-than-human registers and materialities of participation have been attended to and developed across a wide range of feminist technoscience literature. One key reference here is Barad, "Posthumanist Performativity." See also Suchman, "Agencies in Technology Design." Haraway also develops this approach, drawing on Whitehead's notion of "misplaced concreteness," where she works beyond the abstractions of primary and secondary qualities to consider the expanded ways in which the materialities and practices of technoscience are situated. See Haraway, *Modest_Witness*, 269.

84. Stengers, "A Constructivist Reading of Process and Reality."

85. For example, see McCullough, *Ambient Commons*.

86. Latour, *Pandora's Hope*; Stengers, *The Invention of Modern Science*.

1. SENSING AN EXPERIMENTAL FOREST

1. Stengers, "A Constructivist Reading of *Process and Reality*," 109.

2. Ekman, *Throughout*.

3. Akyildiz et al., "A Survey on Sensor Networks"; Pottie and Kaiser, "Wireless Integrated Network Sensors"; Cuff, Hansen, and Kang, "Urban Sensing: Out of the Woods."

4. Whitehead, *Process and Reality*, 88–89.

5. Stengers, "A Constructivist Reading of *Process and Reality*," 99.

6. As N. Katherine Hayles writes in relation to distributed cognition of RFID tags, "When we understand that humans are not the only cognizers who can interpret information and create meaning, we are free to imagine how a world rich in embodied contextual processes might be fashioned to enhance the distributed cognitive systems that surround us and that we ourselves are." While this research is influenced by this important more-than-human media perspective that moves beyond human-centered approaches to computational technology, here I am interested to bring Whitehead into this discussion of distributions of sensing, not in order to emphasize cognition (with its potential connection to consciousness), but rather to expand upon experience as a key way of drawing together multiple sensing entities. See Hayles, "RFID," 69.

7. Whitehead, *Process and Reality*, 20.

8. Ibid., 88.

9. Stengers elaborates on this sense of "tuning" within experiments, and writes, "The idea that experimentation appeals to facts as they are observed by means of experimental appliances only refers to the stabilized end-product of a difficult operation. As Andrew Pickering (1995) marvelously characterized it, in his *Mangle of Practice*, experimenters may well know in advance what they want to achieve—what, for instance, their appliance should detect. However, a long process of tuning will nevertheless be needed, within which nothing will be trusted, neither the human hypothesis nor the observations made. Indeed, the process of tuning works both ways, on human as well as on more-than-human agency, constitutively intertwining a double process of emergence, of a disciplined human agency and of a captured material agency." Stengers, "A Constructivist Reading of *Process and Reality*," 96.

10. Stengers, "The Cosmopolitical Proposal."

11. Benson, *Wired Wilderness*.

12. Estrin, "Reflections on Wireless Sensing Systems."

13. Center for Embedded Networked Sensing (CENS), *Annual Progress Report*.

14. Tolle et al., "A Macroscope in the Redwoods"; Elson and Estrin, "Sensor Networks."

15. This chapter focuses on sensor applications for ecological study. However, this by no means covers the entirety of sensor applications for environmental uses and beyond, including agricultural management and energy saving in buildings. A whole range of automated environments is emerging through sensor applications, some of which are addressed in this study.

16. Szewczyk et al., "An Analysis of a Large Scale Habitat."

17. The U.S. Long Term Ecological Research Network (LTER); National Ecological Observatory Network (NEON).

18. Edwards et al., "Introduction"; Schimel et al., "2011 Science Strategy."

19. Microsoft Research, "SenseWeb." As mentioned in an introductory note, Cosm, which was formerly named Pachube, became a fee-based service in the form of Xively, and

so is now less oriented toward DIY engagements. See also Thingful for a map of devices connected to the Internet.

20. Lehning et al., "Instrumenting the Earth."

21. Nokia, "Sensor Planet"; IBM, "A Smarter Planet"; HP Labs, "Central Nervous System for the Earth (CeNSE)"; Planetary Skin Institute.

22. For a discussion of the multiple actors invested in ubiquitous computing, including governments, see Nold and van Kranenburg, "The Internet of People."

23. While there is not space within this chapter to discuss these developments within sensor systems, for a more extensive discussion of how emerging sensing technologies and practices might be understood as forms of environmentality, see chapter 7 of *Program Earth*, "Citizen Sensing in the Smart and Sustainable City."

24. For a discussion of earlier attempts to program environments with sensors through the prototypical if largely hypothetical technology of smart dust, see Gabrys, "Telepathically Urban."

25. Center for Embedded Network Sensing (CENS), "Annual Progress Report," 3.

26. Mackenzie discusses this issue of how relations or itineraries are drawn through software in *Cutting Code*. See also Fuller, *Behind the Blip*.

27. Center for Embedded Network Sensing (CENS), "Annual Progress Report," 3.

28. Latour, *Science in Action*.

29. Hamilton et al., "New Approaches."

30. Goldman et al., "Distributed Sensing Systems," 3.

31. Rundel et al., "Tansley Review."

32. Center for Embedded Network Sensing (CENS), "Terrestrial Ecology Observing Systems."

33. Center for Embedded Network Sensing (CENS), "Annual Progress Report," 6.

34. Center for Embedded Network Sensing (CENS), "Center for Embedded Networked Sensing," 5.

35. Estrin, "Reflections on Wireless Sensing Systems," 2.

36. Goldman et al., "Distributed Sensing Systems," 12.

37. Elson and Estrin, "Sensor Networks," 2.

38. Ibid., 7.

39. Ibid., 10.

40. Ibid., 7.

41. Inevitably, incompatibilities within data sets are not the only issues with sensors. Given that the embedded sensors are prototypes and tested "in the wild," the devices at times cease to function as intended, whether due to mechanical failure, calibration issues, or bugs in the code. Given that the CENS project has also come to an end, some sensors may cease to function and break down over time. For a more extended discussion on the material processes of electronics as they break down, see Gabrys, *Digital Rubbish*.

42. Bowker, "Biodiversity Datadiversity." For a detailed discussion of data issues that have emerged specifically in relation to the CENS project, see Borgman, Wallis, and Enyedy, "Little Science Confronts the Data Deluge."

43. Roth and Bowen, "Digitizing Lizards."

44. Goldman et al., "Distributed Sensing Systems," 10.

45. Tolle et al., "A Macroscope in the Woods."

46. Goldman et al., "Distributed Sensing Systems," 6.

47. Ibid., 19, emphasis in original.

48. Estrin, "Reflections on Wireless Sensing Systems," 2.

49. Center for Embedded Network Sensing (CENS), "Annual Progress Report," 29.

50. Hyman, Graham, and Hansen, "Imagers as Sensors."

51. Hayles, "RFID," 48–49.

52. Whitehead, *Adventures of Ideas*, 180–81; Whitehead, *Modes of Thought*, 30, 157.

53. Weiser, "The Computer for the 21st Century."

54. Turing, "Intelligent Machinery," 117.

55. Ibid.

56. Whitehead, *Modes of Thought*; Massumi, *Semblance and Event*.

57. Whitehead, *Modes of Thought*, 158; Stengers, *Thinking with Whitehead*, 352–53.

58. Whitehead, *Modes of Thought*, 138.

59. Ibid., 158. And as Stengers writes, "We do not know how a bat, armed with its sonar, or a dog, capable of tracking by smell, perceive 'their' world. We can identify the features they discriminate, but we can only dream of the contrast between 'that which' they perceive and what they are aware of. All we 'know' is that their experience is, like ours, highly interpretative, and that, like ours, it has solved an extraordinarily delicate problem: to give access, in a more or less reliable way, to what it is important to pay attention to." Paying attention is "the interpretative choice from which our experience has issued." Stengers, *Thinking with Whitehead*, 338.

60. Whitehead, *Process and Reality*, 88.

61. Shaviro, *Without Criteria*, 21.

62. While this notion of distributed experience can be found within Whitehead's writings, it also is a prior concept developed by William James in relation to radical empiricism. Adrian Mackenzie takes up the latter concept in relation to wirelessness to discuss how wireless technologies unfold these distributions of experience. See James, *Essays in Radical Empiricism*; Mackenzie, *Wirelessness*.

63. For a more extended discussion on the development of "facticity," see Halewood and Michael, "Being a Sociologist," 34; Stengers, "A Constructivist Reading of *Process and Reality*."

64. Thanks are due to Mike Michael for conversations that have led to these points.

65. Simondon develops the ontogenetic aspects of individuals and experience at length. For a discussion of this work, see Combes, *Gilbert Simondon*.

66. Parisi, "Technoecologies of Sensation and Control." See also Guattari, *Chaosmosis*.

67. Clough, "The New Empiricism," 51.

68. Helmreich, "Intimate Sensing," 130–32.

69. Goodwin, "Seeing in Depth."

70. Foucault, *The Order of Things*; Kittler, "Thinking Colors and/or Machines"; Hayles, "Computing the Human"; Parikka, *Insect Media*; Fuller, "Boxes towards Bananas."

71. Hayles, "RFID."

72. Braidotti, *Transpositions*, 41. See also Guattari, *Chaosmosis*.

73. Braidotti, *Transpositions*, 96–97.

74. Whitehead, *Modes of Thought*; Uexküll, *A Foray*; Canguilhem, *Knowledge of Life*; Foucault, *The Birth of Biopolitics*.

75. Uexküll, *A Foray*, 44–46.

76. Whitehead, *Process and Reality*, 20–21.

77. Whitehead, *Modes of Thought*, 89–90.

78. Whitehead, *Process and Reality,* 58.

79. Shaviro, *Without Criteria,* 25.

80. Whitehead, *Adventures of Ideas,* 238.

81. Foucault, *The Birth of Biopolitics.*

82. Citizen-sensing applications have emerged through the James Reserve sensor research. See Cuff, Hansen, and Kang, "Urban Sensing"; Estrin, "Reflections on Wireless Sensing Systems."

2. FROM MOSS CAM TO SPILLCAM

1. For just one of many examples, see BBC England, "Webcams: Animals."

2. GlobalWarming.House.Gov., "Oil Spill in the Gulf."

3. Simondon, *Du mode d'existence,* 56.

4. For example, see Haraway, "Introduction."

5. This committee was disbanded in 2011. See Sheppard, "Republicans Kill Global Warming Committee."

6. Markey, Edward Markey to Lamar McKay.

7. Interspecies Internet is a project proposed by rock star Peter Gabriel in collaboration with cognitive psychology and animal intelligence researchers, as well as Internet pioneer Vint Cerf. The "user group" for this worldwide network consists of animals in zoos, which are accessed via webcams. See Interspecies Internet.

8. O'Connell, Nichols, and Karanth, eds., *Camera Traps in Animal Ecology.*

9. BBC News, "Secret Life of the Cat." Human-oriented wearable camera technology has also been in use for some time, from Microsoft's SenseCam, which sought to create "a photographic memory for everyone," to the recently discontinued Google Glasses.

10. WildlifeTV, "Wildlife Webcams."

11. The Cornell Lab of Ornithology, Macaulay Library.

12. The Cornell Lab of Ornithology, "All about Birds."

13. As stated by Charles Eldermire, Bird Cams project leader at the Cornell Lab of Ornithology, on "Critter Cams Viewers Becoming Citizen Scientists."

14. Ocean Networks Canada, "Teen Spots Hagfish-Slurping Elephant Seal." The particular discovery in this case was brought forward by a Ukrainian teenage boy who emailed a clip of a hagfish being consumed by an unknown creature in deep underwater spaces. As it turns out, a female elephant seal was the creature eating the hagfish, which had not been previously observed by scientists, as hagfish not only inhabit deep ocean spaces, but were also assumed to be inedible. See http://www.oceannetworks.ca and http://digitalfishers .net for more information on Ocean Networks Canada's citizen-science initiatives.

15. Just a few examples of research that analyzes science and visuality include Haraway, *Modest_Witness*; Daston and Galison, *Objectivity*; Mody, "The Sounds of Science."

16. Mitman, "When Nature Is the Zoo," 139–40.

17. Ibid.

18. Beyond the references mentioned above, Lucy Suchman undertakes a discussion of sociomaterial relations of people and things that engages with visuality as part of its focus. See Suchman, "Reconfigurations."

19. Haraway, *Modest_Witness,* 223; Shapin and Schaffer, *Leviathan and the Air-Pump.*

20. Haraway, "Situated Knowledges."

21. National Geographic, Crittercam.

22. Ibid.

23. Ibid.

24. Donna Haraway, "Crittercam," 249. Haraway draws on Merleau-Ponty in her discussion of "infoldings of the flesh."

25. Ibid.

26. Ibid., 249–50.

27. While some research has argued for the adoption of the concept "emplacement" in comparison to "embodiment," I diverge from this approach as I seek to avoid the fixities that can emerge through denotations and connotations of "place." I am also cautious to avoid appending "environment" to a "mind-body" dyad, thereby maintaining a substantialist approach to sensing. In this sense, Whitehead and Simondon are key thinkers whom I draw on to attempt to work beyond a substantialist understanding of sense (in other words, minds that might decode or sense objects in environment), and instead take up processes such as individuations, concrescences, and subject-superjects of experience to address the particular attachments of sensing entities and environments. For a discussion of emplacement in relation to embodiment, see Howes, "Introduction."

28. Hayles, "RFID," 48.

29. Ibid.

30. Andrejevic, "The Webcam Subculture and the Digital Enclosure."

31. Goldberg, *The Robot in the Garden*.

32. For a collection of discussions on sensory mediation in relation to art and technology, see Jones, *Sensorium*.

33. Hayles, "RFID," 48.

34. Ibid.

35. Whitehead, *Adventures of Ideas*, 179–82.

36. For a discussion of the "sense of place" in relation to rethinking the global–local divide and how this influences environmental imaginations, see Heise, *Sense of Place*. Heise argues that the local is as much an assemblage as the global and does not equate to pure unmediated access to the world.

37. Simondon, *Du mode d'existence*, 50–52.

38. Ibid., 20. As Simondon writes, "L'objet technique un est unité de devenir."

39. Lamarre, "Afterword," 95; See also Simondon, *Du mode d'existence*, 57.

40. Simondon, *Du mode d'existence*, 56. He writes, "Il ne s'agit pas en effet d'un progrès conçu comme marche dans un sens fixé à l'avance, ni d'une humanisation de la nature; ce processus pourrait aussi bien appaître comme une naturalisation de l'homme; entre homme et nature se crée en effet un milieu techno-géographique qui ne devient possible que par l'intelligence de l'homme: l'auto-conditionnement d'un schème par le resultat de son fonctionnement nécessite l'emploi d'une fonction inventive d'anticipation qui ne se trouve ni dans la nature ni dans les objets techniques déjà constitués; c'est une oeuvre de vie de faire ainsi un saut par-dessus la réalité donnée et sa systématique actuelle vers de nouvelles formes qui ne sa maintiennent que parce qu'elles existent toutes ensemble comme un système constitué; quand un nouvel organe apparait dans la série évolutive, il ne se maintient que s'il réalise une convergence systématique et pluri-fonctionnelle."

41. Ibid., 55–56.

42. Ibid., 55. The becoming environmental of technical objects is in this sense normative: it is what is needed to avoid overspecialization and disadaptation, and for a technical object to become "natural."

43. James San Jacinto Mountains Reserve.

44. Rundel et al., "Tansley Review."

45. James San Jacinto Mountains Reserve, "The Moss Cam Project."

46. Mishler, "The Biology of Bryophytes."

47. Hamilton et al., "New Approaches."

48. Hyman, Graham, and Hansen, "Imagers as Sensors."

49. Mishler, "The Biology of Bryophytes."

50. Jasanoff, "Image and Imagination."

51. Haraway, "Situated Knowledges."

52. Whitehead, *Adventures of Ideas*, 178–82.

53. National Commission on the BP Deepwater Horizon Oil Spill and Offshore Drilling, "Deep Water," 167, 187.

54. Markey, Edward Markey to Lamar McKay.

55. Ibid.

56. Jansen and Keilar, "Markey."

57. The Global Language Monitor, "Top Words of 2010."

58. GlobalWarming.House.Gov, "Oil Spill in the Gulf."

59. Jonsson, "BP Live Feed 'Spillcam'."

60. "Live Gulf Oil Spill Cam."

61. Stuever, "BP's Oil Spillcam."

62. Helmreich, "Nature / Culture / Seawater."

63. Woods Hole Oceanographic Institute, "Science in a Time of Crisis."

64. Ibid.

65. Serres, *The Five Senses*.

66. Clough, "The New Empiricism," 51.

67. Roth and Bowen, "Digitizing Lizards."

68. Ibid., 731.

69. Ibid., 721.

70. For a related discussion of environmental sensors, see Gabrys, "Automatic Sensation."

71. Puig de la Bellacasa, "Touching Technologies, Touching Visions."

72. Haraway, "Situated Knowledges," 589.

73. "Critter Cams Viewers."

74. Helmreich, "Intimate Sensing," 149.

75. Combes, *Gilbert Simondon*, 78.

76. Lamarre, "Afterword," 95.

3. ANIMALS AS SENSORS

1. McConnell, "Telemetry in Sea Mammals"; Biuw et al., "Variations in Behavior and Condition."

2. For more information, see Argos; Benson, "One Infrastructure, Many Global Visions."

3. Wilson, Shepard, and Liebsch, "Prying into the Intimate Details."

4. Newton, *Bird Migration*.

5. Pahl et al., "Large Scale Homing in Honeybees."

6. The National Science Foundation, "The Secret Lives of Wild Animals"; Wikelski et al., "Simple Rules Guide Dragonfly Migration."

7. The Arctic Tern Migration Project; Egevang et al., "Tracking of Arctic Terns."

8. Sirtrack, "Sirtrack Working with Emperor Penguin"; Sirtrack, "Happy Feet Transmissions Ceased."

9. Cooke, "Biotelemetry and Biologging"; Johnson, "'Smart Collar' in the Works."

10. Wilson and McMahon, "Measuring Devices on Wild Animals"; Northern Prairie Wildlife Research Center, "A Critique of Wildlife Radio-Tracking"; Scandolara et al., "Impact of Miniaturized Geolocators."

11. For a parallel discussion of biomonitoring, see Kosek, "Ecologies of Empire."

12. Whitehead, *Process and Reality,* 238 and 228–88 passim.

13. Wilson, Shepard, and Liebsch, "Prying into the Intimate Details."

14. May, "Unanswered Questions in Ecology," cited by Wikelski in "Move It, Baby!"

15. Benson, *Wired Wilderness,* 92.

16. Wikelski, "Move It, Baby!"

17. MoveBank; Dodge et al., "The Environmental-Data."

18. Wikelski, "Move It, Baby!"

19. MoveBank, "What Is Animal Tracking?"

20. Wikelski, "Move It, Baby!"

21. Ibid.

22. Ibid.

23. Ibid.

24. Ibid.; Marshall, "Foreword," 4–5.

25. Marshall, "Foreword," 4.

26. Ibid.

27. Ibid., 5.

28. International Cooperation for Animal Research Using Space (ICARUS) Initiative, "Technical Solution"; Wikelski et al., "Going Wild."

29. Pennisi, "Global Tracking of Small Animals."

30. International Cooperation for Animal Research Using Space (ICARUS) Initiative, "Science & Projects."

31. As Pennisi writes, "Because the space station is only 320 kilometers away, the ICARUS system will demand much less energy than ARGOS to send data, reducing battery requirements. As currently designed, most tags for ICARUS will contain a GPS that regularly takes stock of the wearer's location, storing those data for many months if needed. To be most energy efficient, the ICARUS tag will be preprogrammed to turn on only when the space station will pass overhead and will deliver its data only when triggered by the space station." See Pennisi, "Global Tracking of Small Animals."

32. Hinchliffe, "Sensory Biopolitics"; Wikelski et al., "Disaster Alert Mediation Using Nature."

33. Schofield et al., "Novel GPS Tracking."

34. Wilcove, *No Way Home*; Kurvits et al., *Living Planet.*

35. Block et al., "Tracking Apex Marine Predator."

36. Warner Chabot of the Ocean Conservancy, as cited in Bauer, "Tagging Pacific Predators."

37. Wildfowl and Wetlands Trust.

38. BirdsEye Tools for Birders, "BirdLog North America."

39. The Nature Conservancy, "Precision Conservation."

40. The Migratory Connectivity Project, "Citizen Science."

41. Large Pelagics Research Center, "Tag a Tiny Program."

42. Yu, "More Than 300 Sharks In Australia"; Surf Life Saving Western Australia, https://twitter.com/SLSWA; Zoological Society of London, "Eel Conservation"; EpiCollect, Roadkill Garneau. See also OpenScientist, "Mobile Citizen Science Apps!"

43. Allan et al., "A Cost-Effective and Informative Method." An instructional video accompanying this paper is available at http://www.youtube.com/watch?v=UaSvSogrVjw.

44. xClinic Environmental Health Clinic and Living Architecture Lab, "Amphibious Architecture"; Benjamin, Yang, and Jeremijenko, "New Interaction Partners for Environmental Governance."

45. Nitta, "Extreme Green Guerillas."

46. Cornwell and Campbell, "Co-Producing Conservation and Knowledge"; Hurlbert and Liang, "Spatiotemporal Variation in Avian Migration Phenology"; Cooper, Shirk, and Zuckerberg, "The Invisible Prevalence of Citizen Science"; Laughlin et al., "Integrating Information from Geolocators."

47. Wikelski, "Move It, Baby!"

48. Savill et al. *Wytham Woods*; Elton, *The Ecology of Invasions*.

49. University of Oxford, "Wytham Woods."

50. Dyo et al., "WILDSENSING."

51. Dyo et al., "Wildlife and Environmental Monitoring."

52. Dyo et al., "WILDSENSING."

53. Pasztor et al., "Selective Reprogramming."

54. Benson, *Wired Wilderness,* 48.

55. Sea Mammal Research Unit, "Southern Elephant Seals as Oceanographic Samplers."

56. Biuw et al., "Variations in Behavior and Condition."

57. McConnell, "Telemetry in Sea Mammals."

58. For example, see National Oceanic and Atmospheric Administration (NOAA), "Critical New Data."

59. McConnell, "Telemetry in Sea Mammals."

60. Benson, *Wired Wilderness,* 190–93.

61. International Cooperation for Animal Research Using Space (ICARUS) Initiative, "Countdown to ICARUS."

62. Max Planck Institute for Ornithology, "Social Migration in Juvenile White Storks."

63. Rotics, "Comparison of Juvenile and Adult Migration."

64. Brown et al., "Observing the Unwatchable."

65. Rotics, "Comparison of Juvenile and Adult Migration"; Amezian and Khamlichi, "White Stork Satellite-Tracked."

66. Wikelski, "Move It, Baby!"

67. Max Planck Institute for Ornithology, Animal Tracker. http://www.orn.mpg.de/animal_tracker.

68. Wikelski, "Move It, Baby!"; Wald, "Follow That Bird."

69. e-obs.

70. Max Planck Institute for Ornithology, Animal Tracker app.

71. Wikelski, "Move It, Baby!"

72. Wikelski et al., "Disaster Alert Mediation Using Nature."

73. Wikelski, "Move It, Baby!"; Maier, "A Four-Legged Early-Warning System."

74. Canguilhem, *Knowledge of Life*, 109.

75. Combes, *Gilbert Simondon*, 4. As Combes writes, "No individual would be able to exist without a milieu that is its complement, arising simultaneously from the operation of individuation: for this reason, the individual should be seen as but a partial result of the operation bringing it forth. Thus, in a general manner, we may consider individuals as beings that come into existence as so many partial solutions to so many problems of incompatibility between separate levels of being."

76. See also Simondon, *Two Lessons on Animal and Man*.

77. In addition to examples cited above, see Vandenabeele et al., "Excess Baggage for Birds."

78. Wilson and McMahon, "Measuring Devices on Wild Animals."

79. Canguilhem, *Knowledge of Life*, 120.

80. Ibid., 119.

81. This argument resonates with Stengers's discussions in *Cosmopolitics I* and *Cosmopolitics II*.

82. Whitehead, *Process and Reality*, 78.

83. Hayles, *My Mother Was a Computer*, 20.

84. For example, see Bateson, *Steps to an Ecology of Mind*; Negroponte, "SEEK." See also Clarke, "John Lilly."

85. Canguilhem, *Knowledge of Life*, 96.

86. Ibid., 89–90.

87. See Wiener, *The Human Use of Human Beings*. For a parallel discussion on automation, see Stacey and Suchman, "Animation and Automation."

88. Lamarre, "Afterword," 80.

89. Combes, *Gilbert Simondon*, 10; Simondon, *L'individuation à la lumière*, 28.

90. For example, see Stevens, *Sensory Ecology, Behavior, and Evolution*. A related analysis of the informational influences on biology can be found in Keller, "The Body of a New Machine."

91. Technique, however, is one way to approach the open potential of machines, since it is through technique that original engagements unfold, which cannot be limited to one machinic function. See Canguilhem, *Knowledge of Life*, 93.

92. Whitehead, *Process and Reality*, 28. As Whitehead writes, "'becoming' is a creative advance into novelty. It is for this reason that the meaning of the phrase 'the actual world' is relative to the becoming of a definite actual entity which is both novel and actual, relatively to that meaning, and to no other meaning of that phrase. Thus, conversely, each actual entity corresponds to a meaning of 'the actual world' peculiar to itself. . . . An actual world is a nexus; and the actual world of one actual entity sinks to the level of a subordinate nexus in actual worlds beyond that actual entity."

93. Combes, *Gilbert Simondon*, 31.

94. Ibid., 27. As Combes writes, "In effect, a living being, 'in order to exist, needs to be able to continue individualizing by resolving problems in the milieu surrounding it, which is its milieu.' . . . In the analysis proposed by Simondon, perception, for instance, appears as an act of individuation operated by a living being to resolve a conflict into which it has entered with its milieu. In his view, to perceive is not primarily to grasp a form; rather it is the act taking place within an ensemble constituted by the relation between subject and world, through which a subject invents a form and thereby modifies its own structure and

that of the object at the same time: we see only within a system in tension, of which we are a subensemble." See also Simondon, *L'individuation à la lumière,* 264.

95. Simondon, "The Genesis of the Individual," 306.

96. Ibid.

97. See Pritchard, "Thinking with the Animal-Hacker."

98. Stengers, *Cosmopolitics II,* 362.

99. Ibid., 359.

100. Combes, *Gilbert Simondon,* 60; Simondon, *Du mode d'existence,* 152.

101. Tsing, "Empowering Nature," 137.

102. Debaise, "A Philosophy of Interstices," 108.

4. SENSING CLIMATE CHANGE AND EXPRESSING ENVIRONMENTAL CITIZENSHIP

1. The Finnish Society of Bioart, "Environmental Computing."

2. Masco, "Bad Weather."

3. Whitehead, *Modes of Thought,* 5.

4. Coen, *The Earthquake Observers,* 163. See also Aubin, Bigg, and Sibum, *The Heavens on Earth.*

5. Kilpisjärvi Biological Station; Sustaining Arctic Observing Network (SAON); SCANNET; International Network for Terrestrial Research and Monitoring in the Arctic.

6. Arctic Monitoring and Assessment Programme (AMAP), *Report 2010,* 8. This report calls for "a sustained, robust circumpolar monitoring network effective at detecting change and discerning trends over the entire Arctic Region related to a range of environmental stressors including pollutants, climate change and the interaction between them."

7. Anisimov et al., "Polar Regions (Arctic and Antarctic)."

8. Committee on Emerging Research Questions in the Arctic et al., *The Arctic in the Anthropocene,* 12.

9. Cubasch et al., "Introduction."

10. Global Climate Observing System (GCOS), "Essential Climate Variables."

11. Committee on Understanding and Monitoring Abrupt Climate Change and Its Impacts et al., "Abrupt Impacts of Climate Change," 52; Brigham-Grette et al., "Pliocene Warmth," 1421–27.

12. Oceans at MIT with Ron Prinn. "400 ppm CO_2?" http://oceans.mit.edu/featured -stories/5-questions-mits-ron-prinn-400-ppm-threshold.

13. Hartmann et al., "Observations," 166; NASA Jet Propulsion Laboratory, "Orbiting Carbon Observatory 2 (OCO-2)." A number of videos explaining OCO-2 note the importance of gathering constant and global information on CO_2: "We can only manage what we can measure." http://oco.jpl.nasa.gov/galleries/Videos. OCO-2 gathers hundreds of thousands of CO_2 measurements every day by assessing diffracted light levels in the atmosphere. See http://oco.jpl.nasa.gov/observatory/instrument.

14. Stocker et al., *IPCC, 2013;* Scripps Institute of Oceanography, "The Keeling Curve."

15. Hartmann et al., "Observations: Atmosphere and Surface," 167.

16. Stocker et al., *IPCC, 2013,* 11.

17. Hartmann et al., "Observations," 198.

18. CO_2 has now been designated as a pollutant by the EPA. See Supreme Court of the United States, *Massachusetts v. Environmental Protection Agency,* 549 U.S. 497 (April 2, 2007); U.S. Environmental Protection Agency, "Endangerment and Cause." This legislation has

recently been put to work through the "Clean Power Plan" to limit CO_2 levels. See U.S. Environmental Protection Agency, "Carbon Pollution Standards." For an expanded discussion of how CO_2 turns up as a pollutant relative to the environments that it begins to transform, see Gabrys, "Powering the Digital."

19. For instance, see Forest Carbon Project Inventory; Carbon Monitoring for Action (CARMA). For an expanded discussion of climate change in relation to sinks, see Gabrys, "Sink."

20. A resource that lists a number of community-led and traditional ecological knowledge projects in the Arctic is the Committee on Emerging Research Questions in the Arctic et al., *The Arctic in the Anthropocene.*

21. At the time of this fieldwork, a Kilpisjärvi environmental sensor feed was viewable on the Pachube platform, which was later renamed Cosm, and then Xively, as previously mentioned. The Kilpisjärvi feed, which is no longer live but includes historic measurements for light, humidity, temperature, noise, air quality, CO, and CO_2, is now available at https://xively.com/feeds/21544. The Kilpisjärvi webcam can be viewed at http://kilpis5.kilpis.hel sinki.fi/appletvid.html.

22. Forsström et al., "Seasonality of Phytoplankton"; Lemke et al. "Observations."

23. Vaughan et al., "Observations."

24. Prowse et al., "Climate Change Effects."

25. Strangeways, *Measuring the Natural Environment,* 1.

26. Bowker, "Biodiversity Datadiversity."

27. Kohler, *Landscapes and Labscapes,* 118–24. See also Cruikshank, *Do Glaciers Listen?*

28. Whitehead, *Modes of Thought,* 18. As Whitehead writes in relation to the "matter-of-fact," "Consider, for example, the scientific notion of measurement. Can we elucidate the turmoil of Europe by weighing its dictators, its prime ministers, and its editors of newspapers? The idea is absurd, although some relevant information might be obtained. I am not upholding the irrelevance of science. Such a doctrine would be foolish. For example, a daily record of the bodily temperatures of men, above mentioned, might be useful. My point is the incompleteness of the information."

29. Coen, *The Earthquake Observers,* 6, 20; Kohler, *Landscapes and Labscapes,* 122–24.

30. Exchange for Local Observations and Knowledge of the Arctic (ELOKA); Atlas of Community-Based Monitoring in a Changing Arctic. The Arctic Perspective Initiative (API) has worked in collaboration with the Finnish Bioart Society on developing "open authoring, communications and dissemination infrastructures for the circumpolar region" in Kilpisjärvi.

31. On the different ways in which measurement may be operationalized toward much different effects in relation to measure and value, see Verran, "The Changing Lives of Measures and Values." On the ways in which experience and measurement generate speculative registers of evidence and data, see Gabrys, "Pollution Sensing and Fracking."

32. A discussion of the ways in which organisms "incorporate" and sense environments can be found in Gabrys, "Becoming Urban."

33. On the notion of organisms working through the materialities of changing environments, see Gabrys, "Plastic."

34. For more project information, see Preemptive Media, "Area's Immediate Reading (AIR)"; Jeremijenko, "Feral Robotic Dogs."

35. xClinic Environmental Health Clinic and Living Lab, "Amphibious Architecture."

36. Costa, Pigeon Blog; Percifield, dontflush.me; Safecast.

37. Paterson, "Vatnajökull (the sound of)"; Collier, Ray, and Weber, "The Pika Alarm."

38. "Million Trees NYC."

39. Braidotti, *Transpositions*.

40. Ellis and Waterton, "Environmental Citizenship in the Making."

41. Whitehead, *Process and Reality*, 88–89; Whitehead, *Modes of Thought*, 158.

42. Fortun, "Biopolitics and the Informating of Environmentalism."

43. Combes, *Gilbert Simondon*, 5; Simondon, *L'individuation à la lumière*, 32.

44. Simondon, *L'individuation à la lumière*, 32, as translated in Combes, *Gilbert Simondon*, 6.

45. Combes, *Gilbert Simondon*, 32.

46. Simondon, *L'individuation à la lumière*, 253, as translated in Combes, *Gilbert Simondon*, 31.

47. Simondon, *L'individuation à la lumière*, 63, as translated in Combes, *Gilbert Simondon*, 19.

48. Combes, *Gilbert Simondon*, 19.

49. Whitehead, *Process and Reality*, 88.

50. Stengers, "A Constructivist Reading of *Process and Reality*," 103.

51. Stengers, "Introductory Notes on an Ecology of Practices"; Stengers, *Cosmopolitics II*, 349–50.

52. Stengers, *Cosmopolitics II*, 349.

53. Ibid., 350.

54. Braidotti, *Transpositions*, 149–50.

55. Haraway, "Situated Knowledges," 589.

56. For a complementary discussion of radical empiricism in relation to earthly concerns, see Latour, "A Plea for Earthly Sciences."

57. Kim TallBear, "Uppsala 3rd Supradisciplinary Feminist Technoscience Symposium," http://www.kimtallbear.com/homeblog/uppsala-3rd-supradisciplinary-feminist-techno science-symposium-feminist-and-indigenous-intersections-and-approaches-to-technosci ence. See also TallBear, "Why Interspecies Thinking Needs Indigenous Standpoints."

58. On the notion of speculative empiricism, see Debaise, *Un empirisme spéculatif*.

59. Edwards, *A Vast Machine*, 427–39.

60. Stengers, *Thinking with Whitehead*, 252. Stengers also writes (discussing and citing Whitehead's *Process and Reality*, 11), "It pertains to speculative propositions to 'make us feel' what is, in fact, a generality that bears upon every proposition," 267. Generality, in many ways, inheres as an environment in which facts take hold and have relevance.

5. SENSING OCEANS AND GEO-SPECULATING WITH A GARBAGE PATCH

1. Moore et al., "A Comparison of Plastic and Plankton"; Moore and Phillips, *Plastic Ocean*.

2. Ebbesmeyer and Scigliano, *Flotsametrics and the Floating World*, 167.

3. For example, see Carpenter and Smith, "Plastics on the Sargasso Sea Surface."

4. Ebbesmeyer and Scigliano, *Flotsametrics and the Floating World*, 227; Moore and Phillips, *Plastic Ocean*.

5. For example, see National Center for Ecological Analysis and Synthesis, "A Global Map of Human Impacts." Not to be confined to surface views of the ocean, Google Earth

has also added an ocean layer that allows "underwater" views. See http://www.google.com/earth/explore/showcase/ocean.html.

6. National Oceanic and Atmospheric Administration (NOAA), "De-mystifying the 'Great Pacific Garbage Patch.'"

7. For example, see Jordan, *Midway*.

8. Vitaliano, *Legends of the Earth*.

9. Whitehead, *Adventures of Ideas*, 177–82.

10. Whitehead, *Process and Reality*, 89–110.

11. Howell et al., "On North Pacific Circulation"; Pichel et al., "Marine Debris Collects."

12. Lebreton, Greer, and Borrero, "Numerical Modeling of Floating Debris."

13. Howell et al., "On North Pacific Circulation."

14. Ibid.

15. Meanders and eddies are areas where most visible "patches" of garbage emerge, whereas gyres tend to have overall higher concentrations of microplastics that are quite often invisible or difficult to detect. See NOAA, "De-Mystifying the 'Great Pacific Garbage Patch.'"

16. Gregory and Ryan, "Pelagic Plastics"; U.S. EPA, "Marine Debris in the North Pacific."

17. Derraik, "The Pollution of the Marine Environment."

18. Gabrys, "Plastic."

19. As Moore et al. write, "Several limitations restrict our ability to extrapolate our findings of high plastic-to-plankton ratios in the North Pacific central gyre to other areas of the ocean. The North Pacific Ocean is an area of low biological standing stock; plankton populations are many times higher in nearshore areas of the eastern Pacific, where upwelling fuels productivity. . . . Moreover, the eddy effects of the gyre probably serve to retain plastics, whereas plastics may wash up on shore in greater numbers in other areas. Conversely, areas closer to the shore are more likely to receive inputs from land-based runoff and ship loading and unloading activities, whereas a large fraction of the materials observed in this study appear to be remnants of offshore fishing-related activity and shipping traffic." See Moore et al., "A Comparison of Plastic and Plankton."

20. Arthur, Baker, and Bamford, *Proceedings of the International Research Workshop*; Thompson et al., "Lost at Sea"; Thompson et al., "Plastics, the Environment, and Human Health"; Andrady, "Microplastics in the Marine Environment."

21. Gregory, "Environmental Implications."

22. Barnes et al., "Accumulation and Fragmentation"; Takada, "International Pellet Watch"; United Nations Environment Programme, "Marine Litter: A Global Challenge."

23. Kaiser, "The Dirt on Ocean Garbage Patches"; U.S. EPA, "Marine Debris in the North Pacific."

24. Takada, "International Pellet Watch."

25. Boerger et al., "Plastic Ingestion"; Thompson et al., "Plastics, the Environment, and Human Health."

26. Cózar et al., "Plastic Debris in the Open Ocean"; Di Lorenzo et al., "North Pacific Gyre Oscillation"; Goldstein, Rosenberg, and Cheng, "Increased Oceanic Microplastic Debris."

27. As Stengers writes, "The production of the matter of fact that could operate as a reliable witness for the 'adequacy' of an interpretation is always an experimental achievement. As long as this achievement remains a matter of controversy, the putative matter of

fact will remain a matter of collective, demanding, concern." See "A Constructivist Reading of *Process and Reality*," 94. See also Latour, "Why Has Critique Run Out of Steam?"

28. Underwood, *Experiments in Ecology*.

29. Whitehead, *Process and Reality*, 78.

30. Barad, "Posthumanist Performativity."

31. Gabrys, "Speculating with Organisms in the Plastisphere."

32. Gabrys, *Digital Rubbish*; Gabrys, "Plastic."

33. Howell et al., "On North Pacific Circulation."

34. Ebbesmeyer et al., "Tub Toys Orbit the Subarctic Gyre"; Ebbesmeyer and Scigliano, *Flotsametrics and the Floating World*.

35. Madrigal, "Found."

36. Ebbesmeyer, "Using Flotsam to Study Ocean Currents."

37. Ibid.; Law et al., "Plastic Accumulation."

38. Ebbesmeyer writes, "We must develop networks, which remain vigilant to collect this flotsam." See Ebbesmeyer, "Using Flotsam to Study Ocean Currents." This is something Ebbesmeyer has also supported through the efforts of Beachcombers Alert, http://www.beachcombersalert.org.

39. Dohan and Maximenko, "Monitoring Ocean Currents with Satellite Sensors"; International Pacific Research Center, "Tracking Ocean Debris"; SEA-MDI, "Want to Track Marine Debris?"

40. Marine Traffic; Thingful.

41. Interagency Ocean Observation Committee.

42. Gillis, "In the Ocean, Clues to Change."

43. Helmreich, "Intimate Sensing," 148.

44. UK Argo, "Deep Profile Floats."

45. Takada, "International Pellet Watch."

46. Lauro et al., "The Common Oceanographer"; Smerdon, Manning, and Paull, "Crowdsourcing Coastal Oceanographic Data."

47. Protei; National Oceanic and Atmospheric Administration (NOAA), "Marine Debris Monitoring and Assessment Project."

48. WayDownSouth, "Adopt an Argo Float." The Google Earth Argo Application is available at http://argo.jcommops.org/argo.kml. See also EU-Argo RI, "Deployed Floats."

49. Marine Debris Tracker.

50. Schupska, "UGA's Marine Debris Tracker Named.'"

51. Ibid.

52. Llanos, "Average Joe as Oceanographer?"

53. Maximenko, Hafner, and Niiler, "Pathways of Marine Debris."

54. Lumpkin and Pazos, "Measuring Surface Currents with Surface Velocity Program Drifters."

55. Newsroom UNSW, "Our Plastics Will Pollute Oceans for Hundreds of Years."

56. Dohan and Maximenko, "Monitoring Ocean Currents with Satellite Sensors"; Lebreton, Greer, and Borrero, "Numerical Modeling"; Maximenko, Hafner, and Niiler, "Pathways of Marine Debris."

57. Sebille, England, and Froyland, "Origin, Dynamics, and Evolution."

58. Lumpkin, Maximenko, and Pazos, "Evaluating Where and Why Drifters Die."

59. U.S. EPA, "Marine Debris in the North Pacific."

60. Whitehead, *Process and Reality*, 110.

61. Ibid.

62. Stengers, *Thinking with Whitehead*, 223.

63. Ibid., 158.

64. James, *Pragmatism*, 127–64.

65. Whitehead, *Process and Reality*, 103, 105.

66. Ibid., 214–15.

67. Gabrys, "Plastic."

68. Helmreich, "Nature/Culture/Seawater"; Stengers, "A Constructivist Reading of *Process and Reality*."

69. Whitehead, *Adventures of Ideas*, 176.

70. Shaviro, *Without Criteria*, 47. See also Whitehead, *Adventures of Ideas*, 176.

71. Whitehead, *Process and Reality*, 88.

72. Stengers, "A Constructivist Reading of *Process and Reality*," 96.

73. Barthes, "Plastic," 97.

74. Bensaude-Vincent et al., "Matters of Interest."

75. Haraway, *When Species Meet*, 88–93; Barad, *Meeting the Universe Halfway*, 390–93.

76. Stengers, "Including Nonhumans in Political Theory."

6. SENSING AIR AND CREATURING DATA

1. While all of these pollutants affect cardiovascular and pulmonary health, particulate matter (PM) is of particular concern. As the World Health Organization (WHO) notes in a fact sheet on air quality, "PM affects more people than any other pollutant. The major components of PM are sulfate, nitrates, ammonia, sodium chloride, carbon, mineral dust and water. It consists of a complex mixture of solid and liquid particles of organic and inorganic substances suspended in the air. The particles are identified according to their aerodynamic diameter, as either PM (particles with an aerodynamic diameter smaller than 10 μm) or PM (aerodynamic diameter smaller than 2.5 μm). The latter are more dangerous since, when inhaled, they may reach the peripheral regions of the bronchioles, and interfere with gas exchange inside the lungs." See WHO, "Ambient (Outdoor) Air Quality."

2. London Air Quality Network (LAQN).

3. The EU air quality objective (2008) indicates that there should be no more than 40 $\mu m/m^3$ of NO_2 per year. The New Cross Road station (in the borough of Lewisham) recorded 51 $\mu m/m^3$ of NO_2 in 2013. See the London Air Quality Network (LAQN); European Commission, "Air Quality Standards."

4. Preemptive Media, "Area's Immediate Reading."

5. Safecast.

6. Aoki et al., "A Vehicle for Research"; Create Lab, "Speck"; AirCasting. One DIY-guide oriented toward air pollution sensing is Di Justo and Gertz, *Atmospheric Monitoring with Arduino*. Also see ongoing work in the Citizen Sense project in the area of monitoring air pollution.

7. U.S. EPA, "Draft Roadmap," 2. See also Snyder et al., "The Changing Paradigm."

8. Whitehead, *Process and Reality*, 22.

9. Ibid., 7, 16.

10. This notion of experience as constructive functioning has been discussed by Whitehead, *Process and Reality*, 156, and is taken up further by Shaviro in *Without Criteria*, 48.

11. Whitehead, *Modes of Thought*, 114.

12. The U.S. EPA list on "Toxic Air Pollutants" includes 187 hazardous air pollutants, including benzene, dioxin, mercury, and toluene. Some of these substances are carcinogenic, while others can cause serious health effects ranging from respiratory disease, birth defects, and even death. Very few of these substances are monitored regularly at air quality monitoring stations.

13. U.S. EPA, "What Are the Six Common Air Pollutants?"

14. UK Clean Air Act; U.S. EPA, "Clean Air Act"; European Commission, "Directive 2008/50/EC."

15. For example, see Choy, *Ecologies of Comparison*, 139–68.

16. For established limits for common pollutants, see the U.S. AirNow "Air Quality Index"; European Commission, "Air Quality Standards." For a discussion of the ways in which legal disputes become entangled in establishing both the matters of fact and concern of air pollution, see Jasanoff, "Thin Air." For a discussion on how exposure and harm become increasingly difficult to link within newer regimes of chemical living, particularly in relation to indoor air quality, see Murphy, *Sick Building Syndrome*.

17. Ambient PM pollution contributes to 3.2 million deaths annually, and there are increasing levels of heart disease, lung cancer, and cardiopulmonary disease in association with PM 2.5 exposure. See Lim et al., "A Comparative Risk Assessment." The WHO suggests that "exposure to air pollutants is largely beyond the control of individuals and requires action by public authorities at the national, regional, and even international levels." See WHO, "Ambient (Outdoor) Air Quality."

18. Air pollution is increasingly as much a rural environmental problem as an urban one. See Citizen Sense for project work on air pollution and fracking.

19. Bickerstaff and Walker, "The Place(s) of Matter"; Whitehead, *State Science and the Skies;* Irwin, *Citizen Science.*

20. Global Community Monitor. See also Ottinger, "Buckets of Resistance."

21. Lidskog and Sundqvist, *Governing the Air.*

22. Whitehead, *Process and Reality*, 203; Stengers, *Thinking with Whitehead*, 259.

23. U.S. EPA, "Draft Roadmap," 2–5.

24. Gabrys, Pritchard, and Barratt, "'Just Good Enough' Data."

25. Stengers, *Thinking with Whitehead*, 163–64, 518.

26. Ibid., 252.

27. Whitehead, *Process and Reality*, 11.

28. Ibid., 203; Stengers, *Thinking with Whitehead*, 259.

29. Whitehead, *Process and Reality*, 20.

30. As Whitehead notes, "The data upon which the subject passes judgment are themselves components conditioning the character of the judging subject. It follows that any presupposition as to the character of the experiencing subject also implies a general presupposition as to the social environment providing the display for that subject. In other words, a species of subject requires a species of data as its preliminary phase of concrescence. . . . The species of data requisite for the presumed judging subject presupposes an environment of a certain social character." Ibid., 203.

31. Ibid., 20.

32. Whitehead, *Process and Reality,* 233; Stengers, *Thinking with Whitehead,* 275.

33. Whitehead, *Process and Reality,* 155.

34. Ibid., 153–58. As Whitehead notes, the "subjectivist principle" assumes the "datum in the act of experience can be adequately analyzed purely in terms of universals"; and the "sensationalist principle" assumes the "bare subjective entertainment of the datum, devoid of any subjective form of *reception.*" Instead, "the philosophy of organism inverts this analysis" where there are not "objects for knowledge," but rather there are subjects with "experience." Ibid., 156–57.

35. Ibid., 68.

36. For example, see ibid., 219.

37. Bowker, *Memory Practices in the Sciences;* Barry, "Motor Ecology"; Whitehead, *State Science and the Skies;* Stengers, *The Invention of Modern Science.*

38. Bureau of Inverse Technology (BIT), "Feral Robot Engineering."

39. Jeremijenko, "Feral Robotic Dogs."

40. Lane et al. "Public Authoring & Feral Robotics."

41. As mentioned in the introduction to *Program Earth,* these links between information and action are described as "shorter circuits" that enable greater participation. See Bratton and Jeremijenko, "Suspicious Images, Latent Interfaces."

42. Costa, "Reaching the Limits"; Costa, Pigeon Blog.

43. Costa, Pigeon Blog.

44. Preemptive Media, "Area's Immediate Reading."

45. Costa, "Reaching the Limits," 377.

46. Ibid., 379.

47. Braidotti, *Transpositions.*

48. "Air Quality Egg," http://www.airqualityegg.com.

49. DiSalvo et al., "Towards a Public Rhetoric through Participatory Design."

50. For a more extensive discussion of the political economies of participation and maker communities in relation to DIY sensing technologies, see Gabrys, Pritchard, and Calvillo, "Making DIY Air Quality Sensors."

51. Aoki et al., "Common Sense."

52. ClientEarth, "Supreme Court Rules UK Government Is Breaking Air Pollution Laws."

53. Cf. Shaviro's discussion of ethics as unfolding first through aesthetic and affective registers in *The Universe of Things.*

54. Whitehead, *Process and Reality,* 20.

55. Ibid., 23. Whitehead defines ingression as "the particular mode in which the potentiality of an eternal object is realized in a particular actual entity, contributing to the definiteness of that actual entity." Eternal objects, never actual, are figures of pure potential that ingress within actual entities. Eternal objects are then an expression of potential, which nevertheless requires entities to become concrete.

56. Cohen, "Challenges and Benefits of Backyard Science."

57. Whitehead, *Science and the Modern World,* 200; Stengers, *Thinking with Whitehead,* 126–30.

58. Whitehead, *Process and Reality,* 12–13; Stengers, *Thinking with Whitehead,* 265.

59. Whitehead, *Process and Reality,* 88.

60. See Bowker, "Biodiversity Datadiversity"; and Gitelman, *"Raw Data" Is an Oxymoron.*

7. CITIZEN SENSING IN THE SMART AND SUSTAINABLE CITY

1. For example, see Archigram, *A Guide to Archigram;* Forrester, *Urban Dynamics.*

2. A wide range of studies in this area include Batty, "The Computable City"; Castells, *The Informational City*; Droege, *Intelligent Environments;* Gabrys, "Cité Multimedia"; Graham and Marvin, *Splintering Urbanism;* Mitchell, *City of Bits.*

3. Massumi, "National Enterprise Emergency," 155; Foucault, *The Birth of Biopolitics,* 261.

4. Cisco, "Smart + Connected Communities."

5. A study could be written just on the role of white papers within smart city developments. Crafted by industry, universities, and governmental agencies, smart city white papers appear to be a key way in which the imaginings and implementation of these urban developments circulate. The "circulation" of policy as discussed by Jennifer Robinson is part of the way in which cities accumulate multilocated "elsewheres" within projects of urban imagining. See Robinson, "The Spaces of Circulating Knowledge." The documents drafted in support of the Connected Urban Development project are similarly informed by multiple white papers, including Zhen et al., "Cities in Action for Climate Change," which was developed by Cisco, Metropolis, and Connected Urban Development, and gathers together details of eco-actions by and for cities around the world.

6. Foucault, "The Confession of the Flesh."

7. Foucault, *The Birth of Biopolitics.*

8. Allwinkle and Cruickshank, "Creating Smart-er Cities"; European Commission, "Green Digital Charter."

9. Lovink, "The Digital City"; Sassen, *Global Networks, Linked Cities.*

10. Ellison, Burrows, and Parker, "Urban Informatics"; Galloway, "Intimations of Everyday Life"; Graham, *The Cybercities Reader.*

11. Elfrink (for Cisco), "Intelligent Urbanization."

12. Harrison and Abbott Donnelly (for IBM), "A Theory of Smart Cities."

13. Townsend et al., "A Planet of Civic Laboratories," 4.

14. Hollands, "Will the Real Smart City Please Stand Up?"

15. Mackenzie, *Wirelessness.*

16. Borden and Greenfield, "You Are the Smart City."

17. Fuller and Haque, "Urban Versioning System"; Greenfield and Shepard, "Urban Computing and Its Discontents."

18. Foucault, *The Politics of Truth,* 44–45.

19. Foucault, *Security, Territory, Population,* 22–23; Foucault, *The Birth of Biopolitics,* 259–61.

20. Foucault, *The Birth of Biopolitics,* 260.

21. Foucault, *La naissance de la biopolitique;* Foucault, *The Birth of Biopolitics.*

22. Agrawal, *Environmentality.*

23. Luke, "Environmentality as Green Governmentality."

24. While the English version of this passage in *The Birth of Biopolitics* translates this term as "environmentalism," in the French original Foucault uses the term *environnementalité,* which is much closer to conveying the sensing of governmentality distributed through environments, rather than a social movement oriented toward environmental issues. See Foucault, *La naissance de la biopolitique,* 266; Foucault, *The Birth of Biopolitics,* 261.

25. Foucault, *The Birth of Biopolitics,* 259.

26. For example, see Anker, "The Closed World of Ecological Architecture"; Banham, *Architecture of the Well-Tempered Environment.*

27. Massumi, "National Enterprise Emergency."

28. Foucault, *Society Must Be Defended,* 244–45.

29. Ibid., 239–45.

30. Revel, "Identity, Nature, Life," 49–52.

31. See Agamben, *Homo Sacer.*

32. Hayles, "RFID."

33. Foucault, "The Confession of the Flesh," 194.

34. Ibid., 194–95; Agamben, *What Is an Apparatus?*

35. Mitchell and Casalegno, *Connected Sustainable Cities.*

36. Ibid.

37. Ibid., 97.

38. Ibid., 2.

39. Ibid., 58–59.

40. Deleuze, *Negotiations,* 97.

41. Mitchell and Casalegno, *Connected Sustainable Cities,* 98.

42. Gabrys, "Telepathically Urban," 58 and 50–59 passim.

43. Mitchell, *Me++.*

44. Weiser, "The Computer for the 21st Century."

45. Gabrys, "Automatic Sensation"; Hayles, "RFID."

46. Haraway, *Simians, Cyborgs and Women;* see also Taylor, "Technocratic Optimism."

47. Graham, "Software-Sorted Geographies"; Kitchin and Dodge, *Code/Space;* Thrift and French, "The Automatic Production of Space."

48. Gabrys, *Digital Rubbish;* Kittler, "There Is No Software."

49. Mackenzie, "The Performativity of Code," 88.

50. Mitchell and Casalegno, *Connected Sustainable Cities,* 5–6.

51. Gabrys, *Digital Rubbish.*

52. Mackenzie, *Wirelessness.*

53. Connected Urban Development.

54. Ibid.

55. Mitchell and Casalegno, *Connected Sustainable Cities,* 102.

56. Ibid., 89–91.

57. Ibid., 2.

58. Mitchell and Casalegno, *Connected Sustainable Cities,* 101.

59. Townsend et al., "A Planet of Civic Laboratories."

60. Barney, "Politics and Emerging Media." See also Couldry and Powell, "Big Data from the Bottom Up."

61. Mitchell and Casalegno, *Connected Sustainable Cities,* 98.

62. Deleuze, *Negotiations,* 182.

63. Foucault, *The Birth of Biopolitics,* 261.

64. For example, see Wiener, *Cybernetics.*

65. Foucault, *The Birth of Biopolitics,* 246–47.

66. Ibid., 252.

67. For example, see European Commission, "Report of the Meeting of Advisory Group."

68. Deleuze, *Negotiations,* 83–118.

69. Foucault, "Intellectuals and Power," 207.

8. ENGAGING THE IDIOT IN PARTICIPATORY DIGITAL URBANISM

1. SeeClickFix; FixMyStreet.

2. This approach, where computation provides a universal language for addressing problems, can be found in early writings on computing, from Leibniz on to Turing. In *The Universal Computer,* Davis describes Leibniz's dream to achieve a universal means of calculating problems, where even war could be addressed and solved through computation. This classic propensity of computation has been updated and brought into a critical space of analysis in numerous works, not least of which is Morozov's popular text, *To Save Everything, Click Here.* This chapter (as well as the whole of this section on Urban Sensing) is situated in relation to both longer-standing approaches to and more recent critiques of computation as a universal means for solving problems.

3. Stengers, "The Cosmopolitical Proposal."

4. Ibid., 994.

5. Ibid., 995.

6. Ibid.

7. As Mike Michael writes, "The idiot is as much a process as a figure." See Michael, "'What Are We Busy Doing?'"

8. Ibid. For a related discussion of the idiot in relation to twitter, see Hawkins, "Enacting Public Value."

9. For example, see Goriunova, "New Media Idiocy."

10. For example, see Canadian Centre for Architecture (CCA), *Actions; Spontaneous Interventions;* L'atelier d'architecture autogérée; Petrescu, Petcou, and Awan, *Trans-Local-Act;* Public Works; Awan, Schneider, and Till, *Spatial Agency.*

11. For example, see Lydon et al., *Tactical Urbanism, Volume 2;* Ferguson and Urban Drift Projects., *Make_Shift City.*

12. Code for America; Code For Europe; Adopt-a-Hydrant; Adopt-a-Siren; Adopt-a-Drain.

13. Pahlka, "Coding a Better Government."

14. Ibid.

15. Bell, "It's Appy Hour!"

16. Ibid.

17. Hern, "Seven Urban Apps."

18. Bell, "It's Appy Hour!" It bears mentioning that Nesta also provide a summary on how their predictions came to fare, and the civic apps prediction was noted as a "slightly disappointing sector," in that there was neither the explosion in civic apps nor the joining up of these apps to public services in ways that would make them as effective as anticipated. They reflect, "Closer integration of apps with the work of public service providers could help to see more scale, and notably, significant cost savings for cities and their citizens. Local government is traditionally a slow adopter and therefore moving from small experiments to genuine scale is always a challenge." A recurring narrative in many smart city and digital participation initiatives is that it is the slowness and disjointedness of local government that prevents these projects from having the effect they might otherwise have. See Nesta, "2013 Predictions."

19. Mejias, *Off the Network,* 6.

20. There is an extensive and advanced set of literatures that take up issues of digital participation and (free) labor. While there is not room here to discuss this topic, see, for example, Scholz, *Digital Labor.*

21. Goldman et al., "Participatory Sensing."

22. Ibid.

23. Ibid., 14. Emphasis in original.

24. Bell, "It's Appy Hour!"

25. Andrejevic, "Nothing Comes between Me and My CPU," 101–2.

26. Scholz, "What the MySpace Generation Should Know."

27. Stengers, "The Cosmopolitical Proposal," 995.

28. As cited in Davis, *The Universal Computer,* 17.

29. For two examples of urban science initiatives, see City Science; Center for Urban Science + Progress.

30. Fuller and Goffey, "Digital Infrastructures," 317.

31. As Anderson and Pold write, "The city dweller may implement, experience, and use urban scripts, but what happens when she is not satisfied with the foundation of participation in urban public life, the scripts themselves? This experience is a call for action where the city dwellers take control of the urban scripts and become 'writerly' by creating, hacking, and rewriting the city." See "The Scripted Spaces."

32. Sassen, "Open Source Urbanism."

33. Ibid.

34. Ibid.

35. For example, see Fraser's critique of Habermas in "Rethinking the Public Sphere."

36. See Davis, "Hacking as a Civic Duty"; alternatively, Cox, *Speaking Code.*

37. Gray Area Foundation for the Arts, "Urban Prototyping Festival."

38. Ibid.

39. Sustainable Society Network, "UP London-2013."

40. Urban Prototyping London, "Festival Showcase."

41. Jiménez, "The Right to Infrastructure."

42. TechniCity, "Coursera."

43. Deleuze, *Negotiations,* 179

44. See Arnstein, "A Ladder of Citizen Participation."

45. Stengers, "The Cosmopolitical Proposal," 996.

46. Ibid.

47. Deleuze, *Negotiations,* 179–80.

48. Ibid., 181–82.

49. Anderson and Pold, "The Scripted Spaces."

50. Latour, *Pandora's Hope;* Akrich, "The De-Scription of Technical Objects."

51. Suchman, *Human-Machine Reconfigurations.*

52. Andrejevic, "Nothing Comes between Me and My CPU," 102.

53. TechniCity. "Coursera."

54. Deleuze and Guattari, *What Is Philosophy,* 61–62.

55. TechniCity. "Coursera."

56. Mackenzie, *Transductions.*

57. Stengers, "The Cosmopolitical Proposal," 1000.

58. mySociety.

59. See FillThatHole; Citizens Connect.

60. FixMyStreet, "Abandoned Christmas Tree."

61. FixMyStreet, "Council Worker Mowing Rubbish!"

62. FixMyStreet, "Rubbish."

63. IBM Research–Brazil, "Citizen Sensing for Smarter Cities."

64. For representative examples of literature that attends to social media, digital media, and capacities of political engagement, see Boler, *Digital Media and Democracy*; Dean, *Democracy and Other Neoliberal Fantasies*.

65. For example, see Terranova, *Network Culture*; Castells, *Networks of Outrage and Hope*; Rossiter, *Organized Networks*.

66. Fish et al., "Birds of the Internet."

67. Barney, "Politics and Emerging Media."

68. Ratto and Boler, *DIY Citizenship*.

69. Iveson, "Mobile Media," 56.

70. For one example, see Townsend, *Smart Cities*.

71. See *Spontaneous Interventions*; PetaJarkata; Korsgaard and Brynskov, "Prototyping a Smart City"; Cox, "Support of Antisocial Notworking."

72. For a discussion of "versioning" in relation to cities, see Fuller and Haque, *Urban Versioning System*.

73. "The ABC of Tactical Media," a manifesto written by David Garcia and Geert Lovink, states, "Tactical Media are what happens when the cheap 'do it yourself' media, made possible by the revolution in consumer electronics and expanded forms of distribution (from public access cable to the internet) are exploited by groups and individuals who feel aggrieved by or excluded from the wider culture. Tactical media do not just report events, as they are never impartial they always participate and it is this that more than anything separates them from mainstream media."

74. Raley, *Tactical Media*, 28–29. See also Lovink and Rossiter, "Dawn of the Organized Networks."

75. Ibid. Raley draws here on Virno's discussion of virtuosity in *Grammar of the Multitude* (and Virno, in his thinking on collectives, is in turn influenced by Simondon).

76. Stengers, *Thinking with Whitehead*, 17.

77. Stengers, "The Cosmopolitical Proposal," 996.

78. Ibid., 998.

9. DIGITAL INFRASTRUCTURES OF WITHNESS

1. Combes, *Gilbert Simondon*, 12–14; Lamarre, "Afterword," 92–94.

2. This is an interesting point of contrast with Whitehead, who suggests that concrescences are a *less indeterminate* actual entity or actual occasion. Yet with his focus on technology, Simondon draws out the ways in which technical objects may actually give rise to conditions of indeterminateness. In addition to the above, see Simondon, *Du mode d'existence*; Simondon, "Technical Mentality"; Whitehead, *Process and Reality*, 150.

3. Whitehead, *Process and Reality*, 81.

4. Stengers, "The Cosmopolitical Proposal," 996.

5. See Graham, *Disrupted Cities*; and Bennett, "The Agency of Assemblages."

6. Boyle, Yates, and Yeatman, "Urban Sensor Data Streams"; Cambridge Centre for Smart Infrastructure and Construction, "Annual Review 2014"; Turgeman, Alm, and Ratti, "Smart Toilets."

7. Boyle, Yates, and Yeatman, "Urban Sensor Data Streams."

8. Greater London Authority, "Locations of TfL Street Cameras."

9. Ungerleider, "The London Underground"; Microsoft, "Internet of Things."

10. Cambridge Centre for Smart Infrastructure and Construction (CSIC), "Annual Review 2014."

11. City of Westminster, "Parking Bay Sensors."

12. Living PlanIT.

13. Environment Agency, "River and Sea Levels."

14. Environment Agency, "Air Pollution."

15. Monitrain, "Condition Monitoring Protects London's Thames Barrier."

16. Skyview, "London Eye"; Observator Instruments, "Wind Monitoring and Alarm System"; London Eye, "Systems and Controls."

17. Met Office, "Weather Stations"; Met Office, "Weather Observations Website."

18. Boyle, Yates, and Yeatman, "Urban Sensor Data Streams."

19. Living PlanIT. The London City Airport project is in part supported by the Technology Strategy Board (TSB).

20. Future Cities Catapult, "Sensing London."

21. Greater London Authority, "Smart London Vision"; Greater London Authority, "Smart London Plan"; Future Cities Catapult, "Sensing London." The TSB is a UK government-funded body.

22. Greater London Authority, "Smart London Plan," 3.

23. Boyle, Yates, and Yeatman, "Urban Sensor Data Streams," 19. As these authors write, Santander was deemed the smartest city in Europe "largely due to the thousands of sensors—monitoring parking and environmental conditions—deployed as part of a city-scale experiment to prototype typical ICT applications and services that might be found in a future smart city (www.smartsantander.eu). The experiment uses a tiered network infrastructure comprised of thousands of IEEE 802.15.4-enabled sensor nodes, mostly Wasp motes. The Spanish company Libelium (www.libelium.com) produces the devices, targeting embedded monitoring and control systems at the IoT-enabled smart city market."

24. For a discussion of Songdo, see Halpern et al., "Test Bed Urbanism."

25. Singer, "Mission Control, Built for Cities."

26. TechCentral.ie. "Dublin to Become First Fully 'Sensored' City."

27. IBM, "Smarter Cities Challenge."

28. Siemens, "The Crystal."

29. Greater London Authority, "Royal Albert Dock"; Royal Docks; Royal Albert Dock London.

30. In addition to Site IQ, the security system included access cards, an overall security system that automatically armed itself between 10 pm and 6 am, and CCTV cameras. Site IQ's pattern recognition was such that it could discern between a fox, which it would not track, and a human, which it would track. See Siemens, "Siveillance SiteIQ."

31. Miranda and Powell, *Our Urban Future.*

32. Ibid.

33. Ibid., 10.

34. Ibid., 18.

35. Ibid.

36. Siemens, "Creating Cities."

37. Miranda and Powell, *Our Urban Future*, 22.

38. Wiener, *The Human Use of Human Beings*, 22–25.

39. Sustainable Society Network, "UP London 2013."

40. This logic fits with James Beniger's classic description of the cycle of communication technologies developed to control conditions of overload, while also contributing to new conditions of information overload. See Beniger, *The Control Revolution*.

41. Boyle, Yates, and Yeatman, "Urban Sensor Data Streams," 19–20.

42. Kranenburg, "The Internet of Things."

43. Wiener, *Cybernetics*, 5.

44. See CityDashboard. http://citydashboard.org/london.

45. Easterling, "The Action Is the Form," 155.

46. Combes, *Gilbert Simondon*, 15.

47. Simondon, *Du mode d'existence*, 15.

48. Wiener, *Human Use of Human Beings*, 22–27.

49. Simondon, *Du mode d'existence*, 13.

50. Simone, "Infrastructure."

51. Simone, "People as Infrastructure," 407.

52. Mackenzie, *Wirelessness*, 49.

53. Ibid., 96.

54. Ibid., 56.

55. Graham, *Disrupted Cities*, 8.

56. Simone, "Cities of Uncertainty."

57. Weiser, "The Computer for the 21st Century," 94.

58. Ibid., 98.

59. Ibid., 94.

60. Ibid., 99.

61. Ibid., 102.

62. Simondon, *Du mode d'existence*, 55.

63. Ibid.

64. Mackenzie, *Wirelessness*, 20. On withness, see also Munster, *An Aesthesia of Networks*.

65. James, *Pragmatism*, 69. James also suggests that "ordinary empiricism over-emphasizes" withness, whereas his proposal for "radical empiricism" he considered to be "fair to both the unity and the disconnection of experience." See James, *Radical Empiricism*, 47.

66. Whitehead, *Process and Reality*, 62–64, 73, 81.

67. Ibid., 80.

68. Ibid., 79.

69. Whitehead, *Adventures of Ideas*, 69–86.

70. Stengers, *Thinking with Whitehead*, 161–63; cf. Debaise, "A Philosophy of Interstices," 108.

71. Foucault, *The Politics of Truth*, 44–45.

72. Miranda and Powell, *Our Urban Future*.

73. Stengers, *Thinking with Whitehead*, 22, 273. Stengers cites Whitehead, *Process and Reality*, 4.

CONCLUSION

1. Guattari, *Schizoanalytic Cartographies*, 5–13.
2. Ibid., 11–12
3. Weiser, "The Computer for the 21st Century," 104.
4. Simondon, *Du mode d'existence*, 15; Stengers, *Thinking with Whitehead*, 190–92.
5. Massumi, *Semblance and Event*, 173.
6. Combes, *Gilbert Simondon*, 67; Simondon, *Du mode d'existence*, 221.
7. Canguilhem, *Knowledge of Life*, 113. Cf. Braun, "Environmental Issues."
8. Whitehead, *Process and Reality*, 9.
9. Ibid., 35, 42. See also Stengers, *Thinking with Whitehead*, 18.
10. Stengers, *Thinking with Whitehead*, 19.
11. Jasanoff, "Technologies of Humility."
12. Stengers, *Thinking with Whitehead*.
13. Simondon, "Technical Mentality."
14. Stengers, *Thinking with Whitehead*, 17.
15. Ibid.
16. Whitehead, *Modes of Thought*, 49.
17. Stengers, *Thinking with Whitehead*, 213.
18. Combes, *Gilbert Simondon*, 64–65; Stengers, *Thinking with Whitehead*, 164.

BIBLIOGRAPHY

Adopt-a-Drain. http://adoptadrainoakland.com.

Adopt-a-Hydrant. http://www.adoptahydrant.org.

Adopt-a-Siren. http://sirens.honolulu.gov.

Agamben, Giorgio. *Homo Sacer: Sovereign Power and Bare Life.* Translated by Daniel Heller-Roazen. Stanford, Calif.: Stanford University Press, 1998.

———. *What Is an Apparatus?* Translated by David Kishik and Stefan Pedatella. Stanford, Calif.: Stanford University Press, 2009.

Agrawal, Arun. *Environmentality: Technologies of Government and the Making of Subjects.* Durham, N.C.: Duke University Press, 2005.

AirCasting. http://aircasting.org.

AirNow. "Air Quality Index." http://airnow.gov.

Air Quality Egg. http://www.airqualityegg.com.

Akrich, Madeline. "The De-Scription of Technical Objects." In *Shaping Technology / Building Society: Studies in Sociotechnical Change,* edited by Wiebe E. Bijker and John Law, 205–24. Cambridge, Mass.: MIT Press, 1992.

Akyildiz, Ian F., Weilian Su, Yogesh Sankarasubramaniam, and Erdal Cayirci. "A Survey on Sensor Networks." *IEEE Communications Magazine* (August 2002): 102–14.

Allan, Blake M., John P. Y. Arnould, Jennifer K. Martin, and Euan G. Ritchie. "A Cost-Effective and Informative Method of GPS Tracking Wildlife." *Wildlife Research* 40, no. 5 (2013): 345–48.

Allsopp, Michelle, Adam Walters, David Santillo, and Paul Johnston. "Plastic Debris in the World's Oceans." Amsterdam: Greenpeace International, 2006. http://www.greenpeace.org/international/en/publications/reports/plastic_ocean_report.

Allwinkle, Sam, and Peter Cruickshank. "Creating Smart-er Cities: An Overview." *Journal of Urban Technology* 18 (2011): 1–16.

Amezian, Mohamed, and Rachid El Khamlichi. "White Stork Satellite-Tracked by German Ornithologists Found Dead South of Tangier." *Moroccan Birds* (blog). December 11, 2013. http://moroccanbirds.blogspot.com/2013/12/satellite-tracked-white-stork-dead.html.

Anderson, Christian Ulrik, and Søren Pold. "The Scripted Spaces of Urban Ubiquitous Computing: The Experience, Poetics, and Politics of Public Scripted Space." *Fibreculture* 19 (2011): 110–25.

Andrady, Anthony L. "Microplastics in the Marine Environment." *Marine Pollution Bulletin* 62, no. 8 (2011): 1569–605.

Andrejevic, Mark. "Nothing Comes between Me and My CPU: Smart Clothes and Ubiquitous Computing." *Theory, Culture & Society* 22, no. 3 (June 2005): 110–19.

———. "The Webcam Subculture and the Digital Enclosure." In *Mediaspace: Place, Scale, and Culture in a Media Age,* edited by Anna McCarthy and Nick Couldry, 193–208. London: Routledge, 2004.

Anisimov, O. A., D. G. Vaughan, T. V. Callaghan, C. Furgal, H. Marchant, T. D. Prowse, H. Vilhjálmsson, and J. E. Walsh. "Polar Regions (Arctic and Antarctic)." In *Climate Change 2007: Impacts, Adaptation and Vulnerability. Contribution of Working Group II to the Fourth Assessment Report of the Intergovernmental Panel on Climate Change,* edited by M. L. Parry, O. F. Canziani, J. P. Palutikof, P. J. van der Linden, and C. E. Hanson, 653–85. Cambridge: Cambridge University Press, 2007.

Anker, Peter. "The Closed World of Ecological Architecture." *Journal of Architecture* 10 (2005): 527–52.

Aoki, Paul M., R. J. Honicky, Alan Mainwaring, Chris Myers, Eric Paulos, Sushmita Subramanian, and Allison Woodruff. "Common Sense: Mobile Environmental Sensing Platforms to Support Community Action and Citizen Science." In *Adjunct Proceedings UbiComp 2008, 59–60.* Seoul: ACM, 2008.

———. "A Vehicle for Research: Using Street Sweepers to Explore the Landscape of Environmental Community Action." In *CHI '09: Proceedings of the SIGCHI Conference on Human Factors in Computing Systems,* 375–84. New York: ACM, 2009.

Archigram. *A Guide to Archigram, 1961–74.* London: Academy Editions, 1994.

Arctic Monitoring and Assessment Programme (AMAP). *Assessment 2009: Radioactivity in the Arctic.* Oslo: AMAP, 2010.

———. *Report 2010.* Oslo: AMAP: 2010.

Arctic Perspective Initiative. http://arcticperspective.org.

Arctic Tern Migration Project. http://www.arctictern.info.

Argos. http://www.argos-system.org.

Arnstein, Sherry. "A Ladder of Citizen Participation." *Journal of the American Planning Association* 35, no. 4 (July 1969): 216–24.

Arthur, Charles. "Smartphone Explosion in 2014 Will See Ownership in India Pass US." *Guardian.* January 13, 2014. http://www.theguardian.com/technology/2014/jan/13/smartphone-explosion-2014-india-us-china-firefoxos-android.

Arthur, Courtney, Joel Baker, and Holly Bamford, eds. *Proceedings of the International Research Workshop on the Occurrence, Effects and Fate of Microplastic Marine Debris.* Silver Spring, Md.: National Oceanic and Atmospheric Administration. January 2009.

Atlas of Community-Based Monitoring in a Changing Arctic. http://www.arcticcbm.org/index.html.

Aubin, David, Charlotte Bigg, and Otto Sibum, eds. *The Heavens on Earth: Observatories and Astronomy in Nineteenth Century Science and Culture.* Durham, N.C.: Duke University Press, 2010.

Awan, Nishat, Tatjana Schneider, and Jeremy Till. *Spatial Agency: Other Ways of Doing Architecture.* London: Routledge, 2011.

Banham, Reyner. *Architecture of the Well-Tempered Environment: Theory and Design in the First Machine Age.* Chicago: University of Chicago Press, 1984.

Barad, Karen. *Meeting the Universe Halfway: Quantum Physics and the Entanglement of Matter and Meaning.* Durham, N.C.: Duke University Press, 2007.

———. "Posthumanist Performativity: Toward an Understanding of How Matter Comes to Matter." *Signs: Journal of Women in Culture and Society* 28, no. 3 (2003): 801–31.

Barnes, David K. A., Francois Galgani, Richard C. Thompson, and Morton Barlaz. "Accumulation and Fragmentation of Plastic Debris in Global Environments." *Philosophical Transactions of the Royal Society B: Biological Sciences* 364, no. 1526 (2009): 1985–98.

Barney, Darin. "Politics and Emerging Media: The Revenge of Publicity." *Global Media Journal* 1 (2008): 89–106.

Barry, Andrew. "Motor Ecology: The Political Chemistry of Urban Air." Critical Urban Studies: Occasional Papers. London: Centre for Urban and Community Research at Goldsmiths College, 1998.

Barthes, Roland. "Plastic." In *Mythologies,* translated by Annette Lavers, 97–99. 1957. Reprint, New York: Farrar, Straus and Giroux, 1972.

Bateson, Gregory. *Steps to an Ecology of Mind.* Chicago: University of Chicago Press, 1972.

Batty, Michael. "The Computable City." *International Planning Studies* 2 (1995): 155–73.

Bauer, Chris. "Tagging Pacific Predators." Video Story for QUEST Northern California. May 20, 2008. http://science.kqed.org/quest/video/tagging-pacific-predators.

BBC England. "Webcams: Animals." http://www.bbc.co.uk/england/webcams/animals/.

BBC News. "Secret Life of the Cat: What Do Our Feline Companions Get Up To?" http://www.bbc.co.uk/news/science-environment-22567526.

Bell, Haidee. "It's Appy Hour!" Nesta. http://www.nesta.org.uk/news/13-predictions-2013/its-appy-hour.

Bellacasa, María Puig de la. "Touching Technologies, Touching Visions: The Reclaiming of Sensorial Experience and the Politics of Speculative Thinking." *Subjectivity* 28 (2009): 297–315.

Beniger, James R. *The Control Revolution: Technological and Economic Origins of the Information Society.* Cambridge, Mass.: Harvard University Press, 1989.

Benjamin, David, Soo-In Yang, and Natalie Jeremijenko. "New Interaction Partners for Environmental Governance." In *Sentient City: Ubiquitous Computing, Architecture and the Future of Urban Space,* edited by Mark Shepard, 48–63. New York: Architectural League of New York, 2011.

Bennett, Jane. "The Agency of Assemblages and the North American Blackout." *Public Culture* 17, no. 3 (Fall 2005): 445–65.

Bensaude-Vincent, Bernadette, Sacha Loeve, Alfred Nordmann, and Astrid Schwarz. "Matters of Interest: The Objects of Research in Science and Technoscience." *Journal for General Philosophical Science* 42, no. 2 (2011): 365–83.

Benson, Etienne. "One Infrastructure, Many Global Visions: The Commercialization and Diversification of Argos, a Satellite-Based Environmental Surveillance System." *Social Studies of Science* 42, no. 6 (2012): 843–68.

———. *Wired Wilderness: Technologies of Tracking and the Making of Modern Wildlife.* Baltimore, Md.: Johns Hopkins University Press, 2010.

Berkeley, Edmund C. "Edmund C. Berkeley Papers" (1923–1988). University of Minnesota Libraries. Charles Babbage Institute.

Bickerstaff, Karen, and Gordon Walker. "The Place(s) of Matter: Matter out of Place—Public Understandings of Air Pollution." *Progress in Human Geography* 27, no. 1 (2003): 45–67.

BirdsEye Tools for Birders. "BirdLog North America." http://www.birdseyebirding.com.

Biuw, Martin, Lars Boehme, Christophe Guinet, Mark Hindell, Daniel Costa, Jean-Benoît Charrassin, Fabien Roquet, Frédéric Bailleul, Michael Meredith, Sally Thorpe, Yann Tremblay, Birgitte McDonald, Young-Hyang Park, Stephen R. Rintoul, Nathaniel Bindoff, Michael Goebel, Daniel Crocker, Phil Lovell, J. Nicholson, F. Monks, and Michael A. Fedak. "Variations in Behavior and Condition of a Southern Ocean Top Predator in Relation to *In Situ* Oceanographic Conditions." *Proceedings of the National Academy of Sciences* 104, no. 34 (August 21, 2007): 13705–10.

Block, Barbara A., Ian D. Jonsen, Salvador J. Jorgensen, Arliss J. Winship, Scott A. Shaffer, Steven J. Bograd, Elliott Lee Hazen, David G. Foley, G. A. Breed, A.-L. Harrison, James E. Ganong, Alan Swithenbank, Michael Castleton, Heidi Dewar, Bruce R. Mate, George L. Shillinger, K. M. Schaefer, Scott R. Benson, Michael J. Weise, Robert W. Henry, and Daniel P. Costa. "Tracking Apex Marine Predator Movements in a Dynamic Ocean." *Nature* 475 (July 7, 2011): 86–90.

Boerger, Christiana M., Gwendolyn L. Lattin, Shelly L. Moore, and Charles J. Moore. "Plastic Ingestion by Planktivorous Fishes in the North Pacific Central Gyre." *Marine Pollution Bulletin* 60, no. 12 (2010): 2275–78.

Boler, Megan, ed. *Digital Media and Democracy: Tactics in Hard Times*. Cambridge, Mass.: MIT Press, 2008.

Borden, Ed, and Adam Greenfield. "You Are the Smart City." June 30, 2011. https://xively logmein4.wpengine.com/blog/notes/you-are-the-smart-city.

Borgman, Christine L., Jillian C. Wallis, and Noel Enyedy. "Little Science Confronts the Data Deluge: Habitat Ecology, Embedded Sensor Networks, and Digital Libraries." *International Journal on Digital Libraries* 7, no. 1 (2007): 17–30.

Bowker, Geoffrey C. "Biodiversity Datadiversity." *Social Studies of Science* 30, no. 5 (2000): 643–83.

———. "How to Be Universal: Some Cybernetic Strategies, 1943–1970." *Social Studies of Science* 23, no. 1 (1993): 107–27.

———. *Memory Practices in the Sciences*. Cambridge, Mass.: MIT Press, 2005.

Boyle, David E., David C. Yates, and Eric M. Yeatman. "Urban Sensor Data Streams." *Internet Computing, IEEE* 17, no. 6 (2013): 12–20.

Bradley, Joseph, Jeff Loucks, James Macaulay, and Andy Noronha. "Internet of Everything (IoE) Value Index." *Cisco Whitepaper*. 2013. http://internetofeverything.cisco.com/sites/default/files/docs/en/ioe-value-index_Whitepaper.pdf.

Braidotti, Rosi. *Transpositions: On Nomadic Ethics*. Cambridge: Polity, 2006.

Bratton, Benjamin, and Natalie Jeremijenko. "Suspicious Images, Latent Interfaces." In *Situated Technologies Pamphlets 3*. New York: Architectural League of New York, 2008.

Braun, Bruce. "Environmental Issues: Inventive Life." *Progress in Human Geography* 32, no. 5 (2008): 667–79.

Brigham-Grette, Julie, Martin Melles, Pavel Minyuk, Andrei Andreev, Pavel Tarasov, Robert DeConto, Sebastian Koenig, Norbert Nowaczyk, Volker Wennrich, Peter Rosén, Eeva

Haltia, Tim Cook, Catalina Gebhardt, Carsten Meyer-Jacob, Jeff Snyder, Ulrike Herz-schuh. "Pliocene Warmth, Polar Amplification, and Stepped Pleistocene Cooling Recorded in NE Arctic Russia." *Science* 340, no. 21 (June 2013): 1421–27.

Brown, Danielle D., Roland Kays, Martin Wikelski, Rory Wilson, and A. Peter Klimley. "Observing the Unwatchable through Acceleration Logging of Animal Behavior." *Animal Biotelemetry* 1, no. 20 (2013): 1–16.

Bureau of Inverse Technology (BIT). "Feral Robot Engineering." http://www.bureauit.org/feral.

Burke, Jeff, Deborah Estrin, Mark Hansen, Andrew Parker, Nithya Ramanathan, Sasank Reddy, and Mani B. Srivastava. "Participatory Sensing." In *Proceedings of the World Sensor Web Workshop*. Boulder, Colo.: ACM SENSYS, 2006.

Cambridge Centre for Smart Infrastructure and Construction (CSIC). "Annual Review 2014." http://www-smartinfrastructure.eng.cam.ac.uk/news/20140327csicannualreview2014released.

Canadian Centre for Architecture (CCA). "Actions: What You Can Do with the City." http://cca-actions.org.

Canguilhem, Georges. *Knowledge of Life*. Translated by Stefanos Geroulanos and Daniela Ginsburg. 1965. Reprint, New York: Fordham University Press, 2008.

Carbon Monitoring for Action (CARMA). http://carma.org.

Carpenter, Edward J., and K. L. Smith Jr. "Plastics on the Sargasso Sea Surface." *Science* 175, no. 4027 (1972): 1240–41.

Castells, Manuel. *The Informational City: Information Technology, Economic Restructuring, and the Urban-Regional Process*. Oxford: Blackwell, 1989.

———. *Networks of Outrage and Hope: Social Movements in the Internet Age*. Cambridge: Polity, 2012.

Center for Embedded Network Sensing (CENS). University of California, Los Angeles. http://research.cens.ucla.edu.

———. "Annual Progress Report." University of California, Los Angeles. April 30, 2010.

———. "Center for Embedded Networked Sensing." NSF Science and Technology Centers proposal.

———. "Terrestrial Ecology Observing Systems." http://research.cens.ucla.edu/areas/2007/Terrestrial/default.htm.

Center for Urban Science + Progress. http://cusp.nyu.edu.

Chan, Wilson. "IPv6: Internet Protocol Version 6." http://www2.hawaii.edu/~esb/prof/proj/ipv6/wilsonch/IPv6.html.

Choy, Tim. *Ecologies of Comparison: An Ethnography of Endangerment in Hong Kong*. Durham, N.C.: Duke University Press, 2011.

Cisco. "Internet of Everything." http://www.youtube.com/watch?v=M578lU2TGeI.

———. "Smart + Connected Communities." http://www.cisco.com/web/strategy/smart_connected_communities.html.

Citizens Connect. http://www.cityofboston.gov/DoIT/apps/commonwealthconnect.asp.

Citizen Sense. http://citizensense.net.

CityDashboard. http://citydashboard.org/london.

City of Westminster. "Parking Bay Sensors." https://www.westminster.gov.uk/parking-bay-sensors.

City Science. http://cities.media.mit.edu.

Clarke, Bruce. "John Lilly, The Mind of the Dolphin, and Communication Out of Bounds." *communication +1* 3, no. 1 (2014). http://scholarworks.umass.edu/cgi/viewcontent.cgi?article=1025&context=cpo.

ClientEarth. "Supreme Court Rules UK Government Is Breaking Air Pollution Laws." May 1, 2013. http://www.clientearth.org/201305012170/news/press-releases/supreme-court-rules-uk-government-is-breaking-air-pollution-laws-2170.

Clough, Patricia Ticineto. "The New Empiricism: Affect and Sociological Method." *European Journal of Social Theory* 12, no. 1 (2009): 43–61.

Code for America. http://codeforamerica.org.

Code for Europe. http://codeforeurope.net.

Coen, Deborah R. *The Earthquake Observers: Disaster Science from Lisbon to Richter.* Chicago: University of Chicago Press, 2013.

Cohen, Ronald C. "Challenges and Benefits of Backyard Science." *Sensing Change.* Philadelphia, Pa.: Chemical Heritage Foundation. http://sensingchange.chemheritage.org/sensing-change/science/ronald-c-cohen.

Collier, Brian D., Chris Ray, and Shana Weber. "The Pika Alarm." http://www.pikalarm.net.

Combes, Muriel. *Gilbert Simondon and the Philosophy of the Transindividual.* Translated by Thomas Lamarre. 1999. Reprint, Cambridge, Mass.: MIT Press, 2013.

Committee on Emerging Research Questions in the Arctic: Polar Research Board; Division on Earth and Life Studies; National Research Council. *The Arctic in the Anthropocene: Emerging Research Questions.* Washington, D.C.: National Academies Press, 2014.

Committee on Scientific Accomplishments of Earth Observations from Space, National Research Council. *Earth Observations from Space: The First 50 Years of Scientific Achievements.* Washington, D.C.: National Academies Press, 2008.

Committee on Understanding and Monitoring Abrupt Climate Change and Its Impacts; Board on Atmospheric Sciences and Climate; Division on Earth and Life Studies; National Research Council. "Abrupt Impacts of Climate Change: Anticipating Surprises." Washington, D.C.: National Academies Press, 2013.

Connected Sustainable Cities. http://connectedsustainablecities.com.

Connected Urban Development. http://mobile.mit.edu/projects/connected-urban-development; https://web.archive.org/web/20130621085500/http://www.connectedurbandevelopment.org.

———. "Connected Urban Development Visions from MIT's Mobile Experience Lab." Video. https://vimeo.com/6145800.

Cooke, Steven J. "Biotelemetry and Biologging in Endangered Species Research and Animal Conservation: Relevance to Regional, National, and IUCN Red List Threat Assessments." *Endangered Species Research* 4 (2008): 165–85.

Cooper, Caren B., Jennifer Shirk, and Benjamin Zuckerberg. "The Invisible Prevalence of Citizen Science in Global Research: Migratory Birds and Climate Change." *PLoS One* 9, no. 9. (September 2014): e106508.

Corburn, Jason. *Street Science: Community Knowledge and Environmental Health Justice.* Cambridge, Mass.: MIT Press, 2005.

Cornell Lab of Ornithology. "All about Birds." http://cams.allaboutbirds.org.

———. Macaulay Library. http://macaulaylibrary.org.

Cornwell, Myriah L., and Lisa M. Campbell. "Co-Producing Conservation and Knowledge: Citizen-Based Sea Turtle Monitoring in North Carolina, USA." *Social Studies of Science* 42, no. 1 (2012): 101–20.

Cosgrove, Denis. "Contested Global Visions: One-World, Whole-Earth, and the Apollo Space Photographs." *Annals of the Association of American Geographers* 84, no. 2 (1994): 270–94.

Costa, Beatriz da. "Pigeon Blog." http://web.archive.org/web/20120418181942/;http://www.beatrizdacosta.net/pigeonblog.php.

———. "Reaching the Limits." In *Tactical Biopolitics: Art, Activism, and Technoscience*, edited by Beatriz da Costa and Kavita Philip, 365–85. Cambridge, Mass.: MIT Press, 2008.

Couldry, Nick, and Alison Powell. "Big Data from the Bottom Up." *Big Data & Society* 1, no. 2 (July–December 2014): 1–5.

Cox, Geoff. *Speaking Code: Coding as Aesthetic and Political Expression*. Cambridge, Mass.: MIT Press, 2013.

———. "Support of Antisocial Notworking." http://www.anti-thesis.net/?p=869.

Cózar, Andrés, Fidel Echevarría, J. Ignacio González-Gordillo, Xabier Irigoien, Bárbara Úbeda, Santiago Hernández-León, Álvaro T. Palma, Sandra Navarro, Juan García-de-Lomas, Andrea Ruiz, María L. Fernández-de-Puelles, and Carlos M. Duartei. "Plastic Debris in the Open Ocean." *Proceedings of the National Academy of Sciences* 111, no. 28 (July 15, 2014): 10239–44.

Create Lab. Speck. http://www.specksensor.com.

"Critter Cams Viewers Becoming Citizen Scientists." Narrated by Robin Young with Vicki Croke. *Here & Now*. WBUR, July 4, 2014. http://hereandnow.wbur.org/2014/07/04/critter-cams-scientists.

Cruikshank, Julie. *Do Glaciers Listen? Local Knowledge, Colonial Encounters, and Social Imagination*. Vancouver: University of British Columbia Press, 2005.

Cubasch, Ulrich, Donald Wuebbles, Deliang Chen, Maria Cristina Facchini, David Frame, Natalie Mahowald, and Jan-Gunnar Winther. "Introduction." In *IPCC, 2013: Climate Change 2013; The Physical Science Basis. Contribution of Working Group I to the Fifth Assessment Report of the Intergovernmental Panel on Climate Change*, edited by Thomas F. Stocker, Dahe Qin, Gian-Kasper Plattner, Melinda Tignor, Simon K. Allen, Judith Boschung, Alexander Nauels, Yu Xia, Vincent Bex and Pauline M. Midgley, 119–58. Cambridge: Cambridge University Press, 2013.

Cuff, Dana, Mark Hansen, and Jerry Kang. "Urban Sensing: Out of the Woods." *Communications of the Association for Computing Machinery* 51, no. 3 (2008): 24–33.

Daston, Lorraine, and Peter Galison. *Objectivity*. Brooklyn: Zone Books, 2007.

Davis, Martin. *The Universal Computer: The Road from Leibniz to Turing*. New York: W. W. Norton, 2000.

Davis, Paul M. "Hacking as a Civic Duty." March 23, 2012. http://www.shareable.net/blog/hacking-as-a-civic-duty.

Dean, Jodi. *Democracy and Other Neoliberal Fantasies: Communicative Capitalism and Left Politics*. Durham, N.C.: Duke University Press, 2009.

Dean, Jodi, Jon W. Anderson, and Geert Lovink, eds. *Reformatting Politics: Information Technology and Global Civil Society*. London: Routledge, 2006.

Debaise, Didier. "A Philosophy of Interstices: Thinking Subjects and Societies from Whitehead's Philosophy." *Subjectivity* 6, no. 1 (2013): 101–11.

———. *Un empirisme spéculatif: Lecture de Procès et Réalité de Whitehead*. Paris: Vrin, 2006.

Deleuze, Giles. *Negotiations: 1972–1990*. Translated by Martin Joughin. 1990. Reprint, New York: Columbia University Press, 1995.

Deleuze, Giles, and Félix Guattari. *What Is Philosophy?* Translated by Graham Burchell and Hugh Tomlinson. 1991. Reprint, London: Verso, 1994.

Delwiche, Aaron, and Jennifer Jacobs Henderson, eds. *The Participatory Cultures Handbook.* London: Routledge, 2012.

Derraik, José G. B. "The Pollution of the Marine Environment by Plastic Debris: A Review." *Marine Pollution Bulletin* 44, no. 9 (2002): 842–52.

Diederichsen, Diedrich, and Anselm Frank. *The Whole Earth: California and the Disappearance of the Outside.* Berlin: Haus der Kulteren der Welt, 2013.

Di Justo, Patrick, and Emily Gertz. *Atmospheric Monitoring with Arduino.* Sebastopol, Calif.: O'Reilly Media Press, 2013.

Di Lorenzo, Emanuele, Niklas Schneider, Kim M. Cobb, P. J. S. Franks, Kettyah C. Chhak, Arthur J. Miller, James C. McWilliams, Steven J. Bograd, Hernan Arango, Enrique Curchitser, Thomas M. Powell, and Pascal Rivière. "North Pacific Gyre Oscillation Links Ocean Climate and Ecosystem Change." *Geophysical Research Letters* 35 (2008): 1–6.

DiSalvo, Carl, Marti Louw, David Holstius, Illah Nourbakhsh, and Ayca Akin. "Towards a Public Rhetoric through Participatory Design: Critical Engagements and Creative Expression in the Neighborhood Networks Project." *Design Issues* 28, no. 3 (2012): 48–61.

Dodge, Somayeh, Gil Bohrer, Rolf Weinzierl, Sarah C. Davidson, Roland Kays, David Douglas, Sebastian Cruz, Jiawei Han, David Brandes, and Martin Wikelski. "The Environmental-Data Automated Track Annotation (Env-DATA) System: Linking Animal Tracks with Environmental Data." *Movement Ecology* 1, no. 3 (2013): 1–14.

Dohan, Kathleen, and Nikolai Maximenko. "Monitoring Ocean Currents with Satellite Sensors." *Oceanography* 23, no. 4 (2010): 94–103.

Dourish, Paul, and Genevieve Bell. *Divining a Digital Future: Mess and Mythology in Ubiquitous Computing.* Cambridge, Mass.: MIT Press, 2011.

Droege, Peter, ed. *Intelligent Environments: Spatial Aspects of the Information Revolution.* Amsterdam: Elsevier, 1997.

Dyo, Vladimir, Stephen A. Ellwood, David W. MacDonald, Andrew Markham, Cecilia Mascolo, Bence Pásztor, Niki Trigoni, and Ricklef Wohlers. "Wildlife and Environmental Monitoring Using RFID and WSN Technology." In *Proceedings of the 7th ACM Conference on Embedded Networked Sensor Systems* (SenSys 2009). New York, N.Y.: ACM, 2009.

Dyo, Vladimir, Stephen A. Ellwood, David W. MacDonald, Andrew Markham, Niki Trigoni, Ricklef Wohlers, Cecilia Mascolo, Bence Pasztor, Salvatore Scellato, and Kharsim Yousef. "WILDSENSING: Design and Deployment of a Sustainable Sensor Network for Wildlife Monitoring." *ACM Transactions on Embedded Computing Systems* 9, no. 4, (March 2010): 1–33.

Earl, Jennifer, and Katrina Kimport. *Digitally Enabled Social Change: Activism in the Internet Age.* Cambridge, Mass.: MIT Press, 2011.

Easterling, Keller. "The Action Is the Form." In *Sentient Cities: Ubiquitous Computing, Architecture, and the Future of Urban Space,* edited by Mark Shepard, 154–58. Cambridge, Mass.: MIT Press, 2011.

Ebbesmeyer, Curtis. "Using Flotsam to Study Ocean Currents." NASA Ocean Motion and Surface Currents. http://oceanmotion.org/html/gatheringdata/flotsam.htm.

Ebbesmeyer, Curtis, W. James Ingraham Jr., Thomas C. Royer, and Chester E. Grosch, "Tub Toys Orbit the Subarctic Gyre." *EOS, Transactions of the American Geophysical Union* 88, no. 1 (2007): 1–12.

Ebbesmeyer, Curtis, and Eric Scigliano. *Flotsametrics and the Floating World: How One Man's Obsession with Runaway Sneakers and Rubber Ducks Revolutionized Ocean Science.* New York: HarperCollins, 2009.

Edwards, Paul N. *A Vast Machine: Computer Models, Climate Data, and the Politics of Global Warming.* Cambridge, Mass.: MIT Press, 2010.

Edwards, Paul N., Geoffrey C. Bowker, Steven J. Jackson, and Robin Williams. "Introduction: An Agenda for Infrastructure Studies." *Journal of the Association for Information Systems* 10, no. 5 (May 2009): 364–74.

Egevang, Carsten, Iain J. Stenhouse, Richard A. Phillips, Aevar Petersen, James W. Fox, and Janet R. D. Silk. "Tracking of Arctic Terns *Sterna paradisaea* Reveals Longest Animal Migration." *Proceedings of the National Academy of Sciences* 107, no. 5 (February 2, 2010): 1–4.

Ekman, Ulrik, ed. *Throughout: Art and Culture Emerging with Ubiquitous Computing.* Cambridge, Mass.: MIT Press, 2013.

Elfrink, Wim. "Intelligent Urbanization." 2009. http://blogs.cisco.com/news/video.

Elichirigoity, Fernando. *Planet Management: Limits to Growth, Computer Simulation, and the Emergence of Global Spaces.* Evanston, Ill.: Northwestern University Press, 1999.

Ellis, Rebecca, and Clare Waterton. "Environmental Citizenship in the Making: The Participation of Volunteer Naturalists in UK Biological Recording and Biodiversity Policy." *Science and Public Policy* 31, no. 2. (2004): 95–110.

Ellison, Nick, Roger Burrows, and Simon Parker, eds. "Urban Informatics: Software, Cities and the New Cartographies of Knowing Capitalism." *Information, Communication and Society* 10, no. 6 (2007): 785–960.

Elson, Jeremy, and Deborah Estrin. "Sensor Networks: A Bridge to the Physical World." In *Wireless Sensor Networks,* edited by Cauligi S. Raghavendra, Krishna M. Sivalingam, and Taieb Znati, 3–20. Norwell, Mass.: Kluwer, 2004.

Elton, Charles. *The Ecology of Invasions by Animals and Plants.* 1958. Reprint, Chicago: University of Chicago Press, 2000.

Environment Agency (UK). "Air Pollution." http://apps.environment-agency.gov.uk/wiyby/124274.aspx.

———. "River and Sea Levels." http://apps.environment-agency.gov.uk/river-and-sea-levels.

e-obs. http://www.e-obs.de.

EpiCollect. Roadkill Garneau. https://epicollectserver.appspot.com/project.html?name=RoadkillGarneau.

Estrin, Deborah. "Reflections on Wireless Sensing Systems: From Ecosystems to Human Systems." *Radio and Wireless Symposium: IEEE* (January 9–11, 2007): 1–4.

Evans, Dave. "The Internet of Things: How the Next Evolution of the Internet Is Changing Everything." *Cisco Whitepaper.* April 2011. http://www.cisco.com/web/about/ac79/docs/innov/IoT_IBSG_0411FINAL.pdf.

EU-Argo Research Infrastructure (RI). "Deployed Floats." http://www.euro-argo.eu/Main-Achievements/European-Contributions/Floats/Deployments.

European Commission. "Air Quality Standards." http://ec.europa.eu/environment/air/quality/standards.htm.

———. "Directive 2008/50/EC of the European Parliament and of the Council of 21 May 2008 on Ambient Air Quality and Cleaner Air for Europe." *Official Journal of the European Union.* November 6, 2008. L 152/1.

————. "Green Digital Charter." http://ec.europa.eu/information_society/activities/sustainable_growth/green_digital_charter/index_en.htm.

————. "Report of the Meeting of Advisory Group. ICT Infrastructure for Energy-Efficient Buildings and Neighborhoods for Carbon-Neutral Cities." September 16, 2011. http://ec.europa.eu/information_society/activities/sustainable_growth/docs/smart-cities/smart-cities-adv-group_report.pdf.

Exchange for Local Observations and Knowledge of the Arctic (ELOKA). https://eloka-arctic.org.

Feenberg, Andrew, and Darin Barney, eds. *Community in the Digital Age—Philosophy and Practice*. Lanham, Md.: Rowman & Littlefield, 2004.

Ferguson, Francesca, and Urban Drift Projects, eds. *Make_Shift City: Renegotiating the Urban Commons*. Berlin: Jovis, 2014.

FillThatHole. http://www.fillthathole.org.uk.

Finnish Society of Bioart. "Environmental Computing." Field_Notes: A Field Lab for Theory and Practice on Art & Science Work. http://bioartsociety.fi/field_notes/?page_id=8.

Fish, Adam, Luis F. R. Murillo, Lilly Nguyen, Aaron Panofsky, and Christopher M. Kelty. "Birds of the Internet." *Journal of Cultural Economy* 4, no. 2 (2011): 157–87.

FixMyStreet. http://www.fixmystreet.com.

————. "Abandoned Christmas Tree." https://www.fixmystreet.com/report/69857.

————. "Council Worker Mowing Rubbish!" https://www.fixmystreet.com/report/395965.

————. "Rubbish." https://www.fixmystreet.com/report/110531.

Forest Carbon Project Inventory. http://forestcarbonportal.com.

Forrester, Jay W. *Urban Dynamics*. Cambridge, Mass.: MIT Press, 1969.

Forsström, Laura, Sanna Sorvari, Atte Korhola, and Milla Rautio. "Seasonality of Phytoplankton in Subarctic Lake Saanajärvi in NW Finnish Lapland." *Polar Biology* 28, no. 11 (2005): 846–61.

Fortun, Kim. "Biopolitics and the Informating of Environmentalism." In *Lively Capital: Biotechnologies, Ethics, and Governance in Global Markets,* edited by Kaushik Sunder Rajan, 306–26. Durham, N.C.: Duke University Press, 2012.

Fortun, Kim, and Michael Fortun. "Scientific Imaginaries and Ethical Plateaus in Contemporary U.S. Toxicology." *American Anthropologist* 107, no. 1 (2005): 43–54.

Foth, Marcus, Laura Forlano, Christine Satchell, and Martin Gibbs, eds. *From Social Butterfly to Engaged Citizen: Urban Informatics, Social Media, Ubiquitous Computing, and Mobile Technology to Support Citizen Engagement*. Cambridge, Mass.: MIT Press, 2011.

Foucault, Michel. *The Birth of Biopolitics: Lectures at the Collège de France 1978–1979*. Translated by Graham Burchell. 2004. Reprint, New York: Palgrave MacMillan, 2008.

————. "The Confession of the Flesh." In *Power/Knowledge: Selected Interviews and Other Writings 1972–1977*. Translated by Colin Gordon, Leo Marshall, John Mepham, and Kate Soper, 194–228. New York: Vintage Books, 1980.

————. "Intellectuals and Power: A Conversation between Michel Foucault and Gilles Deleuze." In *Language, Counter-Memory, Practice*. Translated by Donald F. Bouchard and Sherry Simon, 205–17. Ithaca, N.Y.: Cornell University Press, 1977.

————. *La naissance de la biopolitique: Cours au Collège de France 1978–1979*. Paris: Éditions du Seuil/Gallimard, 2004.

————. *The Order of Things: An Archaeology of the Human Sciences*. 1970. Reprint, London: Routledge, 1994.

———. *The Politics of Truth*. Translated by Lysa Hochroth and Catherine Porter. Los Angeles: Semiotext(e), 1997.

———. *Security, Territory, Population: Lectures at the Collège de France 1977–1978*. Translated by Graham Burchell. New York: Palgrave Macmillan, 2007.

———. *Society Must Be Defended*. Translated by David Macey. London: Penguin, 2003.

Fraser, Nancy. "Rethinking the Public Sphere: A Contribution to the Critique of Actually Existing Democracy." *Social Text* 25/26 (1990): 56–80.

Fuller, Matthew. *Behind the Blip: Essays on the Culture of Software*. Brooklyn: Autonomedia, 2003.

———. "Boxes towards Bananas: Dispersal, Intelligence and Animal Structures." In *Sentient Cities: Ubiquitous Computing, Architecture, and the Future of Urban Space*, edited by Mark Shepard, 173–81. Cambridge, Mass.: MIT Press, 2011.

———. *Media Ecologies: Materialist Energies in Art and Technoculture*. Cambridge, Mass.: MIT Press, 2005.

Fuller, Matthew, and Andrew Goffey. "Digital Infrastructures and the Machinery of Topological Abstraction." *Theory, Culture & Society* 29, nos. 4–5 (2012): 311–33.

Fuller Matthew, and Usman Haque. "Urban Versioning System." In *Situated Technologies Pamphlets 2*. New York: Architectural League of New York, 2008.

Future Cities Catapult. "Sensing London." https://futurecities.catapult.org.uk/project/-sensing-london/.

Gabrys, Jennifer. "Atmospheres of Communication." In *The Wireless Spectrum: The Politics, Practices and Poetics of Mobile Technologies*, edited by Barbara Crow, Michael Longford and Kim Sawchuk, 46–59. Toronto: University of Toronto Press, 2010.

———. "Automatic Sensation: Environmental Sensors in the Digital City." *Senses and Society* 2, no. 2 (2007): 189–200.

———. "Becoming Urban: Sitework from a Moss-Eye View." *Environment and Planning A* 44, no. 12 (2012): 2922–39.

———. "Cité Multimédia: Noise and Contamination in the Information City." Conference proceedings from "Visual Knowledges," University of Edinburgh, September 17–20, 2003. http://webdb.ucs.ed.ac.uk/malts/other/VKC/dsp-abstract.cfm?ID=84, http://www.jennifergabrys.net/wp-content/uploads/2003/09/Gabrys_InfoCity_VKnowledges.pdf.

———. *Digital Rubbish: A Natural History of Electronics*. Ann Arbor: University of Michigan Press, 2011.

———. "Plastic and the Work of the Biodegradable." In *Accumulation: The Material Politics of Plastic*, edited by Jennifer Gabrys, Gay Hawkins, and Mike Michael, 208–27. London: Routledge, 2013.

———. "Pollution Sensing and Fracking: Practices of Citizen-Based Air-Quality Monitoring." Presentation at the Environment Research Group seminar, University of London, April, 2015.

———. "Powering the Digital: From Energy Ecologies to Electronic Environmentalism." In *Media and the Ecological Crisis*, edited by Richard Maxwell, Jon Raundalen, and Nina Lager Vestberg, 3–18. New York: Routledge, 2014.

———. "Re-Thingifying the Internet of Things." In *Sustainable Media*, edited by Nicole Starosielski and Janet Walker. New York: Routledge, 2016.

———. "Sink: The Dirt of Systems." *Environment and Planning D: Society and Space* 27, no. 4 (2009): 666–81.

———. "Speculating with Organisms in the Plastisphere." In *An Ecosystem of Excess,* edited by Heike Catherina Mertens, 50–61. Berlin: Ernst Schering Foundation, 2014.

———. "Telepathically Urban." In *Circulation and the City: Essays on Urban Culture,* edited by Alexandra Boutros and Will Straw, 48–63. Montreal: McGill-Queen's Press, 2010.

Gabrys, Jennifer, Gay Hawkins, and Mike Michael, eds. *Accumulation: The Material Politics of Plastic.* London: Routledge, 2013.

Gabrys, Jennifer, Helen Pritchard, and Benjamin Barratt. "'Just Good Enough' Data: Recasting Evidence through Citizen-Based Air Pollution Monitoring."

Gabrys, Jennifer, Helen Pritchard, and Nerea Calvillo. "Making DIY Air Sensors and Materializing Digital Citizenship."

Galloway, Anne. "Intimations of Everyday Life: Ubiquitous Computing and the City." *Cultural Studies* 18, nos. 2–3 (2004): 384–408.

Garcia, David, and Geert Lovink. "The ABC of Tactical Media." nettime. May 16, 1997. http://www.nettime.org/Lists-Archives/nettime-l-9705/msg00096.html.

Gertz, Emily, and Patrick Di Justo. *Environmental Monitoring with Arduino: Building Simple Devices to Collect Data about the World around Us.* Beijing: O'Reilly Maker Press, 2012.

Gillis, Justin. "In the Ocean, Clues to Change." *New York Times.* August 11, 2014. http://www.nytimes.com/2014/08/12/science/in-the-ocean-clues-to-change.html?_r=0.

Gitelman, Lisa, ed. *"Raw Data" Is an Oxymoron.* Cambridge, Mass.: MIT Press, 2013.

Global Climate Observing System (GCOS). "Essential Climate Variables." http://www.wmo.int/pages/prog/gcos/index.php?name=EssentialClimateVariables.

Global Community Monitor. http://www.gcmonitor.org.

Global Drifter Program. Satellite-Tracked Surface Drifting Buoys. http://www.aoml.noaa.gov/phod/dac.

Group on Earth Observations (GEO). Global Earth Observation System of Systems (GEOSS) Portal. http://www.geoportal.org/web/guest/geo_home_stp.

Global Language Monitor. "Top Words of 2010." November 27, 2010. http://www.languagemonitor.com/new-words/top-words-of-2010.

GlobalWarming.House.Gov. "Oil Spill in the Gulf—Live Cam." BP Oil Spill Live Video Feed. http://www.markey.senate.gov/GlobalWarming/spillcam.html.

Goldberg, Ken, ed. *The Robot in the Garden: Telerobotics and Telepistemology in the Age of the Internet.* Cambridge, Mass.: MIT Press, 2000.

Goldman, Jeffrey, Katie Shilton, Jeffrey A. Burke, Deborah Estrin, Mark Hansen, Nithya Ramanathan, Sasank Reddy, Vids Samanta, Mani Srivastava, and Ruth West. "Participatory Sensing: A Citizen-Powered Approach to Illuminating the Patterns That Shape Our World." Washington, D.C.: Woodrow Wilson International Center for Scholars, 2009.

Goldman, Jeffrey, Nithya Ramanathan, Richard Ambrose, David A. Caron, Deborah Estrin, Jason C. Fisher, Robert Gilbert, Mark H. Hansen, Thomas C. Harmon, Jennifer Jay, William J. Kaiser, Gaurav S. Sukhatme, and Yu-Chong Tai. "Distributed Sensing Systems for Water Quality Assessment and Management." Washington, D.C.: Woodrow Wilson International Center for Scholars, Foresight and Governance Project, February 2007.

Goldstein, Miriam C., Marci Rosenberg, and Lanna Cheng. "Increased Oceanic Microplastic Debris Enhances Oviposition in an Endemic Pelagic Insect." *Biology Letters* 8, no. 5 (2012): 817–20.

Goodchild, Michael F. "Citizens as Sensors: The World of Volunteered Geography." *Geo-Journal* 69 (2007): 211–21.

Goodwin, Charles. "Seeing in Depth." *Social Studies of Science* 25, no. 2 (1995): 237–74.

Google Glass. http://www.google.com/glass.

Goriunova, Olga. "New Media Idiocy." *Convergence* 19, no. 2 (May 2013): 223–35.

Graham, Stephen, ed. *The Cybercities Reader.* London: Routledge, 2004.

———. *Disrupted Cities: When Infrastructure Fails.* New York: Routledge, 2010.

———. "Software-Sorted Geographies." *Progress in Human Geography* 29 (2005): 562–80.

Graham, Stephen, and Simon Marvin. *Splintering Urbanism: Networked Infrastructures, Technological Mobilities and the Urban Condition.* London: Routledge, 2001.

Gray Area Foundation for the Arts. "Urban Prototyping Festival to Transform Public Space through Citizen Experiments." October 11, 2012. http://grayarea.org/press/urban-proto typing-festival-to-transform-public-space-through-citizen-experiments.

Greater London Authority. "Locations of TfL Street Cameras." http://80.79.211.134/data store/applications/locations-tfl-street-cameras.

———. "Royal Albert Dock." http://www.london.gov.uk/priorities/housing-land/land -assets/royal-albert-dock.

———. Smart London Plan. http://www.london.gov.uk/sites/default/files/smart_lon don_plan.pdf.

———. Smart London Vision. https://www.london.gov.uk/priorities/business-economy /vision-and-strategy/smart-london.

Greenfield, Adam, and Mark Shepard. "Urban Computing and Its Discontents." In *Situated Technologies Pamphlets 1.* New York: Architectural League of New York, 2007.

Gregory, Murray R. "Environmental Implications of Plastic Debris in Marine Settings—Entanglement, Ingestion, Smothering, Hangers-On, Hitch-Hiking and Alien Invasions." *Philosophical Transactions of the Royal Society B: Biological Sciences* 364, no. 1526 (2009): 2013–25.

Gregory, Murray R., and Peter G. Ryan. "Pelagic Plastics and Other Seaborne Persistent Synthetic Debris: A Review of Southern Hemisphere Perspectives." In *Marine Debris,* edited by James M. Coe and Donald B. Rogers, 49–66. New York: Springer, 1997.

Gross, Neil. "The Earth Will Don an Electronic Skin." *Business Week.* August 30, 1999. http://www.businessweek.com/1999/99_35/b3644024.htm.

Guattari, Felix. *Chaosmosis: An Ethico-Aesthetic Paradigm.* Translated by Paul Bains and Julian Pefanis. Bloomington: Indiana University Press, 1995.

———. *Schizoanalytic Cartographies.* Translated by Andrew Goffey. 1989. Reprint, London: Bloomsbury, 2013.

———. *The Three Ecologies.* Translated by Ian Pindar and Paul Sutton. London: Athlone, 2000.

Halewood, Michael, and Mike Michael. "Being a Sociologist and Becoming a Whiteheadian: Toward a Concrescent Methodology." *Theory, Culture & Society* 25, no. 4 (2008): 31–56.

Halpern, Orit, Jesse LeCavalier, Nerea Calvillo, and Wolfgang Pietsch. "Test-Bed Urbanism." *Public Culture* 25, no. 2 (2013): 272–306.

Hamilton, Michael P., Eric A. Graham, Philip W. Rundel, Michael F. Allen, William Kaiser, Mark H. Hansen, and Deborah L. Estrin. "New Approaches in Embedded Networked Sensing for Terrestrial Ecological Observatories." *Environmental Engineering Science* 24, no. 2 (2007): 192–204.

Haraway, Donna. "Crittercam: Compounding Eyes in Naturecultures." In *When Species Meet*, 249–263. Minneapolis: University of Minnesota Press, 2008.

———. "The High Cost of Information in Post–World War II Evolutionary Biology: Ergonomics, Semiotics, and the Sociobiology of Communication Systems." *Philosophical Forum* 13 (Winter-Spring 1981–1982): 244–78.

———. "Introduction: The Persistence of Vision." In *Primate Visions: Gender, Race, and Nature in the World of Modern Science*. New York: Routledge, 1989.

———. *Modest_Witness@Second_Millennium. FemaleMan©_Meets_Onco-Mouse*™. New York: Routledge, 1997.

———. *Simians, Cyborgs and Women: The Reinvention of Nature*. New York: Routledge, 1991.

———. "Situated Knowledges: The Science Question in Feminism and the Privilege of Partial Perspective." *Feminist Studies* 14, no. 3 (1988): 575–99.

———. *When Species Meet*. Minneapolis: University of Minnesota Press, 2008.

Harrison, Colin, and Ian Abbott Donnelly. "A Theory of Smart Cities." *Proceedings of the 55th Annual Meeting of the International Society for the Systems Sciences (ISSS)*. 2011. http://journals.isss.org/index.php/proceedings55th/article/view/1703.

Hartmann, Dennis L., Albert M. G. Klein Tank, Matilde Rusticucci, Lisa V. Alexander, Stefan Brönnimann, Yassine Abdul-Rahman Charabi, Frank J. Dentener, Edward J. Dlugokencky, David R. Easterling, Alexey Kaplan, Brian J. Soden, Peter W. Thorne, Martin Wild, and Panmao Zhai. "Observations: Atmosphere and Surface." In *IPCC, 2013: Climate Change 2013; The Physical Science Basis. Contribution of Working Group I to the Fifth Assessment Report of the Intergovernmental Panel on Climate Change*, edited by Thomas F. Stocker, Dahe Qin, Gian-Kasper Plattner, Melinda Tignor, Simon K. Allen, Judith Boschung, Alexander Nauels, Yu Xia, Vincent Bex and Pauline M. Midgley, 159–254. Cambridge: Cambridge University Press, 2013.

Harvey, David, and Donna Haraway. "Nature, Politics, and Possibilities: A Debate and Discussion with David Harvey and Donna Haraway." *Environment and Planning D: Society and Space* 13, no. 5 (1995): 507–27.

Hawkins, Gay. "Enacting Public Value on the ABC's *Q and A*: From Normative to Performative Approaches." *Media International Australia* 146 (2013): 82–92.

Hayles, N. Katherine. "Computing the Human." *Theory, Culture & Society* 22, no. 1 (2005): 131–51.

———. *My Mother Was a Computer: Digital Subjects and Literary Texts*. Chicago: University of Chicago Press, 2005.

———. "RFID: Human Agency and Meaning in Information-Intensive Environments." *Theory, Culture & Society* 26, nos. 2–3 (2009): 47–72.

Heise, Ursula K. *Sense of Place and Sense of Planet: The Environmental Imagination of the Global*. Oxford: Oxford University Press, 2008.

Helmers, Carl. "Creating Sensors: The Story of a New Publication." *Sensors* (January 1984): 18–19.

Helmreich, Stefan. *Alien Ocean: Anthropological Voyages in Microbial Seas*. Berkeley: University of California Press, 2009.

———. "Intimate Sensing." In *Simulation and Its Discontents*, edited by Sherry Turkle, 129–50. Cambridge, Mass.: MIT Press, 2009.

———. "Nature/Culture/Seawater." *American Anthropologist* 113 (2011): 132–44.

Hern, Alex. "Seven Urban Apps Guaranteed to Improve the Quality of City Life." *Guardian.* March 11, 2014. http://www.theguardian.com/cities/2014/mar/11/city-apps-should-not-be-without.

Hinchliffe, Steve. "Sensory Biopolitics: Knowing Birds and a Politics of Life." In *Sentient Creatures,* edited by Kristin Asdal, Tone Druglitrø, and Steve Hinchliffe. May 2014, draft chapter. http://www.biosecurity-borderlands.org/research.

Höhler, Sabine. "'Spaceship Earth': Envisioning Human Habitats in the Environmental Age." *GHI Bulletin,* 42 (Spring 2008): 65–85.

Hollands, Robert G. "Will the Real Smart City Please Stand Up?" *City* 12, no. 3 (2008): 303–20.

Hörl, Erich, ed. *The Ecological Paradigm: Perspectives of a General Ecology of Media and Technology.* New York: Fordham University Press, 2014.

Horowitz, Andrew. "Domestic Communications Satellites." *Radical Software* 2, no. 5 (Winter 1973): 36–40.

Howell, Evan A., Steven J. Bograd, Carey Morishige, Michael P. Seki, and Jeffrey J. Polovina. "On North Pacific Circulation and Associated Marine Debris Concentration." *Marine Pollution Bulletin* 65, nos. 1–3 (2012): 16–22.

Howes, David. "Introduction: Empires of the Senses." In *Empire of the Senses: The Sensual Culture Reader,* edited by David Howes, 1–17. Oxford: Berg, 2005.

HP Labs. "Central Nervous System for the Earth (CeNSE)." http://www.hp.com/hpinfo/globalcitizenship/environment/techgallery/cense.html.

Hsu, V., Joseph M. Kahn, and Kris S. J. Pister. "Wireless Communications for Smart Dust." *Electronics Research Laboratory Memorandum Number M98/2.* January 30, 1998.

Hurlbert, Allen H., and Zhongfei Liang. "Spatiotemporal Variation in Avian Migration Phenology: Citizen Science Reveals Effects of Climate Change." *PLoS One* 7, no. 2 (February 2012): e31662.

Hyman, Josh, Eric Graham, and Mark Hansen. "Imagers as Sensors: Correlating Plant CO_2 Uptake with Digital Visible-Light Imagery." *Proceedings of the 4th International Workshop on Data Management for Sensor Networks.* Vienna, Austria. September 23–28, 2007.

IBM. "Citizen Sensing for Smarter Cities." IBM Research-Brazil. 2013. http://researcher.watson.ibm.com/researcher/files/br-fernandokoch/IBM-CitizenSensor%28web%29.pdf.

———. "The Internet of Things." IBMSocialMedia. http://www.youtube.com/watch?v=sfEbMV295Kk.

———. "Smarter Cities Challenge." http://smartercitieschallenge.org/index.html.

———. "A Smarter Planet." http://www.ibm.com/smarterplanet.

Intel. "What Does The Internet of Things Mean?" March 12, 2014. http://www.youtube.com/watch?v=Q3ur8wzzhBU.

Interagency Ocean Observation Committee (IOOC). http://www.iooc.us/ocean-observations/platforms.

International Cooperation for Animal Research Using Space (ICARUS) Initiative. "Countdown to ICARUS." http://icarusinitiative.org.

———. "Science & Projects." http://icarusinitiative.org/science-projects.

———. "Technical Solution." Max Planck Institute. November 15, 2011. http://icarusinitiative.org/technical-solution.

International Network for Terrestrial Research and Monitoring in the Arctic. http://www.eu-interact.org.

International Pacific Research Center (IPRC). "Tracking Ocean Debris." *Newsletter of the International Pacific Research Center* 8, no. 2 (2008): 14–16. http://iprc.soest.hawaii.edu/newsletters/iprc_climate_vol8_no2.pdf.

Interspecies Internet. http://interspecies-internet.org.

Irwin, Alan. *Citizen Science: A Study of People, Expertise, and Sustainable Development.* London: Routledge, 1995.

Irwin, Alan, and Mike Michael. *Science, Social Theory and Public Knowledge.* Maidenhead: Open University Press, 2003.

Iveson, Kurt. "Mobile Media and the Strategies of Urban Citizenship: Control, Responsibilization, Politicization." In *From Social Butterfly to Engaged Citizen: Urban Informatics, Social Media, Ubiquitous Computing, and Mobile Technology to Support Citizen Engagement,* edited by Marcus Foth, Laura Forlano, Christine Satchell, and Martin Gibbs, 55–70. Cambridge, Mass.: MIT Press, 2011.

James, William. *Essays in Radical Empiricism.* 1912. Reprint, Lincoln: University of Nebraska Press, 1996.

———. *Pragmatism.* 1907. Reprint, Cambridge: Cambridge University Press, 2014.

James San Jacinto Mountains Reserve. http://www.jamesreserve.edu.

———. "The Moss Cam Project: The Star Moss, *Tortula princeps,* a Desiccation-Tolerant Plant." http://www.jamesreserve.edu/mosscam (site discontinued).

Jansen, Lesa, and Brianna Keilar. "Markey: Spillcam Was Game-Changer in BP Disaster Response." CNN. April 20, 2011. http://edition.cnn.com/2011/POLITICS/04/20/markey.bp.

Jasanoff, Sheila. "Image and Imagination: The Formation of Global Environmental Consciousness." In *Changing the Atmosphere: Expert Knowledge and Environmental Governance,* edited by Clark A. Miller and Paul N. Edwards, 309–37. Cambridge, Mass.: MIT Press, 2001.

———. "Technologies of Humility: Citizen Participation in Governing Science." *Minerva* 41 (2003): 223–44.

———. "Thin Air." In *Débordements: Mélanges Offerts à Michel Callon,* edited by Madeleine Akrich, Yannick Barthe, Fabian Muniesa, and Philippe Mustar, 191–202. Paris: Presses des Mines, 2010.

Jenkins, Henry, and David Thorburn, eds. *Democracy and New Media.* Cambridge, Mass.: MIT Press, 2003.

Jeremijenko, Natalie. "Feral Robotic Dogs." https://www.nyu.edu/projects/xdesign/feralrobots.

Jimenez, Alberto Corsín. "The Right to Infrastructure: A Prototype for Open Source Urbanism." *Environment and Planning D: Society and Space* 32, no. 2 (2014): 342–62.

Johnson, Kirk. "'Smart Collar' in the Works to Manage Wildlife Better." *New York Times.* August 29, 2011. http://www.nytimes.com/2011/08/30/us/30collars.html?_r=1&hp.

Jones, Caroline A., ed. *Sensorium: Embodied Experience, Technology, and Contemporary Art.* Cambridge, Mass.: MIT Press, 2006.

Jonsson, Patrik. "BP Live Feed 'Spillcam' Now Offers 12 Views of Oil Spill." *Christian Science Monitor.* June 3, 2010. http://www.csmonitor.com/USA/2010/0603/BP-live-feed-spillcam-now-offers-12-views-of-oil-spill.

Jordan, Chris. Midway: Message from the Gyre (2009-). http://www.chrisjordan.com/gallery/midway.

Kaiser, Jocelyn. "The Dirt on Ocean Garbage Patches." *Science* 328, no. 5985 (2010): 1506.

Keller, Evelyn Fox. "The Body of a New Machine: Situating the Organism between Telegraph and Computers." In *Refiguring Life: Changing Metaphors of Twentieth-Century Biology,* 79–91. New York: Columbia University Press, 1995.

Kilpisjärvi Biological Station. http://www.helsinki.fi/kilpis/english/index.htm.

Kitchin, Rob, and Martin Dodge. *Code/Space: Software and Everyday Life.* Cambridge, Mass.: MIT Press, 2011.

Kittler, Friedrich A. "There Is No Software." *CTHEORY.* October 18, 1995. http://www.ctheory.net/articles.aspx?id=74.

———. "Thinking Colors and/or Machines." *Theory, Culture & Society* 23, nos. 7–8 (2006): 39–50.

Kohler, Robert E. *Landscapes and Labscapes: Exploring the Lab-Field Border in Biology.* Chicago: University of Chicago Press, 2002.

Korsgaard, Henrik, and Martin Brynskov. "Prototyping a Smart City." DC8 Workshop, 6th International Conference on Communities and Technologies (C&T 6). Munich, Germany. June 30, 2013. https://pure.au.dk/ws/files/70459732/DC8_6_Korsgaard_Brynskov.pdf.

Kosek, Jake. "Ecologies of Empire: On the New Uses of the Honeybee." *Cultural Anthropology* 25, no. 4 (2010): 650–78.

Kranenburg, Rob van. "The Internet of Things: A Critique of Ambient Technology and the All-Seeing Network of RFID Network Notebooks." Amsterdam: Institute of Network Cultures, 2008.

Kurvits, Tiina, Christian Nellemann, Björn Alfthan, Aline Kühl, Peter Prokosch, Melanie Virtue, and Janet F. Skaalvik, eds. *Living Planet: Connected Planet—Preventing the End of the World's Wildlife Migrations through Ecological Networks.* United Nations Environment Programme, GRID-Arendal, 2011.

Lamarre, Thomas. "Afterword: Humans and Machines." In Muriel Combes, *Gilbert Simondon and the Philosophy of the Transindividual,* translated by Thomas Lamarre, 79–108. Cambridge, Mass.: MIT Press, 2013.

Lane, Giles, Camilla Brueton, George Roussos, Natalie Jeremijenko, George Papamarkos, Dima Diall, Dimitris Airantzis, and Karen Martin. "Public Authoring & Feral Robotics." *Cultural Snapshot* 11 (March 2006). http://proboscis.org.uk/publications/SNAPSHOTS_feralrobots.pdf.

Large Pelagics Research Center. "Tag a Tiny Program." http://www.tunalab.org/tagatiny.htm.

L'atelier d'architecture autogérée. http://www.urbantactics.org.

Latour, Bruno. *Pandora's Hope: Essays on the Reality of Science Studies.* Cambridge, Mass.: Harvard University Press, 1999.

———. "A Plea for Earthly Sciences." In *New Social Connections: Sociology's Subjects and Objects,* edited by Judith Burnett, Syd Jeffers, and Graham Thomas, 72–84. London: Palgrave Macmillan, 2010.

———. *Science in Action: How to Follow Scientists and Engineers through Society.* Cambridge, Mass.: Harvard University Press, 1987.

———. "Why Has Critique Run Out of Steam? From Matters of Fact to Matters of Concern." *Critical Inquiry* 30 (2004): 225–48.

Laughlin, Andrew J., Caz M. Taylor, David W. Bradley, Dayna LeClair, Robert G. Clark, Russell D. Dawson, Peter O. Dunn, Andrew Horn, Marty Leonard, Daniel R. Sheldon, Dave Shutler, Linda A. Whittingham, David W. Winkler, and D. Ryan Norris. "Integrating Information from Geolocators, Weather Radar, and Citizen Science to Uncover a Key Stopover Area of an Aerial Insectivore." *Auk* 130, no. 2 (2013): 230–39.

Lauro, Federico M., Svend Jacob Senstius, Jay Cullen, Russell Neches, Rachelle M. Jensen, Mark V. Brown, Aaron E. Darling, Michael Givskov, Diane McDougald, Ron Hoeke, Martin Ostrowski, Gayle K. Philip, Ian T. Paulsen, and Joseph J. Grzymski. "The Common Oceanographer: Crowdsourcing the Collection of Oceanographic Data." *PLoS Biology* 12, no. 9 (September 9, 2014): 1–5.

Law, Kara Lavender, Skye Morét-Ferguson, Nikolai A. Maximenko, Giora Proskurowski, Emily E. Peacock, Jan Hafner, and Christopher M. Reddy. "Plastic Accumulation in the North Atlantic Subtropical Gyre." *Science* 329, no. 5996 (2010): 1185–88.

Lebreton, Laurent C. M., S. D. Greer, and Jose Carlos Borrero. "Numerical Modeling of Floating Debris in the World's Oceans." *Marine Pollution Bulletin* 64, no. 3 (2012): 653–61.

Lehning, Michael, Nicholas Dawes, Mathias Bavay, Marc Parlange, Suman Nath, and Feng Zhao. "Instrumenting the Earth: Next-Generation Sensor Networks and Environmental Science." In *The Fourth Paradigm: Data-Intensive Scientific Discovery*. Microsoft Research, 2009.

Lemke, Peter, Jiawen Ren, Richard B. Alley, Ian Allison, Jorge Carrasco, Gregory Flato, Yoshiyuki Fujii, Georg Kaser, Robert Thomas, and Tingjun Zhang. "Observations: Changes in Snow, Ice and Frozen Ground." In *Climate Change 2007: The Physical Science Basis; Contribution of Working Group I to the Fourth Assessment Report of the Intergovernmental Panel on Climate Change*, edited by S. Solomon, D. Qin, M. Manning, Z. Chen, M. Marquis, K. B. Averyt, M. Tignor, and H. L. Miller, 337–83. Cambridge: Cambridge University Press, 2007.

Lidskog, Rolf, and Göran Sundqvist, eds. *Governing the Air: The Dynamics of Science, Policy, and Citizen Interaction*. Cambridge, Mass.: MIT Press, 2011.

Lim, Stephen, et al. "A Comparative Risk Assessment of Burden of Disease and Injury Attributable to 67 Risk Factors and Risk Factor Clusters in 21 Regions, 1990–2010: A Systematic Analysis for the Global Burden of Disease Study 2010." *Lancet* 380, no. 9859 (December 15, 2012): 2224–60.

"Live Gulf Oil Spill Cam: Watch Video Feed of BP Spill." *Huffington Post*. May 20, 2010. http://www.huffingtonpost.com/2010/05/20/live-gulf-oil-spill-video-feed_n_583682.html.

Living PlanIt. http://www.living-planit.com.

Llanos, Miguel. "Average Joe as Oceanographer? Crowdsourcing Goes to Sea (with Caveats)." NBC News. September 13, 2014. http://www.nbcnews.com/science/environment/average-joe-oceanographer-crowdsourcing-goes-sea-caveats-n202411.

London Air Quality Network (LAQN). http://www.londonair.org.uk.

London Eye. "Systems and Controls." http://www.londoneye.com/LearningAndDiscovery/Education/TeacherResource/OnlineResource/systems/systpup.pdf (site discontinued).

Lovink, Geert. "The Digital City: Metaphor and Community." In *Dark Fiber: Tracking Critical Internet Culture*, 42–67. Cambridge, Mass.: MIT Press, 2002.

———. *Networks without a Cause: A Critique of Social Media*. Cambridge: Polity, 2012.

Lovink, Geert, and Ned Rossiter, "Dawn of the Organized Networks." *Fibreculture Journal* 5 (2005). http://five.fibreculturejournal.org/fcj-029-dawn-of-the-organised-networks.

Luke, Timothy W. "Environmentality as Green Governmentality." In *Discourses of the Environment*, edited by Éric Darier, 121–51. Oxford: Blackwell, 1999.

Lumpkin, Rick, Nikolai Maximenko, and Mayra Pazos. "Evaluating Where and Why Drifters Die." *Journal of Atmospheric and Oceanic Technology* 29, no. 2 (2012): 300–308.

Lumpkin, Rick, and Mayra Pazos. "Measuring Surface Currents with Surface Velocity Program Drifters: The Instrument, Its Data, and Some Recent Results." In *Lagrangian Analysis and Prediction of Coastal and Ocean Dynamics (LAPCOD)*, edited by Annalisa Griffa, A. D. Kirwan, Arthur J. Mariano, Tamary Ozgokmen, and Thomas Rossby, 39–67. Cambridge: Cambridge University Press, 2007.

Lydon, Mike, Dan Bartman, Tony Garcia, Russ Preston, and Ronald Woudstra. *Tactical Urbanism, Volume 2*. http://issuu.com/streetplanscollaborative/docs/tactical_urbanism_vol_2_final.

Mack, Pamela. *Viewing the Earth: The Social Construction of the Landsat Satellite System*. Cambridge, Mass.: MIT Press, 1990.

Mackenzie, Adrian. *Cutting Code: Software and Sociality*. New York: Peter Lang, 2006.

———. "The Performativity of Code: Software and Cultures of Circulation." *Theory, Culture & Society* 22, no. 1 (2005): 71–92.

———. *Transductions: Bodies and Machines at Speed*. London: Continuum, 2002.

———. *Wirelessness: Radical Empiricism in Network Cultures*. Cambridge, Mass.: MIT Press, 2010.

Madrigal, Alexis C. "Found: World's Oldest Message in a Bottle, Part of 1914 Citizen-Science Experiment." *Atlantic*. September 5, 2012. http://www.theatlantic.com/technology/archive/2012/09/found-worlds-oldest-message-in-a-bottle-part-of-1914-citizen-science-experiment/261981.

Maier, Elke. "A Four-Legged Early-Warning System." *MaxPlanckResearch* 2 (2014): 58–63.

Marine Debris Tracker. http://www.marinedebris.engr.uga.edu.

Marine Traffic. http://www.marinetraffic.com.

Markey, Edward J. "Edward Markey to Lamar McKay." Washington, D.C: House Committee on Energy and Commerce, 111th Congress, May 19, 2010.

Marshall, Greg. "Foreword: Advances in Animal-Borne Imaging." *Marine Technology Society Journal* 41, no. 4 (Winter 2007/2008): 4–5.

Martin, Reinhold. "Environment c. 1973." *Grey Room* 14 (Winter 2004): 78–101.

Masco, Joseph. "Bad Weather: On Planetary Crisis." *Social Studies of Science* 40, no. 1 (February 2010): 7–40.

Massumi, Brian. "National Enterprise Emergency: Steps toward an Ecology of Powers." *Theory, Culture & Society* 26, no. 6 (2009): 153–85.

———. *Semblance and Event: Activist Philosophy and the Occurrent Arts*. Cambridge, Mass.: MIT Press, 2011.

Maximenko, Nikolai, Jan Hafner, and Peter Niiler. "Pathways of Marine Debris Derived from Trajectories of Lagrangian Drifters." *Marine Pollution Bulletin* 65, nos. 1–3 (2012): 51–62.

Max Planck Institute for Ornithology. Animal Tracker app. https://itunes.apple.com/us/app/animal-tracker/id813629500?mt=8&ign-mpt=uo%3D4.

———. Animal Tracker. http://www.orn.mpg.de/animal_tracker.

———. "Social Migration in Juvenile White Storks." https://www.orn.mpg.de/3297372/Social_Migration_in_Juvenile_White_Storks.

May, Robert. "Unanswered Questions in Ecology." *Philosophical Transactions of the Royal Society of London B* 354 (1999): 1951–59.

McConnell, Bernie. "Telemetry in Sea Mammals." Presentation at New Techniques in Mammal Research. London: Mammal Society Symposium, Zoological Society of London. November 26–27, 2010.

McCullough, Malcolm. *Ambient Commons: Attention in the Age of Embodied Information.* Cambridge, Mass.: MIT Press, 2013.

McLuhan, Marshall. "At the Moment of Sputnik the Planet Became a Global Theater in Which There Are No Spectators but Only Actors." *Journal of Communication* 24, no. 1 (1974): 48–58.

Mejias, Ulises Ali. *Off the Network: Disrupting the Digital World.* Minneapolis: University of Minnesota Press, 2013.

Met Office. "Weather Observations Website." http://wow.metoffice.gov.uk.

———. "Weather Stations." http://www.metoffice.gov.uk/learning/science/first-steps/observations/weather-stations.

Michael, Mike. "'What Are We Busy Doing? Engaging the Idiot.'" *Science, Technology & Human Values* 37, no. 5 (September 2012): 528–54.

Microsoft. "Internet of Things." http://www.microsoft.com/windowsembedded/en-gb/internet-of-things.aspx.

Microsoft Research. "SenseCam." http://research.microsoft.com/en-us/um/cambridge/projects/sensecam.

———. "SenseWeb." http://research.microsoft.com/en-us/projects/senseweb.

Migratory Connectivity Project. "Citizen Science." http://www.migratoryconnectivityproject.org/citizen-science.

Million Trees NYC. http://www.milliontreesnyc.org.

Miranda, Pedro Pires de, and Martin Powell, eds. *Our Urban Future: The Crystal; A Sustainable Cities Initiative by Siemens.* Dublin: Booklink, 2013.

Mishler, Brent D. "The Biology of Bryophytes with Special Reference to Water." *Fremontia* 31, no. 3 (2003): 34–38.

Mitchell, William J. *City of Bits: Space, Place, and the Infobahn.* Cambridge, Mass.: MIT Press, 1995.

———. *Me++: The Cyborg Self and the Networked City.* Cambridge, Mass.: MIT Press, 2003.

Mitchell, William J., and Federico Casalegno. *Connected Sustainable Cities.* Cambridge, Mass.: MIT Mobile Experience Lab Publishing, 2008. http://connectedsustainablecities.com/downloads/connected_sustainable_cities.pdf.

Mitman, Gregg. "When Nature Is the Zoo: Vision and Power in the Art and Science of Natural History." *Osiris* 11 (1996): 117–43.

Mody, Cyrus. "The Sounds of Science: Listening to Laboratory Practice." *Science, Technology, & Human Values* 30, no. 2 (2005): 175–98.

Monitrain. "Condition Monitoring Protects London's Thames Barrier." http://www.monitran.com/products/mtn2200ic-mtnm2200ic.

Moore, Charles J., Shelly L. Moore, Molly K. Leecaster, and Steve B. Weisberg. "A Comparison of Plastic and Plankton in the North Pacific Central Gyre." *Marine Pollution Bulletin* 42, no. 12 (2001): 1297–1300.

Moore, Charles J., and Cassandra Phillips. *Plastic Ocean: How a Sea Captain's Chance Discovery Launched a Determined Quest to Save the Oceans.* New York: Avery, 2011.

Morozov, Evgeny. *To Save Everything, Click Here: Technology, Solutionism, and the Urge to Fix Problems That Don't Exist.* London: Allen Lane, 2013.

MoveBank. http://www.movebank.org.

———. "What Is Animal Tracking?" http://www.movebank.org/node/857.

Muir, John. *My First Summer in the Sierra.* 1911. Reprint, Boston: Houghton Mifflin Harcourt, 1998.

Munster, Anna. *An Aesthesia of Networks: Conjunctive Experience in Art and Technology.* Cambridge, Mass.: MIT Press, 2013.

Murphy, Michelle. *Sick Building Syndrome and the Problem of Uncertainty.* Durham, N.C.: Duke University Press, 2006.

mySociety. http://www.mysociety.org.

NASA Jet Propulsion Laboratory. "Orbiting Carbon Observatory 2 (OCO-2)." http://oco.jpl.nasa.gov.

National Center for Ecological Analysis and Synthesis (NCEAS). "A Global Map of Human Impacts to Marine Ecosystems." http://www.nceas.ucsb.edu/globalmarine.

National Commission on the BP Deepwater Horizon Oil Spill and Offshore Drilling. "Deep Water: The Gulf Oil Disaster." Report to the President, January 2011.

National Ecological Observatory Network (NEON). http://www.neoninc.org.

National Geographic. Crittercam. http://animals.nationalgeographic.com/animals/crittercam-about.

National Oceanic and Atmospheric Administration (NOAA). "Critical New Data to be Added to the Integrated Ocean Observing System in 2011." May 16, 2011. http://oceanservice.noaa.gov/news/weeklynews/may11/ioostagging.html.

———. "De-Mystifying the 'Great Pacific Garbage Patch.'" http://marinedebris.noaa.gov/info/patch.html.

———. "Marine Debris Monitoring and Assessment Project." http://marinedebris.noaa.gov/research/marine-debris-monitoring-and-assessment-project.

National Science Foundation. "The Secret Lives of Wild Animals." http://www.nsf.gov/news/special_reports/animals/dragonfly.jsp.

Nature Conservancy. "Precision Conservation." http://www.conserveca.org/our-stories/all/7-spotlight/132-precision-conservation#.U_dUlojjm9c.

Negroponte, Nicholas, with the Architecture Machine Group. "Seek." In *Software: Information Technology: Its New Meaning for Art,* curated by Jack Burnham. New York: Jewish Museum, 1970.

Nesta. "2013 Predictions: How Did We Do?" http://www.nesta.org.uk/news/13-predictions-2013/2013-predictions-how-did-we-do.

Neumann, John von. "First Draft of a Report on the EDVAC." Contract No. W-670-ORD-4926 between the United States Army Ordnance Department and the University of Pennsylvania. Moore School of Electrical Engineering: University of Pennsylvania, June 30, 1945.

Newsroom UNSW (University of New South Wales, Australia). "Our Plastics Will Pollute Oceans for Hundreds of Years." January 9, 2013. http://newsroom.unsw.edu.au/news/science/our-plastics-will-pollute-oceans-hundreds-years.

Newton, Ian. *Bird Migration.* London: Collins, 2010.

New York Times. "Sputnik's Legacy." September 24, 2007. http://www.nytimes.com/video/science/space/1194817120962/sputnik-s-legacy.html.

Nitta, Michiko. "Extreme Green Guerillas." 2007. http://www.michikonitta.co.uk.

Nokia. "Sensor Planet." http://research.nokia.com/page/232 (site discontinued).

Nold, Christian, and Rob van Kranenburg. "The Internet of People for a Post-Oil World." In *Situated Technologies Pamphlets 8.* New York: Architectural League of New York, 2011.

Northern Prairie Wildlife Research Center. "A Critique of Wildlife Radio-Tracking and Its Use in National Parks." August 3, 2006. http://pubs.er.usgs.gov/publication/93895.

Observator Instruments. "Wind Monitoring and Alarm System." http://www.observator meteohydro.com/cms/uploads/documenten/licentie_1/document_34.pdf.

Ocean Networks Canada. "Teen Spots Hagfish-Slurping Elephant Seal." http://www.you tube.com/watch?v=nzMB8jqioVo.

Oceans at MIT with Ron Prinn. "400 ppm CO_2? Add Other GHGs, and It's Equivalent to 478 ppm." http://oceans.mit.edu/featured-stories/5-questions-mits-ron-prinn-400-ppm -threshold.

O'Connell, Allan F., James D. Nichols, K. Ullas Karanth, eds. *Camera Traps in Animal Ecology: Methods and Analyses.* Tokyo: Springer, 2011.

Odum, Eugene. *Ecology: The Link between the Natural and the Social Sciences.* 1963. Reprint, London: Holt Rinehart and Winston, 1975.

Odum, Howard T. *Environment, Power and Society.* New York: Wiley-Interscience, 1971.

OpenScientist. "Mobile Citizen Science Apps!" http://www.openscientist.org/p/citizen -science-for-your-phone.html.

OpenStreetMap. http://www.openstreetmap.org.

O'Sullivan, Dan, and Tom Igoe. *Sensing and Controlling the Physical World with Sensors.* Boston: Thompson Course Technology, 2004.

Ottinger, Gwen. "Buckets of Resistance: Standards and the Effectiveness of Citizen Science." *Science, Technology, & Human Values* 35, no. 2 (2010): 244–70.

Pahl, Mario, Hong Zhu, Jürgen Tautz, and Shaowu Zhang. "Large Scale Homing in Honeybees." *PLoS ONE* 6, no. 5 (May 2011): e19669.

Pahlka, Jennifer. "Coding a Better Government." *TED.* February 2012. http://www.ted.com /talks/jennifer_pahlka_coding_a_better_government?language=en.

Parikka, Jussi. *Insect Media: An Archaeology of Animals and Technology.* Minneapolis: University of Minnesota Press, 2010.

Parisi, Luciana. "Technoecologies of Sensation and Control." In *Deleuze and Guattari and Ecology,* edited by Bernd Herzogenrath, 182–200. Basingstoke: Palgrave Macmillan, 2009.

Parks, Lisa. *Cultures in Orbit: Satellites and the Televisual.* Durham, N.C.: Duke University Press, 2005.

Pasztor, Bence, Luca Mottola, Cecilia Mascolo, Gian Pietro Picco, Stephen Ellwood, and David Macdonald. "Selective Reprogramming of Mobile Sensor Networks through Social Community Detection." In *Proceedings of 7th European Conference on Wireless Sensor Networks* (EWSN2010). Coimbra: Springer, 2010.

Paterson, Katie. "Vatnajökull (the Sound of)." http://www.katiepaterson.org/vatnajokull.

Paulos, Eric, R. J. Honicky, and Ben Hooker. "Citizen Science: Enabling Participatory Urbanism." In *Handbook of Research on Urban Informatics: The Practice and Promise of the Real-Time City,* edited by Marcus Foth, 414–36. Hershey, Pa.: IGI Global, 2009.

Pennisi, Elizabeth. "Global Tracking of Small Animals Gains Momentum." *Science* 334, no. 6059 (November 25, 2011): 1042.

Percifield, Leif. dontflush.me. http://dontflush.me.

PetaJakarta. http://petajakarta.org.

Petrescu, Doina, Constantin Petcou, and Nishat Awan. *Trans-Local-Act: Cultural Practices Within and Across.* Paris: aaa/peprav, 2010.

Pichel, William G., James H. Churnside, Timothy S. Veenstra, David G. Foley, Karen S. Friedman, Russell E. Brainard, Jeremy B. Nicoll, Quanan Zheng, and Pablo Clemente-Colon. "Marine Debris Collects within the North Pacific Subtropical Convergence Zone." *Marine Pollution Bulletin* 54 (2007): 1207–11.

Pickering, Andrew. *The Mangle of Practice: Time, Agency, and Science.* Chicago: University of Chicago Press, 1995.

Planetary Skin Institute. http://www.planetaryskin.org.

Plastics Europe. "Plastics—The Facts: An Analysis of European Plastics Production, Demand and Waste Data for 2011." http://www.plasticseurope.org.

Poole, Robert. *Earthrise: How Man First Saw the Earth.* New Haven, Conn.: Yale University Press, 2008.

Postscapes. "A Brief History of the Internet of Things." http://postscapes.com/internet-of -things-history.

Pottie, Gregory J., and William J. Kaiser. "Wireless Integrated Network Sensors." *Communications of the ACM* 43, no. 5 (2000): 51–58.

Preemptive Media. "Area's Immediate Reading (AIR)." http://www.pm-air.net/index.php.

Pritchard, Helen. "Thinking with the Animal-Hacker: Articulation in Ecologies of Earth Observation." *APRJA.* 2013. http://www.aprja.net/?p=990.

Protei. http://scoutbots.com.

Prowse, Terry D., Frederick J. Wrona, James D. Reist, John J. Gibson, John E. Hobbie, Lucie M. J. Lévesque, and Warwick F. Vincent. "Climate Change Effects on Hydroecology of Arctic Freshwater Ecosystems." *Ambio* 35, no. 7 (November 2006): 347–58.

Public Works. http://www.publicworksgroup.net.

Rainie, Lee, and Barry Wellman. *Networked: The New Social Operating System.* Cambridge, Mass.: MIT Press, 2012.

Raley, Rita. *Tactical Media.* Minneapolis: University of Minnesota Press, 2009.

Ratto, Matt, and Megan Boler. *DIY Citizenship: Critical Making and Social Media.* Cambridge, Mass.: MIT Press, 2014.

Revel, Judith. "Identity, Nature, Life: Three Biopolitical Deconstructions." *Theory, Culture & Society* 26, no. 6 (2009): 45–54.

Robinson, Jennifer. "The Spaces of Circulating Knowledge: City Strategies and Global Urban Governmentality." In *Mobile Urbanism Cities and Policymaking in the Global Age,* edited by Eugene McCann and Kevin Ward, 15–40. Minneapolis: University of Minnesota Press, 2011.

Rossiter, Ned. *Organized Networks: Media Theory, Creative Labor, New Institutions.* Amsterdam: NAI, 2006.

Roth, Wolff-Michael, and G. Michael Bowen. "Digitizing Lizards: The Topology of 'Vision' in Ecological Fieldwork." *Social Studies of Science* 29, no. 5 (1999): 719–64.

Rotics, Shay. "Comparison of Juvenile and Adult Migration in White Storks (*Ciconia ciconia*) with Implications on Survival." Presentation at "Symposium on Animal Movement and the Environment." May 5, 2014. http://www.youtube.com/watch?v=zhe07jTtM60.

Royal Albert Dock London. "Funding and Investment." http://abp-london.co.uk/business -centre/funding-and-investment.

Royal Docks. http://www.royaldocks.london.

Rundel, Philip W., Eric A. Graham, Michael F. Allen, and Jason C. Fisher. "Tansley Review: Environmental Sensor Networks in Ecological Research." *New Phytologist* 182 (2009): 589–607.

Ryan, Peter G., Charles J. Moore, Jan A. van Franeker, and Coleen Moloney. "Monitoring the Abundance of Plastic Debris in the Marine Environment." *Philosophical Transactions of the Royal Society B: Biological Sciences* 364, no. 1526 (2009): 1999–2012.

Safecast. http://blog.safecast.org.

Sassen, Saskia, ed. *Global Networks, Linked Cities*. London: Routledge, 2002.

———. "Open Source Urbanism." domus. June 29, 2011. http://www.domusweb.it/en/op-ed/2011/06/29/open-source-urbanism.html.

Savill, P. S., C. M. Perrins, K. J. Kirby, and N. Fisher. *Wytham Woods: Oxford's Ecological Laboratory*. Oxford: Oxford University Press, 2010.

Scandolara, Chiara, Diego Rubolini, Roberto Ambrosini, Manuela Caprioli, Steffen Hahn, Felix Liechti, Andrea Romano, Maria Romano, Beatrice Sicurella, and Nicola Saino. "Impact of Miniaturized Geolocators on Barn Swallow *Hirundo rustica* Fitness Traits." *Journal of Avian Biology* 45, no. 5 (September 2014): 417–23.

SCANNET: A Circumarctic Network of Terrestrial Field Bases. http://www.scannet.nu/content/view/43/140.

Schiele, Edwin. "From Your Shopping Cart to the Ocean Garbage Patch." NASA Ocean Motion and Surface Currents. http://oceanmotion.org/html/impact/garbagepatch.htm.

Schimel, Dave, Michael Keller, Steve Berukoff, Becky Kao, Hank Loescher, Heather Powell, Tom Kampe, Dave Moore, Wendy Gram, Dave Barnett, Rachel Gallery, Cara Gibson, Keli Goodman, Courtney Meier, Stephanie Parker, Lou Pitelka, Yuri Springer, Kate Thibault, and Ryan Utz. "2011 Science Strategy: Enabling Continental-Scale Ecological Forecasting." National Ecological Observatory Network (NEON), 2011.

Schofield, Gail, Charles M. Bishop, Grant MacLean, Peter Brown, Martyn Baker, Kostas A. Katselidis, Panayotis Dimopoulos, John D. Pantis, and Graeme C. Hays, "Novel GPS Tracking of Sea Turtles as a Tool for Conservation Management." *Journal of Experimental Marine Biology and Ecology* 347 (2007): 58–68.

Scholz, Trebor, ed. *Digital Labor: The Internet as Playground and Factory*. New York: Routledge, 2012.

———. "What the MySpace Generation Should Know about Working for Free." April 3, 2007. http://collectivate.net/journalisms/2007/4/3/what-the-myspace-generation-should-know-about-working-for-free.html (site discontinued).

Schupska, Stephanie. "UGA's Marine Debris Tracker Named in Apple's 'Apps We Can't Live Without.'" June 5, 2014. http://news.uga.edu/releases/article/marine-debris-tracker-apple-apps-we-cant-live-without-0614.

Scottish Sensor Systems Centre. http://sensorsystems.org.uk.

Scripps Institute of Oceanography. "The Keeling Curve." https://scripps.ucsd.edu/programs/keelingcurve.

Sea Mammal Research Unit. "Southern Elephant Seals as Oceanographic Samplers." http://biology.st-andrews.ac.uk/seaos.

Sebille, Erik van, Matthew H. England, and Gary Froyland. "Origin, Dynamics, and Evolution of Ocean Garbage Patches from Observed Surface Drifters." *Environmental Research Letters* 7, no. 4 (2012): doi:10.1088/1748–9326/7/4/044040.

SeeClickFix. http://en.seeclickfix.com.

Serres, Michel. *The Five Senses: A Philosophy of Mingled Bodies.* Translated by Margaret Sankey and Peter Cowley. 1985. Reprint, London: Continuum, 2008.

———. *Hermes: Literature, Science, Philosophy.* Edited by Josué V. Harari and David F. Bell. Baltimore: Johns Hopkins University Press, 1982.

Shapin, Steven, and Simon Schaffer. *Leviathan and the Air-Pump: Hobbes, Boyle, and the Experimental Life.* Princeton, N.J.: Princeton University Press, 1985.

Shaviro, Steven. *The Universe of Things: On Speculative Realism.* Minneapolis: University of Minnesota Press, 2014.

———. *Without Criteria: Kant, Whitehead, Deleuze, and Aesthetics.* Cambridge, Mass.: MIT Press, 2009.

Sheppard, Kate. "Republicans Kill Global Warming Committee." *Guardian.* January 6, 2011. http://www.theguardian.com/environment/2011/jan/06/republicans-kill-global-warming-committee.

Siemens. "Creating Cities." http://www.thecrystal.org/exhibition.html.

———. "The Crystal." http://www.thecrystal.org.

———. "Siveillance SiteIQ—A Better Way to View Automated Video Surveillance." http://w3.siemens.co.uk/buildingtechnologies/uk/en/security/product-portfolio/siveillance-site-iq/pages/siveillance-site-iq.aspx.

Silva, Adriana de Souza e, and Jordan Frith. *Mobile Interfaces in Public Spaces: Locational Privacy, Control, and Urban Sociability.* London: Routledge, 2012.

Simondon, Gilbert. *Du mode d'existence des objets techniques.* Editions Aubier, 1958.

———. "The Genesis of the Individual." Translated by Mark Cohen and Sanford Kwinter. In *Incorporations,* edited by Jonathan Crary and Sanford Kwinter, 296–319. New York: Zone Books, 1992.

———. *L'individuation à la lumière des notions de forme et d'information.* Grenoble: Éditions Jérome Millon, 2005.

———. "Technical Mentality." Translated by Arne De Boever. In *Gilbert Simondon: Being and Technology,* edited by Arne De Boever, Alex Murra, Jon Roffe, and Ashley Woodward, 1–15. Edinburgh: Edinburgh University Press, 2012.

———. *Two Lessons on Animal and Man.* Translated by Drew S. Burk. 2004. Reprint, Minneapolis: Univocal Publishing, 2011.

Simone, AbdouMaliq. "Cities of Uncertainty: Jakarta, the Urban Majority, and Inventive Political Technologies." *Theory, Culture & Society* 30, nos. 7–8 (December 2013): 243–63.

———. "Infrastructure: Introductory Commentary by AbdouMaliq Simone." November 26, 2012. http://www.culanth.org/curated_collections/11-infrastructure/discussions/12-infrastructure-introductory-commentary-by-abdoumaliq-simone.

———. "People as Infrastructure: Intersecting Fragments in Johannesburg." *Public Culture* 16, no. 3 (2004): 407–29.

Singer, Natasha. "Mission Control, Built for Cities: I.B.M. Takes 'Smarter Cities' Concept to Rio de Janeiro." *New York Times.* March 3, 2012. http://www.nytimes.com/2012/03/04/business/ibm-takes-smarter-cities-concept-to-rio-de-janeiro.html?pagewanted=all&_r=0.

Sirtrack. "Happy Feet Transmissions Ceased." September 12, 2011. http://www.sirtrack.com/index.php/news/news/21-happy-feet-transmissions-ceased.

———. "Sirtrack Working with Emperor Penguin 'Happy Feet.'" http://www.sirtrack.com/index.php/news/case-studies/70-sirtrack-working-with-emperor-penguin-qhappy-feetq.

Skyview. "London Eye." http://www.skyview.co.uk/acatalog/London_Eye.html.

Smerdon, Andrew M., James P. Manning, and Elizabeth M. Paull. "Crowdsourcing Coastal Oceanographic Data: A New Wireless Instrument for Fishing Boats to Gather Seabed Temperature." http://www.nefsc.noaa.gov/epd/ocean/MainPage/lob/Aquatec_poster.pdf.

Snyder, Emily G., Timothy H. Watkins, Paul A. Solomon, Eben D. Thoma, Ronald W. Williams, Gayle S. W. Hagler, David Shelow, David A. Hindin, Vasu J. Kilaru, and Peter W. Preuss. "The Changing Paradigm of Air Pollution Monitoring." *Environmental Science and Technology* 47 (2013): 11369–77.

Southeast Atlantic Marine Debris Initiative (SEA-MDI). "Want to Track Marine Debris? There's an App for That." http://sea-mdi.engr.uga.edu/?p=155.

Spontaneous Interventions. http://www.spontaneousinterventions.org/interventions.

Stacey, Jackie, and Lucy Suchman. "Animation and Automation—The Liveliness and Labors of Bodies and Machines." *Body and Society* 18, no. 1 (March 2012): 1–46.

Stengers, Isabelle. "A Constructivist Reading of Process and Reality." *Theory, Culture & Society* 25, no. 4 (2008): 91–110.

———. "The Cosmopolitical Proposal." In *Making Things Public: Atmospheres of Democracy,* edited by Bruno Latour and Peter Weibel, 994–1003. Cambridge, Mass.: MIT Press, 2005.

———. *Cosmopolitics.* Translated by Robert Bononno. 1997. Minneapolis: University of Minnesota Press, 2010.

———. *Cosmopolitics II.* Translated by Robert Bononno. 1997. Reprint, Minneapolis: University of Minnesota Press, 2011.

———. "Including Nonhumans in Political Theory: Opening Pandora's Box?" In *Political Matter: Technoscience, Democracy and Public Life,* edited by Bruce Braun and Sarah Whatmore, 3–34. Minneapolis: University of Minnesota Press, 2010.

———. "Introductory Notes on an Ecology of Practices." *Cultural Studies Review* 11, no. 1 (2005): 183–96.

———. *The Invention of Modern Science.* Minneapolis, University of Minnesota Press, 2000.

———. *Thinking with Whitehead: A Free and Wild Creation of Concepts.* Translated by Michael Chase. 2002. Reprint, Cambridge: Harvard University Press, 2011.

Stevens, Martin. *Sensory Ecology, Behavior and Evolution.* Oxford: Oxford University Press, 2013.

Stocker, Thomas F., Dahe Qin, Gian-Kasper Plattner, Melinda Tignor, Simon K. Allen, Judith Boschung, Alexander Nauels, Yu Xia, Vincent Bex, and Pauline M. Midgley, eds. *IPCC, 2013: Climate Change 2013; The Physical Science Basis. Contribution of Working Group I to the Fifth Assessment Report of the Intergovernmental Panel on Climate Change.* Cambridge: Cambridge University Press, 2013.

Strangeways, Ian. *Measuring the Natural Environment.* 2000. Reprint, Cambridge: Cambridge University Press, 2003.

Stuever, Hank. "BP's Oil Spillcam: A Horror Movie about the Gulf That's Deeply Compelling." *Washington Post.* May 26, 2010. http://www.washingtonpost.com/wp-dyn/content/article/2010/05/25/AR2010052505047.html.

Suchman, Lucy. "Agencies in Technology Design: Feminist Reconfigurations." Digital Cultures: Participation—Empowerment—Diversity. Online Proceedings of the 5th European

Symposium on Gender & ICT, University of Bremen. March 5, 2009. http://www.infor
matik.uni-bremen.de/soteg/gict2009/proceedings/GICT2009_Suchman.pdf.

———. *Human-Machine Reconfigurations: Plans and Situated Actions.* 2nd ed. Cambridge: Cambridge University Press, 2007.

———. "Reconfigurations." In *Human-Machine Reconfigurations: Plans and Situated Actions,* 259–86. 2nd ed. Cambridge: Cambridge University Press, 2007.

———. "Subject Objects." *Feminist Theory* 12, no. 2 (2011): 119–45.

Supreme Court of the United States. *Massachusetts v. Environmental Protection Agency.* 549 U.S. 497. April 2, 2007.

Surf Life Saving Western Australia. https://twitter.com/SLSWA.

Sustainable Society Network. "UP London 2013." http://uplondon.org.

Sustaining Arctic Observing Network (SAON). http://arcticobserving.org.

Szewczyk, Robert, Alan Mainwaring, Joseph Polastre, Joseph Anderson, and David Culler. "An Analysis of a Large Scale Habitat Monitoring Application." In *Sensys'04: Proceedings of the Second ACM Conference on Embedded Networked Sensor Systems,* 214–26. New York: ACM Press, 2004.

Takada, Hideshige. "International Pellet Watch: Studies of the Magnitude and Spatial Variation of Chemical Risks Associated with Environmental Plastics." In *Accumulation: The Material Politics of Plastic,* edited by Jennifer Gabrys, Gay Hawkins, and Mike Michael, 184–207. London: Routledge, 2013.

TallBear, Kim. "Uppsala 3rd Supradisciplinary Feminist Technoscience Symposium: Feminist and Indigenous Intersections and Approaches to Technoscience." October 17, 2013. http://www.kimtallbear.com/homeblog/uppsala-3rd-supradisciplinary-feminist-tech noscience-symposium-feminist-and-indigenous-intersections-and-approaches-to-techno science.

———. "Why Interspecies Thinking Needs Indigenous Standpoints." Fieldsights—Theorizing the Contemporary, *Cultural Anthropology Online.* April 24, 2011. http://culanth .org/fieldsights/260-why-interspecies-thinking-needs-indigenous-standpoints.

Taylor, Peter. "Technocratic Optimism: H. T. Odum and the Partial Transformation of Ecological Metaphor after World War II." *Journal of the History of Biology* 21, no. 2 (Summer 1988): 213–44.

TechCentral.ie. "Dublin to Become First Fully 'Sensored' City." April 1, 2014. http://www .techcentral.ie/dublin-to-become-first-fully-sensored-city.

TechniCity. Coursera. http://www.coursera.org/course/techcity.

Terranova, Tiziana. *Network Culture: Politics for the Information Age.* London: Pluto Press, 2004.

Thingful. https://thingful.net.

Thomas, Daniel. "Smartphones Set to Become Even Smarter." *Financial Times.* January 3, 2014. http://www.ft.com/cms/s/0/210864c6-72f9-11e3-b05b-00144feabdc0.html#axzz3ihi UWZgT.

Thompson, Richard C., Charles J. Moore, Frederick S. vom Saal, and Shanna H. Swan. "Plastics, the Environment, and Human Health: Current Consensus and Future Trends." *Philosophical Transactions of the Royal Society B: Biological Sciences* 364, no. 1526 (2009): 2153–66.

Thompson, Richard C., Ylva Olsen, Richard P. Mitchell, Anthony Davis, Steven J. Rowland, Anthony W. G. John, Daniel McGonigle, and Andrea E. Russell. "Lost at Sea: Where Does All the Plastic Go?" *Science* 304, no. 5672 (2004): 838.

Thompson, Richard C., Shanna H. Swan, Charles J. Moore, and Frederick S. vom Saal. "Our Plastic Age." *Philosophical Transactions of the Royal Society B: Biological Sciences* 364, no. 1526 (2009): 1973–76.

Thrift, Nigel, and Shaun French. "The Automatic Production of Space." *Transactions of the Institute of British Geographers* 27, no. 3 (2002): 309–35.

Tikhonravov, M. K. "The Creation of the First Artificial Earth Satellite: Some Historical Details." *Journal of the British Interplanetary Society* 47, no. 5 (May 1994): 191–94.

Tolle, Gilman, Joseph Polastre, Robert Szewczyk, David Culler, Neil Turner, Kevin Tu, Stephen Burgess, Todd Dawson, Phil Buonadonna, David Gay, and Wei Hong. "A Macroscope in the Redwoods." In *SenSys'05: Proceedings of the Third International Conference on Embedded Networked Sensor Systems,* 51–63. New York: ACM Press, 2005.

Townsend, Anthony. *Smart Cities: Big Data, Civic Hackers, and the Quest for a New Utopia.* New York: W. W. Norton. 2013.

Townsend, Anthony, Rachel Maguire, Mike Liebhold, and Matthias Crawford. "A Planet of Civic Laboratories: The Future of Cities, Information, and Inclusion." Palo Alto, Calif.: Institute for the Future and the Rockefeller Foundation, 2010.

Tsing, Anna Lowenhaupt. "Empowering Nature; or, Some Gleanings in Bee Culture." In *Naturalizing Power: Essays in Feminist Cultural Analysis,* edited by Sylvia Junko Yanagisako and Carol Lowery Delaney, 113–44. New York: Routledge, 1995.

Turgeman, Yaniv J., Eric Alm, and Carlo Ratti. "Smart Toilets and Sewer Sensors Are Coming." *Wired.* March 21, 2014. http://www.wired.co.uk/magazine/archive/2014/03/ideas-bank/yaniv-j-turgeman.

Turing, Alan. "Intelligent Machinery." In *Mechanical Intelligence: Collected Works of A. M. Turing,* edited by Darrel C. Ince, 107–27. 1948. Reprint, Amsterdam: North Holland, 1992.

Turner, Fred. *From Counterculture to Cyberculture: Stewart Brand, the Whole Earth Network, and the Rise of Digital Utopianism.* Chicago: University Of Chicago Press, 2006.

Uexküll, Jacob von. *A Foray into the Worlds of Animals and Humans.* Translated by Joseph D. O'Neil. 1934. Reprint, Minneapolis: University of Minnesota Press, 2010.

UK Argo. "Deep Profile Floats." http://www.ukargo.net/about/technology/deep_profile_floats.

UK Clean Air Act, 1956. 4 & 5 Eliz. 2 chapter 52. London: HMSO, 1956.

Underwood, A. J. *Experiments in Ecology: Their Logical Design and Interpretation Using Analysis of Variance.* Cambridge: Cambridge University Press. 1997.

Ungerleider, Neal. "The London Underground Has Its Own Internet of Things." *Fast Company.* May 8, 2014. http://www.fastcolabs.com/3030367/the-london-underground-has-its-own-internet-of-things.

United Nations Environment Programme. "Marine Litter: An Analytical Overview." Nairobi: UNEP, 2005. http://www.unep.org/regionalseas/marinelitter/publications/docs/anl_oview.pdf.

———. "Marine Litter: A Global Challenge." Nairobi: UNEP, 2009. http://www.unep.org/pdf/unep_marine_litter-a_global_challenge.pdf.

United Nations Environment Programme Yearbook. "Plastic Debris in the Ocean." Nairobi: UNEP, 2011. http://www.unep.org/yearbook/2011/pdfs/plastic_debris_in_the_ocean.pdf.

University of Oxford. "Wytham Woods." http://www.wytham.ox.ac.uk.

Urban Prototyping London. "Festival Showcase." 2013. http://uplondon.net/wp-content/uploads/2013/05/UP-London-Street-Exposition.pdf.

U.S. EPA. "Carbon Pollution Standards for New, Modified and Reconstructed Power Plants." http://www2.epa.gov/carbon-pollution-standards.

———. "Clean Air Act." http://www.epa.gov/air/caa/amendments.html.

———. "Draft Roadmap for Next Generation Air Monitoring." March 8, 2013. http://www.eunetair.it/cost/newsroom/03-US-EPA_Roadmap_NGAM-March2013.pdf.

———. "Endangerment and Cause or Contribute Findings for Greenhouse Gases under Section 202(a) of the Clean Air Act." http://www3.epa.gov/climatechange/endangerment.

———. "Marine Debris in the North Pacific. A Summary of Existing Information and Identification of Data Gaps." EPA-909-R-11–006. San Francisco, Calif.: EPA, 2011.

———. "Toxic Air Pollutants." http://www3.epa.gov/airtrends/agtrnd95/tap.html.

———. "What Are the Six Common Air Pollutants?" http://www3.epa.gov/airquality/urbanair.

U.S. Long Term Ecological Research Network (LTER). http://www.lternet.edu.

Vandenabeele, Sylvie P., Edward Grundy, Michael I. Friswell, Adam Grogan, Stephen C. Votier, and Rory P. Wilson. "Excess Baggage for Birds: Inappropriate Placement of Tags on Gannets Changes Flight Patterns." *PLoS One* 9, no. 3 (March 2014): e92657.

Vaughan, David G., Josefino C. Comiso, Ian Allison, Jorge Carrasco, Georg Kaser, Ronald Kwok, Philip Mote, Tavi Murray, Frank Paul, Jiawen Ren, Eric Rignot, Olga Solomina, Konrad Steffen, and Tingjun Zhang. "Observations: Cryosphere. In *Climate Change 2013: The Physical Science Basis; Working Group I Contribution to the Fifth Assessment Report of the Intergovernmental Panel on Climate Change,* edited by Thomas F. Stocker, Dahe Qin, Gian-Kasper Plattner, Melinda Tignor, Simon K. Allen, Judith Boschung, Alexander Nauels, Yu Xia, Vincent Bex, and Pauline M. Midgley, 317–82. Cambridge: Cambridge University Press, 2013.

Verran, Helen. "The Changing Lives of Measures and Values: From Center Stage in the Fading 'Disciplinary' Society to Pervasive Background Instrument in the Emergent 'Control' Society." *Sociological Review* 59, special issue, "Measure and Value," edited by Lisa Adkins and Celia Lury (2012): 60–72.

Virno, Paolo. *A Grammar of the Multitude.* Los Angeles: Semiotext(e), 2004.

Vitaliano, Dorothy B. *Legends of the Earth.* Bloomington: Indiana University Press, 1973.

Wald, Chelsea. "Follow That Bird: Real-Time Data on Migrating Birds, Coming to a Phone Near You." June 20, 2014. http://www.earthtouchnews.com/discoveries/innovation/follow-that-bird-real-time-data-on-migrating-birds-coming-to-a-phone-near-you.

WayDownSouth. "Adopt an Argo Float." http://argofloats.wikispaces.com/Adopt+an+Argo+Float.

Weiser, Mark. "The Computer for the 21st Century." *Scientific American* 265, no. 3 (1991): 94–104.

Welsh, Matt. "Sensor Networks, Circa 1967." January 21, 2010. http://matt-welsh.blogspot.co.uk/2010/01/sensor-networks-circa-1967.html.

Whitehead, Alfred N. *Adventures of Ideas.* 1933. Reprint, New York: The Free Press, 1967.

———. *Modes of Thought.* 1938. Reprint, New York: The Free Press, 1966.

———. *Process and Reality.* 1929. Reprint, New York: The Free Press, 1985.

———. *Science and the Modern World.* 1925. Reprint, New York: The Free Press, 1967.

Whitehead, Mark. *State Science and the Skies: Governmentalities of the British Atmosphere.* Malden, Mass.: Wiley-Blackwell, 2009.

Wiener, Norbert. *Cybernetics; or, Control and Communication in the Animal and the Machine.* New York: John Wiley, 1948.

———. *The Human Use of Human Beings.* Boston: Houghton Mifflin, 1954.

Wikelski, Martin. "Move It, Baby!" Keynote presentation at Symposium on Animal Movement and the Environment. May 5, 2014. https://www.youtube.com/watch?v=PxtJAX QQU40&index=2&list=PLxu—qUEe7JoCMayKHK-Ryi_tcq1RxTA4.

Wikelski, Martin, Roland W. Kays, N. Jeremy Kasdin, Kasper Thorup, James A. Smith, and George W. Swenson. "Going Wild: What a Global Small-Animal Tracking System Could Do for Experimental Biologists." *Journal of Experimental Biology* 210 (2007): 181–86.

Wikelski Martin, David Moskowitz, James S. Adelman, Jim Cochran J, David S. Wilcove, and Michael L. May. "Simple Rules Guide Dragonfly Migration." *Biology Letters* 2, no. 3 (2006): 325–29.

Wikelski, Martin, Uschi Müller, Wolfgang Arne Heidrich, and Franz Xaver Kümmeth. "Disaster Alert Mediation Using Nature." (May 8, 2012 patent filing date; November 14, 2013 publication date), WO 2013167661 A2.

Wilcove, David S. *No Way Home: The Decline of the World's Great Animal Migrations.* Washington, D.C.: Island Press, 2008.

Wildfowl and Wetlands Trust. http://www.wwt.org.uk.

WildlifeTV. "Wildlife Webcams." http://www.wildlifetv.co.uk/webcams.

Wilson, Jon S., ed. *Sensor Technology Handbook.* Amsterdam: Elsevier, 2005.

Wilson, Rory P., and Clive R. McMahon. "Measuring Devices on Wild Animals: What Constitutes Acceptable Practice?" *Frontiers in Ecology and the Environment* 4, no. 3 (2006): 147–54.

Wilson, Rory P., E. L. C. Shepard, and N. Liebsch. "Prying into the Intimate Details of Animal Lives: Use of a Daily Diary on Animals." *Endangered Species Research* 4 (2008): 123–37.

Woods Hole Oceanographic Institute. "Science in a Time of Crisis." http://www.whoi.edu/deepwaterhorizon.

World Health Organization (WHO). "Ambient (Outdoor) Air Quality and Health." Fact Sheet No. 313 (March 2014). http://www.who.int/mediacentre/factsheets/fs313/en.

Wynne, Brian. "May the Sheep Graze Safely? A Reflexive View of the Expert-Lay Knowledge Divide." In *Risk, Environment and Modernity: Towards a New Ecology,* edited by Scott Lash, Bronislaw Szerszynski, and Brian Wynne, 44–83. London: Sage, 1996.

xClinic Environmental Health Clinic and Living Architecture Lab. "Amphibious Architecture." http://www.environmentalhealthclinic.net/amphibiousarchitecture.

Xively. https://xively.com.

Yu, Alan. "More Than 300 Sharks In Australia Are Now on Twitter." *All Things Considered.* National Public Radio. January 1, 2014. http://www.npr.org/blogs/alltechconsidered/2013/12/31/258670211/more-than-300-sharks-in-australia-are-now-on-twitter.

Zhen, Zhao, et al. "Climate Change: Cities in Action." Montreal: Metropolis and Cisco in partnership with Connected Urban Development, 2009. http://old.metropolis.org/sites/default/files/publications/2009/Climate-Change-Cities-in-Action_0.pdf.

Zoological Society of London. "Eel Conservation." http://www.zsl.org/conservation/regions/uk-europe/eel-conservation.

INDEX

Page numbers in italics refer to photographs and other illustrations.

(continued from page ii)

JENNIFER GABRYS is reader in the Department of Sociology, Goldsmiths, University of London, and principal investigator of the Citizen Sense research project. She is author of *Digital Rubbish: A Natural History of Electronics* and coeditor of *Accumulation: The Material Politics of Plastic.*